Electrochemistry

Volume 14

Specialist Periodical Reports Online

Application for Free Access to Electronic Chapters

Customers purchasing a print volume are now entitled to free site-wide access to the electronic version of that title. (For the definition of a site, please consult **www.rsc.org/subagree**)

To apply for free access, please contact **ejournals@rsc.org** with proof of purchase (a scanned copy is acceptable) containing details of:

- **Title of book**
- **Volume number**
- **ISBN**
- **Address of your organisation**

By providing these details you will be agreeing to the terms of the Electronic Information Licence Agreement at

www.rsc.org/subagree

ROYAL SOCIETY OF **CHEMISTRY**

A Specialist Periodical Report

Electrochemistry

Volume 14

Editors
Craig Banks, *Manchester Metropolitan University, UK*
Steven McIntosh, *Lehigh University, Bethlehem PA, USA*

Authors
N. Alonso-Vante, *University of Poitiers, France*
S. D. Bhat, *CSIR Madras Complex, Chennai, India*
Mary Clare Sison Escaño, *University of Fukai, Japan*
Sadia Kabir, *University of New Mexico, Albuquerque, NM, USA*
Y. Luo, *University of Poitiers, France*
Javier Parrondo, *Washington University in St. Louis, MO, USA*
Vijay Ramani, *Washington University in St. Louis, MO, USA*
A. K. Sahu, *CSIR Madras Complex, Chennai, India*
Alexey Serov, *University of New Mexico, Albuquerque, NM, USA*
P. Sridhar, *Mesha Energy Solutions Pvt. Ltd., Bengaluru, India*
Lihui Wang, *Illinois Institute of Technology, Chicago, IL, USA*
Zhongyang Wang, *Washington University in St. Louis, MO, USA*

ROYAL SOCIETY OF CHEMISTRY

THE QUEEN'S AWARDS FOR ENTERPRISE: INTERNATIONAL TRADE 2013

Print ISBN: 978-1-78262-114-0
PDF eISBN: 978-1-78262-272-7
ePUB eISBN: 978-1-78801-150-1
ISSN: 0305-9979
DOI: 10.1039/9781782622727

A catalogue record for this book is available from the British Library

Published by The Royal Society of Chemistry,
Thomas Graham House, Science Park, Milton Road,
Cambridge CB4 0WF, UK

Registered Charity Number 207890

For further information see our web site at www.rsc.org

Printed in the United Kingdom by CPI Group (UK) Ltd, Croydon,
CR0 4YY, UK

Preface

DOI: 10.1039/9781782622727-FP005

We are happy to introduce volume 14 of *Specialist Periodical Reports in Electrochemistry* which presents comprehensive and critical reviews with a thematic discussion of recent progress in electrolytes and electrocatalysts for fuel cells. In this volume, Escaño provides an in-depth review of recent density functional theory studies of borohydride electro-oxidation on metal electrodes that provide new insights the composition and structural dependence of this critical reaction for direct borohydride fuel cells. Continuing with the theme of emerging types of fuel cell, Ramani and co-workers review the most recent advances in anion exchange membranes with insights into both the most promising compositions and identifying the critical material requirements. This is complemented by the excellent review of anode electrocatalysts for direct alcohol anion exchange membrane fuel cells authored by Kabir and Serov. They focus their review on the critical need to understand how the complex interactions between structure, composition and physico-chemical properties of the catalysts dictate the electrode performance. The goal of direct alcohol utilization within fuel cells is further discussed by Sridhar and co-authors who discuss the recent progress in developing both polymer electrolytes and electrocatalysts for proton conducting direct alcohol fuel cells. Finally, Luo and Alonso-Vante have authored an insightful discussion of the application of metal organic frameworks (MOFs) in fuel cell electrocatalysis where they can play the role of electrocatalytic centers, electrocatalyst supports, or as electrocatalyst precursors.

Steven McIntosh

Author biographies

DOI: 10.1039/9781782622727-FP006

Nicolas Alonso-Vante is Professor at the University of Poitiers. He has developed materials based on transition metal and chalcogenides in metallic and semiconducting form in Germany and France for photoelectrocatalytic processes. He has pioneered the use of non-platinum and non-precious metal catalysts center chalcogenides, using facile chemical routes, in order to understand, *e.g.*, the complex process of the oxygen reduction reaction process, in low temperature fuel cells. He has received the awards of the Alexander von Humboldt foundation fellowship, of the National Polytechnic Institute-Mexico as a R&D distinguished graduate, and of the Mexican Council of Technology SNI-III recognition as a Mexican researcher working outside Mexico.

Dr Santoshkumar D. Bhat is a Senior Scientist at CSIR-Central Electrochemical Research Institute, Madras Unit, CSIR Madras Complex Chennai, India. His research interests are on polymer membrane electrolytes for fuel cells. In particular, he has developed new polymer electrolytes for restricting the methanol cross-over in direct methanol fuel cells. He has also worked on new electrocatalyst support for oxygen reduction reaction in polymer electrolyte fuel cells. He has contributed towards the development of low temperature fuel cell stacks for stationary applications. He is also interested in developing wick based humidification techniques for polymer electrolyte fuel cells. He has 60 international publications and 2 patents to his credit.

Dr Mary Clare Escaño is currently a tenure-track assistant professor at University of Fukui, Japan. She received her PhD in 2012 from the Department of Applied Physics, Osaka University, Japan. Dr Escaño was a Young Scientist Fellow of the Japan Society for the Promotion of Science (JSPS) for three years and is a multi-awarded early career researcher. She holds two masters degrees in experimental and theoretical applied physics and has published more than 30 original articles in high reputable journals. Her research interest involves interdisciplinary studies bridging magnetism and catalysis on metals.

Sadia Kabir is a PhD candidate in the Department of Chemical and Biological Engineering at the University of New Mexico. In 2013, she received her Masters in Chemical Engineering from UNM while working on characterizing the structure–property relationships of nitrogen-functionalized nanomaterials using a synergistic combination of computational and spectroscopic techniques. Currently, Sadia's research is primarily focused on developing and integrating hierarchically structured nanocomposite materials for fuel cells applications; she has 7 peer-reviewed journal publications in the field. Sadia is also involved in several energy-conversion related projects in collaboration with Los Alamos National Laboratory, Daihatsu, and the University of Utah.

Yun Luo, born in China in 1984, Yun Luo obtained her MS of analytical chemistry at Sichuan University in 2005 on fluorescent probe, received her PhD in 2012 on porous and luminescent Ln-based MOFs in INSA-Rennes, France. From 2012 to 2013, she worked as post-doc on photo-electroreduction of CO_2 and silicon surface grafting at University of Rennes 1. In 2013, she joined Prof. N. Alonso-Vante's group at the University of Poitiers for a European project (Cathcat) on nanomaterials for oxygen reduction reaction and low-temperature fuel cell. Her research interests are focused on coordination polymers, solar fuels and electrocatalysis.

Javier Parrondo is a Research Scientist in the Department of Energy, Environmental and Chemical Engineering at Washington University in Saint Louis, Missouri. He received both his BSc and his PhD in Chemical Engineering from the University of Oviedo, Spain. He was working as Assistant Professor at the University of the Basque Country in Bilbao, Spain and as Research Associate at Illinois Institute of Technology, Chicago, Illinois. He is interested in the development of new catalysts and anion exchange membranes for alkaline membrane fuel cells, solid-state alkaline water electrolyzers and redox flow batteries.

Vijay Ramani holds the Roma B. and Raymond H. Wittcoff Professorship in the Department of Energy Environmental and Chemical Engineering at Washington University in St. Louis, and concurrently serves as the Director of the Center for Solar Energy and Energy Storage at Washington University. His research interests lie at the confluence of electrochemical engineering, materials science, and renewable energy technologies. Current research directions in his group include multi-functional electrolyte and electrocatalyst materials for electrochemical systems, mitigating component degradation in electrochemical devices, and *in situ* diagnostics to probe electrochemical systems. NSF (NSF CAREER), ONR (ONR Young Investigator Award), and DOE have funded his research. Vijay has a PhD from the University of Connecticut, Storrs, and a BE from Annamalai University, India, both in Chemical Engineering.

Dr Akhila Kumar Sahu, obtained his PhD degree in Chemistry from the University of Madras in 2009. He joined Central Electrochemical Research Institute-Madras Unit as a Senior Research Fellow In 2005, and became Research Associate in the same Institute during 2006–2008 and devoted his research work in the field of Fuel Cell Technologies. In 2008, he joined as Scientist, at Central Electrochemical Research Institute-Madras Unit, CSIR Madras Complex, Chennai and continued the research on fuel cell area. He has about 52 research publications in peer reviewed journals and 3 international patents to his credit. His present research activities include development of composite membranes for PEFCs operation at elevated temperature and reduced relative humidity, new membrane materials for mitigating methanol cross-over in DMFCs, Synthesis of nano-porous/mesoporous carbon materials for fuel cell electrodes, Pt based and non-precious metal catalysts for efficient oxygen reduction reaction (ORR) for fuel cell applications.

Alexey Serov (MS in Inorganic Chemistry from Moscow State University, Russia and PhD in Physical Chemistry from University of Bern, Switzerland) is a Research Associate Professor of Chemical & Biological Engineering Department at the University of New Mexico. Dr Serov's research focuses on materials design and characterization for energy applications and includes development of electrocatalyst for fuel cells, electrolyzers and lithium-ion batteries. He is the author of more than 85 peer-reviewed articles and 65 issued patents (USA and international).

Dr Sridhar Parthasarathi, Chief Scientist, M/s Mesha Energy Solutions Pvt. Ltd. has more than three decades of experience in the area of Electrochemical Energy Generation and Storage and has been working with M/s Mesha Energy Solutions Pvt. Ltd., Bengaluru, India since August, 2014. He was involved in the development of Polymer Electrolyte Membrane based fuel cell materials, cell, stack and system in various projects funded by Ministry of New and Renewable Energy, Council of Scientific and Industrial Research, Govt. of India. He has guided students for their PhD and MS degrees. He has more than 100 publications; majority of them in Polymer Electrolyte Membrane based fuel cells. He is a Fellow of Royal Society of Chemistry.

Lihui Wang, PhD candidate in Illinois Institute of Technology (Department of Chemical and Biological Engineering), received her BSc (2011) from Hainan University (China) and MSc (2013) from Illinois Institute of Technology. Her research focuses on investigating different kinds of cations to stabilize polysulfone (PSF) and polyphenylene oxide (PPO) based anion exchange membranes (AEMs), characterizing on AEMs and testing the synthesized new AEMs to alkaline fuel cell, water electrolyzer and redox flow battery.

Zhongyang Wang, PhD candidate in department of Energy, Environmental and Chemical Engineering at Washington University in Saint Louis. His research interest lies in alkaline polymer electrolyte in energy storage and conversion devices such as flow battery, fuel cell and water electrolyzer. Research projects include: reinforced triblock copolymer membranes for flow batteries and bipolar membranes for direct borohydride fuel cells.

CONTENTS

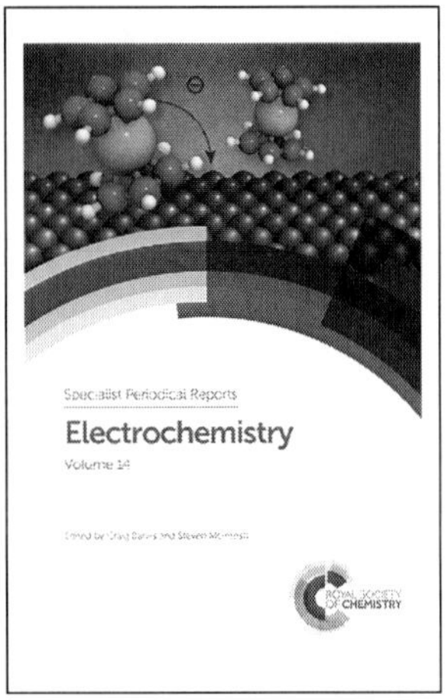

Cover

Borohydride electro-oxidation on metal electrodes: structure, composition and solvent effects from DFT

Mary Clare Sison Escaño

DOI: 10.1039/9781782622727-00001

1 Introduction

An efficient borohydride (BH_4^-) electro-oxidation in alkaline media facilitates rapid development of direct borohydride fuel cells (DBFC) as power sources for portable devices and other applications.[1–7] While there have been reviews on the experimental and theoretical studies of BH_4^- electro-oxidation,[2,3,8] there has been no reports that consolidate, compare systematically and update electro-oxidation of borohydride on vast structures and composition of metals and metal alloys. Such focused review is very insightful in terms of formation of new ideas for design and development of better catalysts. Theoretical investigations are mostly conducted using density functional theory (DFT). Beyond-DFT has also been used and will also be discussed.

One of the parameters in evaluating the effectiveness of an anode catalyst is the resulting initial oxidative adsorption structure of BH_4^-. The theoretical anodic reaction of direct BH_4^- oxidation in a DBFC can be described as:

$$BH_4^- + 8OH^- \rightarrow BO_2^- + 6H_2O + 8e^- \quad E^\circ_{298\,K} = -1.24\,V\ vs.\ SHE \tag{1}$$

Furthermore, depending on the anode catalyst, pH, temperature and residence time on the anode, a thermo-catalytic hydrolysis reaction can also occur, generating $H_{2,(g)}$:

$$BH_4^- + 4H_2O \rightarrow B(OH)_4^- + 4H_2. \tag{2}$$

The *in situ* $H_{2(g)}$ evolution complicates the anode operation by generating a mixed potential and gas shielding the anode surface, thereby hampering the BH_4^- mass transfer and adsorption and lowering the effective ionic conductivity in the anode compartment. Ideally, an effective electrocatalyst should promote reaction (1) over reaction (2). Promoting (1) implies a more controlled interaction of BH_4^- and H_2O species with the anode catalyst. To discuss this in more detail, the initial step of oxidation is considered without solvation:

$$BH_4^- + \ast(\text{metal site}) \rightarrow BH_y^\ast + (4-y)H^\ast + e^- \tag{3}$$

and with solvation effects:

$$BH_{4aq}^- + H_2O^\ast \rightarrow BH_y^\ast + (4-y)H^\ast + H_2O_{aq} + e^- \tag{4}$$

University of Fukui, 3-9-1 Bunkyo, Fukui 910-8507, Japan.
E-mail: mcescano@u-fukui.ac.jp

Eqn (3) indicates that a borohydride ion "sees" a metal site, adsorbs and releases electron due to the potential difference. There are four ways by which the ion adsorbs depending on the catalyst: as a molecule, $y = 4$ (BH_4^*) or as a dissociated structure, $y = 1$–3 (*i.e.* $BH^* + 3H^*$, $BH_2^* + 2H^*$, $BH_3^* + H^*$). Same is the case for eqn (4), however the formation of either of these structures depends on the catalyst and the interaction with water (or solvent). It can be noted that a molecular adsorption structure (BH_4^*) can inhibit fast formation of H_2 gas making the catalyst selective for direct pathway (1), on the other hand, an almost completely dissociated BH_4^- ($BH^* + 3H^*$) can promote H_2 gas evolution and hence leading to reaction pathway (2), and the partially dissociated structures ($BH_2^* + 2H^*$, $BH_3^* + H^*$) can be somewhere in between. Moreover, the strength of the adsorption of the borohydride especially for the molecularly adsorbed and partially adsorbed ones can impact shifts in the overpotentials (*e.g.* weak adsorption → can contribute to higher overpotential; strong adsorption → can contribute to lower overpotential). In this article, the initial oxidative adsorption structures and energies of borohydride with and without solvation effects obtained by DFT across many structures of metal anode catalysts are systematically compared.

2 Pure metal surfaces

A Same metal facet: (111)

A borohydride is a trigonal molecule (Fig. 1 inset) with a calculated B–H bond distance of 1.25 Å and H–B–H angle of 109°, in excellent agreement with experiment.[9]

Early theoretical studies on borohydride electro-oxidation tackled the adsorption of borohydride *via* eqn (3) on the (111) facet of 3d, 4d and 5d metals.[10–13] The (111) surface of the metal is shown in Fig. 1, along with

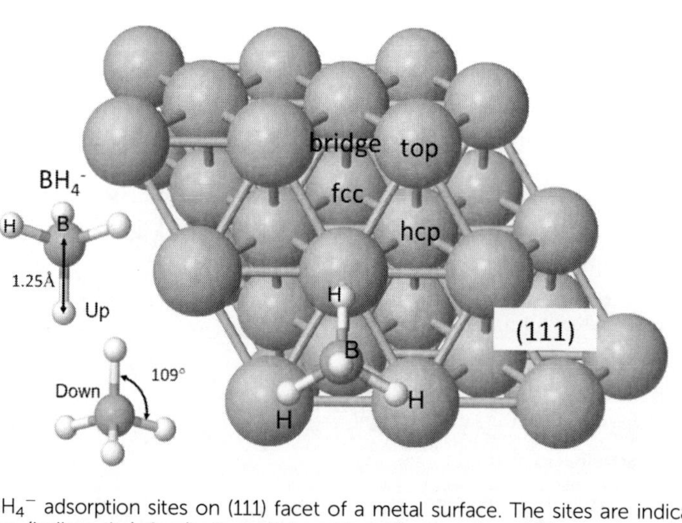

Fig. 1 BH_4^- adsorption sites on (111) facet of a metal surface. The sites are indicated as bridge, hcp (hollow site), fcc (hollow site) and top. The gas phase BH_4^- in trigonal form is shown on the left. (Reprinted with permission from M. C. S. Escaño, E. Gyenge, R. L. Arevalo and H. Kasai, Reactivity Descriptors for Borohydride Interaction with Metal Surfaces, *J. Phys. Chem. C*, 2011, **115**(40), 19883. Copyright 2011 American Chemical Society.)

Table 1 Adsorption structures and energies of borohydride on 3d, 4d and 5d metals as taken from ref. 10 and 11. The surface unit cell used is (3×3). The same adsorption structures for Au and Pt are noted in ref. 12 and 13.

Metals	Adsorption structure	Adsorption energy (eV)
5d[10]		
Os	BH_4*	−4.59
Ir	BH_3* + H*	−4.56
Pt	BH_3* + H*[12]	−4.56
Au	BH_4*[13]	−1.96
4d[10]		
Ru	BH_4*	−4.37
Rh	BH_4*	−4.01
Pd	BH* + 3H*	−3.63
Ag	BH_4*	−2.31
3d[11]		
Cr	BH_3* + H*	−7.14
Mn	BH_2* + 2H*	−4.93
Fe	BH_2* + 2H*	−5.36
Co	BH_4*	−3.55
Ni	BH_4*	−3.50
Cu	BH_4*	−2.87

the adsorption site (∗) labelled as bridge, hcp-hollow, fcc-hollow or top. The adsorption site is defined as the location of the boron atom, which is the center of the trigonal molecule. Table 1 lists the adsorption structure and energies depending on the transition metal.

The adsorption energies are referenced to isolated metal surface and the gas phase borohydride ion. Generally, there is an increasing trend on the adsorption energies as one goes from left to right within the same period of the periodic table. For the 3d transition metals, stronger adsorption leads to partially dissociated structures. However, the adsorption structures for 4d and 5d metals do not really correlate well with the adsorption energies. That is, a strong adsorption does not imply dissociated structures. For instance, in osmium (Os), the adsorption energy is strong but a molecular adsorption structure is noted. The adsorption structures for 4d and 5d transition metals are depicted in Fig. 2 and for the 3d metals, the structures are shown in Fig. 3.

B Different metal surface facets

The relationship between the adsorption structure and energies with the different facets of metal surfaces is investigated in Os.[14] The facets considered, the adsorption structures and the adsorption energies are given in Table 2. Figures 4, 5 and 6 depict the (0001), $(10\bar{1}0)$ and $(1\bar{1}00)$ facets of Os and the adsorption structures, respectively. (0001) is the basal plane of the Os hcp bulk lattice, while the $(10\bar{1}0)$ and $(1\bar{1}00)$ are the prism planes. It is well known that (0001) (Fig. 4) is *akin* to the fcc (111) surface except for the stacking of the atomic layers. On the other hand, $(10\bar{1}0)$ and $(1\bar{1}00)$ have characteristic rows and trenches (Figs. 5–6). These two surfaces differ mainly in the position of the trench atoms. For

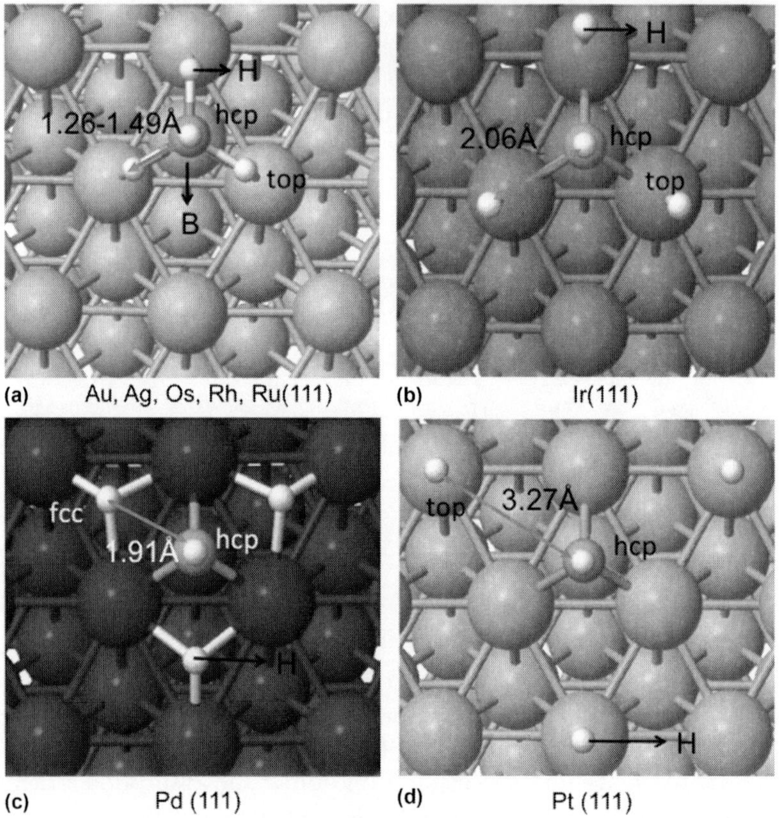

Fig. 2 BH_4^- adsorption structures on (111) facet of 4d and 5d metal surfaces: (a) BH_4* and (b)–(d) BH_3* + H*. The boron (B) sites, the hydrogen (H) atom sites and the B–H distances are indicated. (Reprinted with permission from M. C. S. Escaño, E. Gyenge, R. L. Arevalo and H. Kasai, Reactivity Descriptors for Borohydride Interaction with Metal Surfaces, *J. Phys. Chem. C*, 2011, **115**(40), 19883. Copyright 2011 American Chemical Society.)

instance, in the $(10\bar{1}0)$, a trench atom is directly under the fourfold hollow site (h1) while in the $(1\bar{1}00)$, it is under the bridge site (b2).

Based on the Table 2, it can be noted that the adsorption energy increases depending on how "open" the facets are. The $(10\bar{1}0)$ and $(1\bar{1}00)$ facets contain rows and trenches that expose the second layer atoms (hence "open"), thereby facilitating greater conformation and interaction with the borohydride than the (0001), which does not have these structures. The "b" distance, which is the distance between the parallel rows, is greater in $(1\bar{1}00)$ than in the $(10\bar{1}0)$, making the former more "open". Thus, the adsorption energy follows this order: $(0001) < (10\bar{1}0) < (1\bar{1}00)$. In terms of the adsorption structure, it is observed that within the same kind of metal, the structure can be correlated with the order of the adsorption energies: $BH_4{}^* < BH_2{}^* + 2H^* < BH^* + 3H^*$. For (0001), the lateral and vertical interlayer distances are generally not changed with respect to the relevant atoms' distance from the adsorption site, in contrast to the that of $(10\bar{1}0)$ and $(1\bar{1}00)$. The adsorption of the molecule on the open surfaces involves binding or interaction with the atoms in the second layer as can be seen in Figs. 5 and 6.

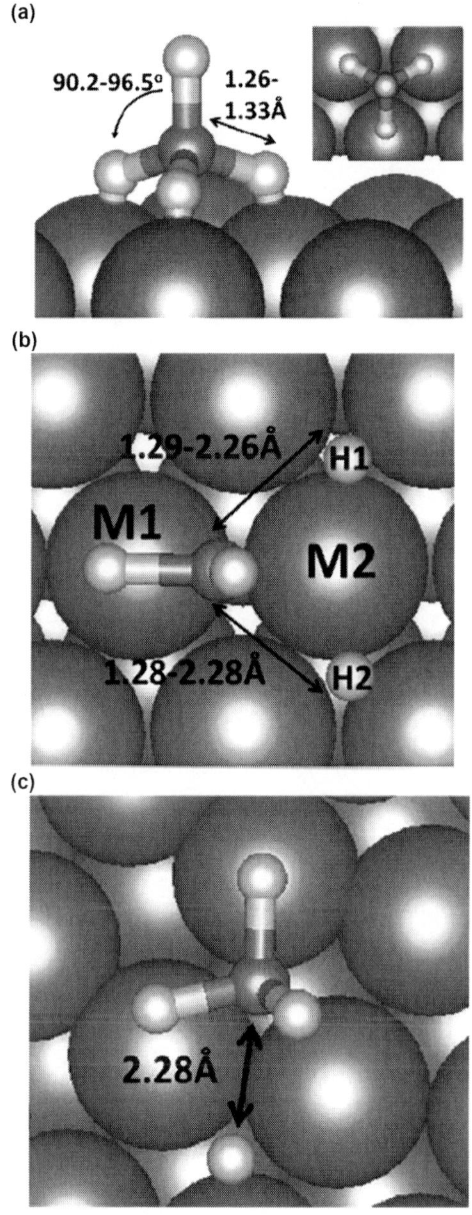

Fig. 3 BH_4^- adsorption structures on (111) facet of 3d metal surfaces: (a) BH_4^* for Co, Ni and Cu; (b) $BH_2^* + 2H^*$ for Mn and Fe; and (c) $BH_3^* + H^*$ for Cr. The boron (B) sites, the hydrogen (H) atom sites and the B–H distances and angles are shown. M1 and M2 are two metal atoms that bind with B differently. (Reprinted from R. L. Arevalo, M. C. S. Escaño, E. Gyenge and H. Kasai, A theoretical study of the structure and stability of borohydride on 3d transition metals, *Surf. Sci.*, **606**, 1954–1959, Copyright 2012 with permission from Elsevier.)

3 Metal alloy surfaces

A Same composition but different types of alloying metal

Because of the development of anode catalysts that are both selective and cost-effective, the metal alloys have been studied theoretically.[15,16]

Table 2 Adsorption structures and energies of borohydride depending on the facets of Os.[14]

Facet	Adsorption structure	Adsorption energy (eV)
(0001)	$BH_4{}^*$	−4.96
$(10\bar{1}0)$	$BH_2{}^* + 2H^*$	−5.64
$(1\bar{1}00)$	$BH^* + 3H^*$	−6.72

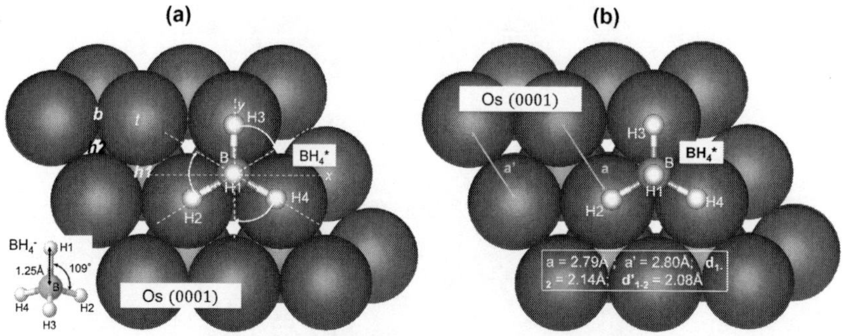

Fig. 4 $BH_4{}^-$ adsorption structures on (0001) facet of Os. Initial configuration considered and the most favored adsorption structure are shown in (a) and (b), respectively. The trigonal gas phase molecule is shown in (a) inset. In (a), the adsorption site for the center of the molecule (B) is indicated as *b* (for bridge), *t* (for top) and *h1*, *h2* (for two types of hollow sites). The atomic distances in the surface as a result of adsorption are given in (b). d_{1-2} and d'_{1-2} are the first interlayer distances corresponding to a and a′ lateral distances. The a and a′ are the distance near and far from the adsorption site, respectively. (Reproduced from ref. 14 with permission from The Royal Society of Chemistry.)

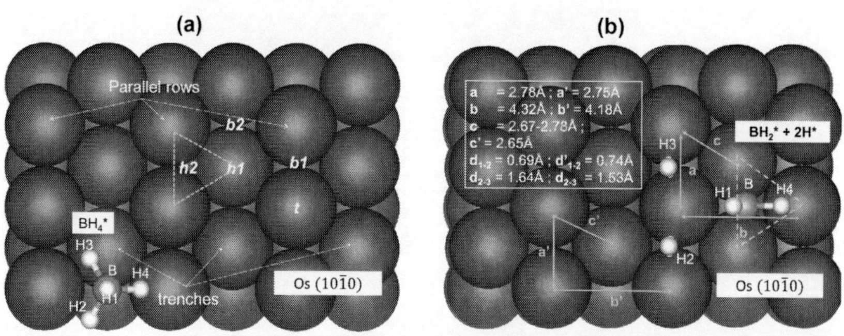

Fig. 5 $BH_4{}^-$ adsorption structures on $(10\bar{1}0)$ facet of Os. Initial configuration considered and the most favored adsorption structure are shown in (a) and (b), respectively. In (a), the adsorption site for the center of the molecule (B) is indicated as *b1* or *b2* (for bridges), *t* (for top) and *h1*, *h2* (for two types of hollow sites). The triangular section shows that the atoms in the vertices belong to different surface layers. The atomic distances in the surface as a result of adsorption are given in (b). d_{1-2} and d'_{1-2} are the first interlayer distances corresponding to a and a′ lateral distances. The difference between these lateral distances is whether they are far or near the adsorption site. d_{2-3} and d'_{2-3} are the second interlayer distances. (Reproduced from ref. 14 with permission from The Royal Society of Chemistry.)

In ref. 15, metal alloys based on Au, which are of Au_3M type (M is a 3d transition metal = Cr, Mn, Fe, Co and Ni), are investigated. Table 3 lists the adsorption structure and the adsorption energies on Au_3M (111) surfaces. The initial oxidative adsorption structures are depicted in Fig. 7.

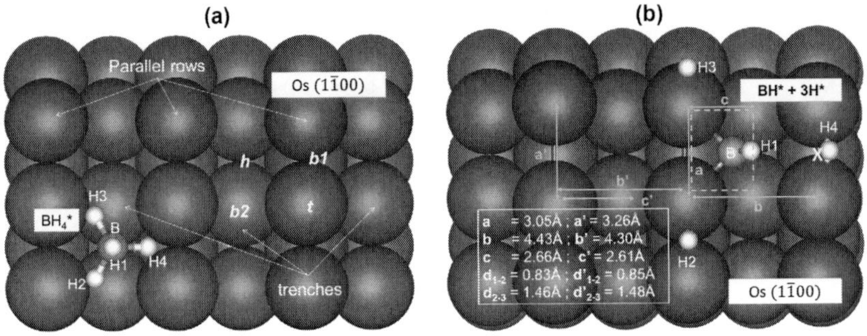

Fig. 6 BH_4^- adsorption structures on (1$\bar{1}$00) facet of Os. Initial configuration considered and the most favored adsorption structure are shown in (a) and (b), respectively. In (a), the adsorption site for the center of the molecule (B) is indicated as *b*1 or *b*2 (for bridges), *t* (for top) and *h* (for hollow site). The atomic distances in the surface as a result of adsorption are given in (b). d_{1-2} and d'_{1-2} are the first interlayer distances (vertical) corresponding to a and a′ lateral distances. The difference between these lateral distances is whether they are far or near the adsorption site. d_{2-3} and d'_{2-3} are the second interlayer distances. (Reproduced from ref. 14 with permission from The Royal Society of Chemistry.)

Table 3 Adsorption structures and energies of borohydride on Au_3M surface alloys and on Au.[15] (2×2) surface unit cell is used.

$Au_3M(111)$	Adsorption structure	Adsorption energy (eV)
Au_3Cr	BH_4*	−2.31
Au_3Mn	BH_4*	−2.19
Au_3Fe	BH_4*	−2.34
Au_3Co	BH_4*	−2.41
Au_3Ni	BH_4*	−2.16
Au	BH_4*	−1.83

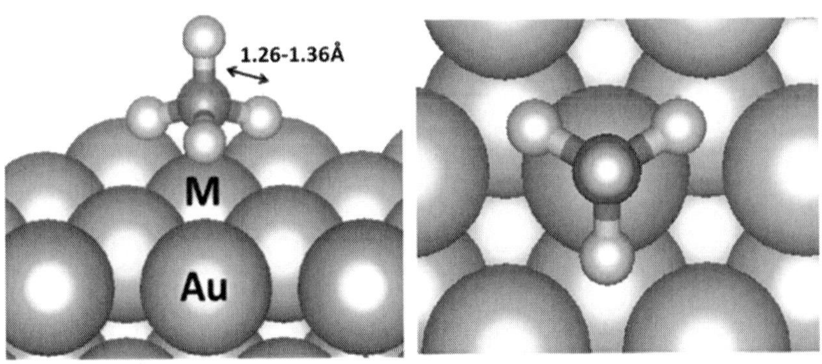

Fig. 7 BH_4^- adsorption structure on (111) facet of Au_3M metal alloy surfaces: (left) side view and (b) top view. Au and M (= Cr, Mn, Fe, Co, Ni) atoms are labelled. For the molecule, the center atom is boron and the smaller ones are hydrogen. (Reproduced from ref. 15 with permission from The Royal Society of Chemistry.)

It can be noted that for Au, alloying does not change the adsorption structure of borohydride. It remains molecular with a slight elongation of three B–H bonds. The molecule prefers the M site for the adsorption and the adsorption energy increases in the following

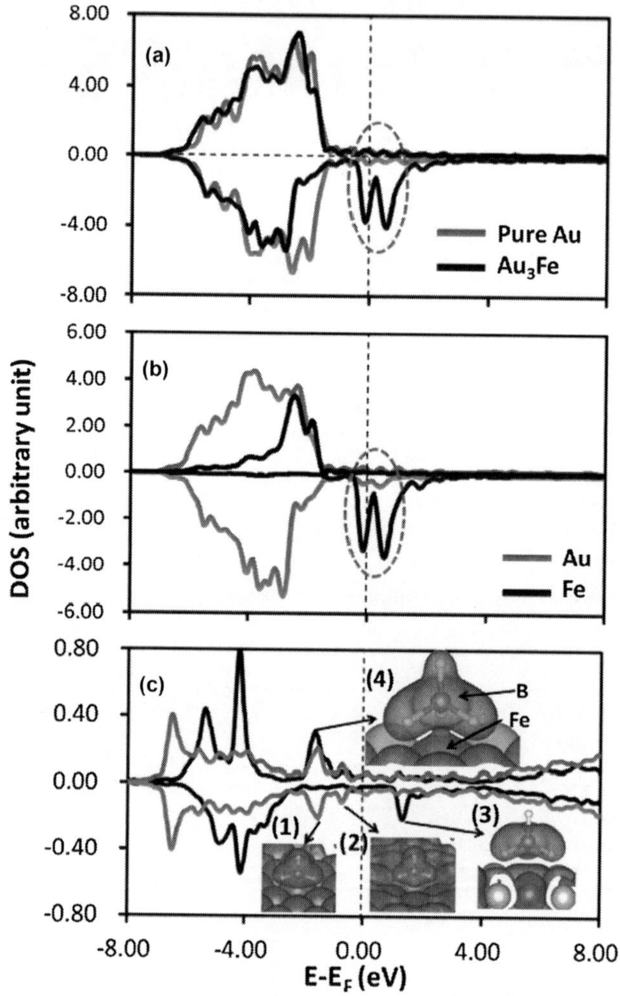

Fig. 8 Density of states (DOS) on (a) pure Au and Au_3Fe; (b) local DOS on Au and Fe atoms and (c) partial charge densities corresponding to the DOS peaks near the Fermi level (E_F). (Reproduced from ref. 15 with permission from The Royal Society of Chemistry.)

order: Au < Au_3Ni < Au_3Mn < Au_3Cr < Au_3Fe < Au_3Co. The mechanism of the alloying effect is explained using the density of states (DOS) of the metal and the molecule as shown in Fig. 8. Without the M atom, in this case Fe, the DOS of the Au is fully occupied (Fig. 8(a)). Upon alloying with Fe, a spin-down DOS forms at and above the Fermi level (E_F). The local DOS shown in Fig. 8(b) indicates that these states originate from the Fe, explaining the preference of the borohydride on the M site. Figure 8(c) further clarifies the bonding interactions *via* partial charge density plots corresponding to DOS peaks near the E_F.

B Same alloying metal but different compositions

Other metal alloys have also been considered for instance, Pd–Ir.[16] In this case, the alloying component is not a 3d transition metal but a 5d. Pd is a

Table 4 Adsorption structures and energies of borohydride on (111) surface of Pd and Pd–Ir alloys. The energies are free energies of adsorption that takes into account the electrochemical potential difference (−0.5V *vs.* SHE) and the solvated borohydride ion.[16] The adsorption structure for Pd(111) is obtained from ref. 10.

Composition	Adsorption structure	Free energy (eV)
$Pd_2Ir_1(111)$	BH_4*	−1.12
$Pd_2Ir_2(111)$	BH_4*	−0.89
Pd(111)	BH* + 3H*[10]	−1.77

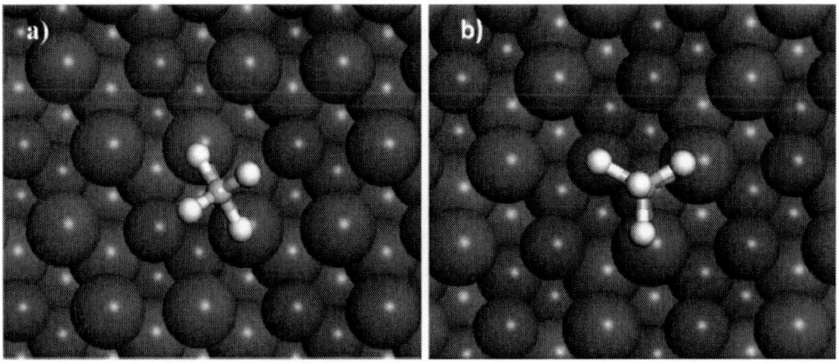

Fig. 9 BH_4^- adsorption structure on (a) $Pd_2Ir_1(111)$ and (b) $Pd_2Ir_2(111)$ alloys. Medium-sized atoms are Ir atoms and the small ones are boron and hydrogen atoms. (Reprinted from I. Merino-Jimenez, M. J. Janik, P. de Leon and F. C. Walsh, Pd–Ir alloy as an anode material for borohydride oxidation, *J. Power Sources*, **269**, 498–508, Copyright 2014, with permission from Elsevier.)

4d metal. Two types of composition are studied: $Pd_2Ir_1(111)$ and $Pd_2Ir_2(111)$. The adsorption structure and energies are shown in Table 4. Figure 9 depicts the preferred adsorption structures.

Alloying the Pd with Ir, increases the free energy of the adsorption (*i.e.* more positive shift in the free energy). Such energy also increases upon increase in the composition of the alloying metal. The structures on the Pd–Ir alloys are shown in Fig. 9. Although both structures are molecular, the orientation of the B–H bonds and the position of the center B atom are different. For $Pd_2Ir_1(111)$, the boron is at the bridge site with two H atoms spanning towards Pd top sites and the other two H atoms towards the Ir top sites. Thus, the molecule conforms with the surface *via* two H atoms. In the $Pd_2Ir_2(111)$ (greater Ir composition), the molecule conforms with the surface *via* three H atoms spanning towards two Pd top sites and one Ir top site. This is the typical conformation of the borohydride on some 4d and 5d metals.

4 Pure metal surfaces – solvated

A Effect of water in Au(111) and Pt(111)

To consider the effect of the water as the solvent, the Gibbs free energy, ΔG_{ads}, for the *initial* oxidative adsorption of the BH_4^- based on eqn (4) is derived to draw the thermodynamics of the reaction. There are two ways by which this can be done, the vacuum slab method[17,18] and the double-reference method.[19,20] Mainly, the difference between the two methods is the inclusion of water molecules in the model (or supercell). In the former, as the name suggests, there are no water molecules that interact with the borohydride within the supercell and the effect of solvent is integrated *via* the free energies of H_2O adsorption (H_2O^*) and the aqueous H_2O_{aq} (please see eqn (4)) alone, which are computed in separate supercells. The latter on the otherhand, includes water molecules within the supercell and so the effect of solvent is integrated *via* the usual free energies of H_2O adsorption (H_2O^*) and the aqueous H_2O_{aq} as in the above method *plus* the borohydride–water interactions. The reader is suggested to refer to the given sources above for details. Figure 10 best captures the solvent effects using the double reference method on Au and Pt.[12]

The re-orientation of the adsorbed borohydride occurs on Au. Two B–H bonds, instead of one, point away from the metal surface. There is negligible change in the orientation of the borohydride on Pt. It can be noted that for molecularly adsorbed structure (BH_4^*), the effect of the interaction with water molecules in the orientation of borohydride can be

Fig. 10 BH_4^- adsorption structure on Au(111) (top panels) and on Pt(111) (bottom panels) with and without water molecules in the supercell. (Reprinted from G. Rostamikia and M. J. Janik, First principles mechanistic study of borohydride oxidation over the Pt(111) surface, *Electrochim. Acta*, **55**, 1175–1183, Copyright 2010, with permission from Elsevier.)

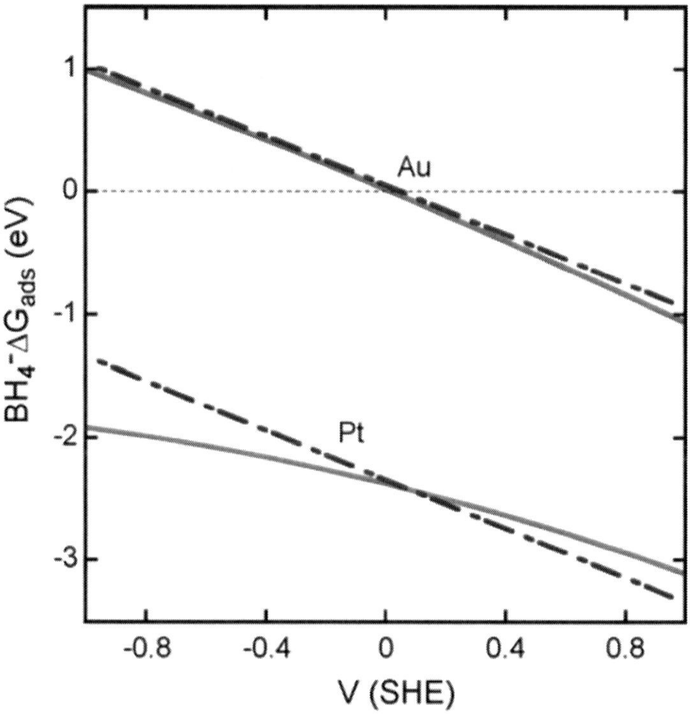

Fig. 11 ΔG_{ads} as a function of electrode potential, V (SHE) on Au(111) and on Pt(111) obtained using vacuum slab method (dashed line) and double reference method (solid line). (Reprinted from G. Rostamikia and M. J. Janik, First principles mechanistic study of borohydride oxidation over the Pt(111) surface, *Electrochim. Acta*, **55**, 1175–1183, Copyright 2010, with permission from Elsevier.)

observed. In terms of energetics (ΔG_{ads}), Fig. 11 shows that for Pt, the ΔG_{ads} as a function of electrode potential, V (SHE) is steeper when the vacuum slab method is used (dashed line) than when the double reference method is used (solid line). This is attributed to the dipole moment or polarizability effects.[21] Basically, there is not much difference between the two methods for Au. Now, when the energies of the adsorption of borohydride in unsolvated (Table 1 of Section 1) and the solvated (Fig. 11) models of Au and Pt are compared, it can be noted that the trends are actually well-preserved.

B Water on different facets of metal surface

As discussed above the adsorption of water (H_2O^*) on the metal surface can bring about shifts in ΔG_{ads}. The study of the differences in the H_2O^* structure on various facets of metal surfaces is worthwhile. Due to its interesting interaction of borohydride with Os as discussed previously, the H_2O^* structure is obtained on this metal. The (0001), $(10\bar{1}0)$ and $(1\bar{1}00)$ surfaces of Os are used. Extensive adsorption configurations are evaluated, that is using the flat, up and down orientation of the molecule as well as its translation and rotation on the surface (please see Fig. 12). For instance, on (0001), for each water orientation (up and down), the molecule is rotated in-plane such that the O–H bond "parallel" to the

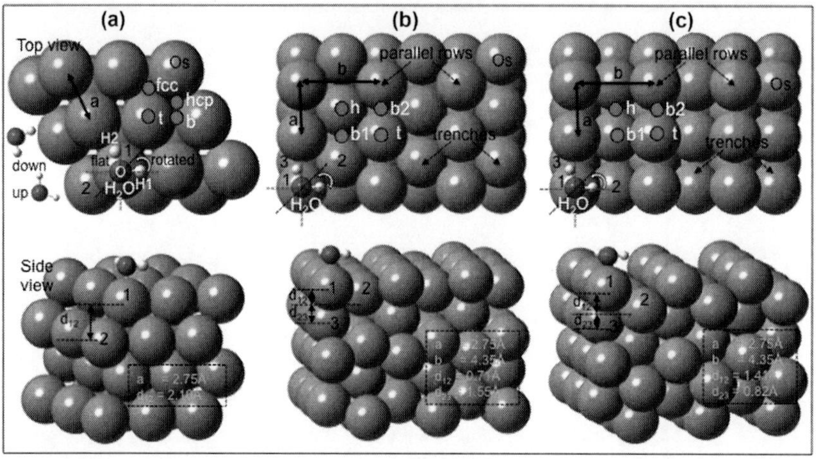

Fig. 12 (a), (b) and (c) show the top and side views of the (0001), (10$\bar{1}$0) and (1$\bar{1}$00) slabs, respectively. The lateral distances are shown on the top panels and the interlayer distances are depicted in the bottom panels. The water orientations considered (flat, down and up) are shown in Fig. 12(a) top panel. The potential energy scanning to obtain the most favorable configuration on the surfaces are depicted *via* planar rotation (top panels) per orientation of the molecule, and translation *via* shifts to several high symmetric sites: for (0001) → hollow sites (fcc/hcp), bridge (*b*) and top (*t*); for (10$\bar{1}$0) and (1$\bar{1}$00) → hollow site (*h*), top (*t*) and bridge sites (*b*1 and *b*2). The molecule and the Os atoms are labeled accordingly.

surface can point towards different directions. Also, the O atom position is shifted to other high symmetric sites (*i.e.* translation). Thus, each adsorption configuration can be identified by X1–X2–X3, where X1 indicates the molecule orientation (flat (f), one H down (d) or one H up (u)), X2 is the position of the oxygen atom (hollow (h), top (t), bridge (b)) and X3 is the direction where the H expands as a result of planar rotation (towards the bridge (b) or hollow site (h)). Hence, *f-t-b* means a flat orientation with O atom on top site and the O–H1 bond (see Fig. 12(a) for the H1 atom) pointing towards the bridge. A total of 24 adsorption configurations on (0001) facet are considered. The adsorption energy per configuration is calculated based on the following:

$$E_{\mathrm{ads}} = E_{\mathrm{T}} - (E_{\mathrm{g}} + E_{\mathrm{s}}) \tag{5}$$

where E_{T}, E_{g} and E_{s} are the total energy of adsorbed system, isolated molecule and the metal surface (or slab), respectively. Table 5 lists the adsorption energies on (0001).

The most stable H_2O^* structure originated from the *f-t-h* initial configuration with an adsorption energy of −0.484 eV. Figure 13(a) depicts the final adsorption structure. The oxygen atom is on the top site and the O–H bonds are almost parallel to the surface. Both O–H bonds have the same lengths and the O–H–O angle is ~105.495°. The O–Os distance is 2.302 Å. A small change in the interlayer distance, d_{12} of ~0.029 Å of the surface upon H_2O adsorption can be noted. Water monomer adsorption on metal surfaces has been studied in other metals previously.[22–25] Although a direct comparison cannot be made due to the different metals used and the size of the supercell, in general, it can be noted that

Table 5 Adsorption energies of water monomer on Os(0001) at various configurations. The configurations are labeled X1-X2-X3, where X1 indicates the molecule orientation (flat (*f*), one H down (*d*) or one H up (*u*)), X2 is the position of the oxygen atom as shown in Fig. 12(a) and X3 is the direction where the H atom points as a result of planar rotation (towards the bridge (*b*) or hollow site (*h*)). Bold form indicates the most stable adsorption. Initial configurations that resulted to the same most stable adsorption energies do not necessarily indicate the same final states. Due to the rounding off of the figures, the difference can not be shown. The very minimal difference in the structure comes from the very small angle of in-plane rotation of the molecule.

Configuration	Adsorption energy (eV)	Configuration	Adsorption energy, E_{ads} (eV)
f-t-b	−0.484	*d-b-h*	−0.481
f-b-b	−0.481	*d-fcc-h*	−0.484
f-fcc-b	−0.480	*d-hcp-h*	−0.477
f-hcp-b	−0.484	*u-t-b*	−0.477
f-t-h	**−0.484**	*u-b-b*	−0.467
f-b-h	−0.095	*u-fcc-b*	−0.483
f-fcc-h	−0.471	*u-hcp-b*	−0.480
f-hcp-h	−0.483	*u-t-h*	−0.481
d-t-b	−0.053	*u-b-h*	−0.060
d-b-b	−0.073	*u-fcc-h*	−0.479
d-fcc-b	−0.482	*u-hcp-h*	−0.477
d-hcp-b	−0.481		
d-t-h	−0.056		

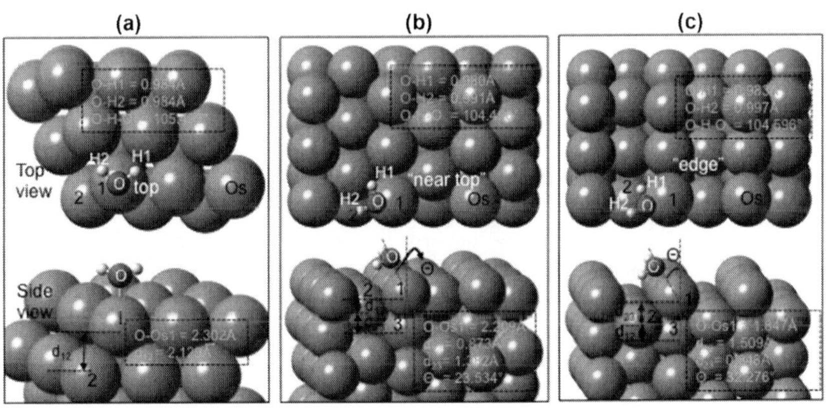

Fig. 13 (a), (b) and (c) show the top and side views of the most stable adsorption state of H_2O on (0001), (10$\bar{1}$0) and (1$\bar{1}$00), respectively. Relevant distances and angles for the molecule are shown in the top panels while those for the slab are given in the bottom panels. The molecule and the Os atoms are labeled accordingly.

the O–Os, O–H distances and the H–O–H angle are comparable to those reported in the literature for the fcc (111) metals. For Au(111), the literature reports a slightly weaker binding of water on the surface.[22]

Next, for the (10$\bar{1}$0), Fig. 12(b) shows the symmetric sites. Here, there are two bridge sites (*b*1, *b*2) corresponding to the *a* and *b* lateral distances. The same three orientations of H_2O, in-plane rotations, and configuration notations in (0001) are also employed in (10$\bar{1}$0). Table 6 gives the adsorption energies. The most stable site originated from the *d-b1-h* initial adsorption configuration. The adsorption energy

Table 6 Adsorption energies of water monomer on Os(10$\bar{1}$0) at various configurations. The configurations are labeled X1–X2–X3, where X1 indicates the molecule orientation (flat (*f*), one H down (*d*) or one H up (*u*)), X2 is the position of the oxygen atom as shown in Fig. 12(b) and X3 is the direction where the H expands as a result of planar rotation (towards the bridge (*b*) or hollow site (*h*)). Bold form indicates the most stable adsorption.

Configuration	Adsorption energy (eV)	Configuration	Adsorption energy (eV)
f-t-b	−0.608	***d-b1-h***	**−0.660**
f-b1-b	−0.592	*d-b2-h*	−0.001
f-b2-b	−0.592	*d-h-h*	−0.646
f-h-b	−0.199	*u-t-b*	−0.454
f-t-h	−0.598	*u-b1-b*	−0.491
f-b1-h	−0.149	*u-b2-b*	−0.321
f-b2-h	−0.599	*u-h-b*	−0.357
f-h-h	−0.199	*u-t-h*	−0.596
d-t-b	−0.159	*u-b1-h*	−0.648
d-b1-b	−0.201	*u-b2-h*	−0.629
d-b2-b	−0.214	*u-h-h*	−0.247
d-h-b	−0.229		
d-t-h	−0.578		

Table 7 Adsorption structures and energies of borohydride on different facets of Os-214.[14] The letters indicate the binding site on a single facet and are depicted in Fig. 17.

Facet		Adsorption structure	Adsorption energy (eV)
(0001)	(a)	BH_4*	−5.05
	(b)	BH_4*	−5.33
	(c)	BH_4*	−5.15
(10$\bar{1}$0)	(d)	BH_2* + 2H*	−5.71
	(e)	BH_4*	−5.41
	(f)	BH_3* + H*	−5.66
(1$\bar{1}$00)	(g)	BH* + 3H*	−5.39
	(h)	BH_2* + 2H*	−5.61

is −0.660 eV. Figure 13(b) depicts the final adsorption state. The molecule is in bent configuration with the O–Os axis forming a 23.53° angle with the vertical axis (see Fig. 13(b), bottom panel). One of the O–H bonds (O–H2) points towards the *b*1 site and is longer than the O–H1 bond. This is because the H2 atom is lower (or closer to the surface) than the H1. This bent structure is due to the "near top" adsorption, in contrast to (0001) where the molecule directly sits on the top site. The O–Os distance in (10$\bar{1}$0) is shorter by 0.093 Å as compared to that of the (0001). We note that this bending of water molecule towards the trench resulted in a stronger adsorption.

Similar trend can be observed for (1$\bar{1}$00). Table 7 lists the adsorption energies and the final state of the molecule is shown in Fig. 13(c). This stable state originated from the *d-b1-h* configuration with an adsorption energy of −0.944 eV. The water molecule is more bent as compared to the previous two cases. The deviation of the O–Os axis from the vertical axis is 32.27° (see Fig. 13(c), bottom panel) and in the same way as in (10$\bar{1}$0), one O–H bond (*i.e.* O–H2) is lower (or closer to the surface) than the other (*i.e.* O–H1). Because of the larger deviation of the molecule from the top site, the O–Os vertical distance is much shorter (1.847 Å) and the O–H2 bond is longer (0.997 Å), all leading to a much stronger adsorption energy. This configuration of the H_2O molecule on $(1\bar{1}00)$ is termed "edge site" adsorption. We can recall that the trench atom in $(1\bar{1}00)$ is directly under the *b*1 site, creating a stronger interaction with the H2 atom of the H_2O molecule than that of the $(10\bar{1}0)$. Based on the above, we clearly see the role of the trench atom and its contribution to the adsorption of water monomer on Os surfaces. The interaction of H_2O with other metals having the same trenches and rows can be predicted from this study on Os.

Next, what is the effect of the water interaction with Os surface on the initial oxidative adsorption of borohydride? Using eqn (4) and the vacuum slab model, the ΔG_{ads} as a function of electrode potential, U (SHE) for the different facets is shown in Fig. 14. First, it can be noted that the initial borohydride oxidation is favorable across potentials of interest for all facets, that is, the lines are below 0. The favorability of the initial oxidative adsorption of borohydride on Os surfaces increases in the following order: $(0001) < (10\bar{1}0) < (1\bar{1}00)$. This suggests that the overpotentials for BH_4^- oxidation on the more open surfaces are lower. However, as a result of BH_4^- interaction with such surfaces (recall Figs. 5 and 6), larger hydrogen atom coverage can be expected, which in turn can also promote H_2 evolution. The investigation of the competing reactions is worthwhile. Lastly, in Fig. 14, comparison of the ΔG_{ads} *vs.* U_{SHE} with H_2O (thick lines) and without H_2O effects (thin lines), the absence of adsorbed H_2O shifts the Gibbs free energy of the oxidative BH_4^- adsorption downward. Thus, without the solvent effects, the adsorption energies obtained for the borohydride on metals are generally higher.

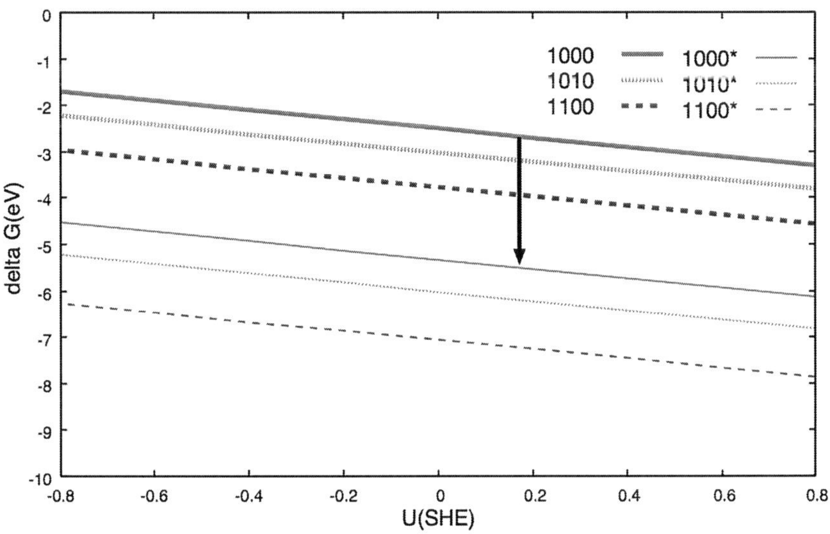

Fig. 14 The ΔG_{ads} with respect to the electrode potential U_{SHE} for (0001), $(10\bar{1}0)$ and $(1\bar{1}00)$ including H_2O effects (thick lines). Thinner lines give the ΔG_{ads} *vs.* U_{SHE} without the H_2O effects and are labelled as (0001)*, $(10\bar{1}0)$* and $(1\bar{1}00)$*.

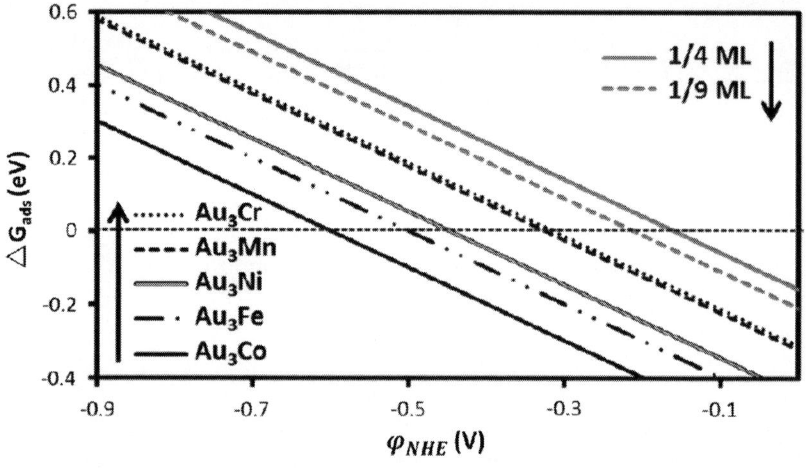

Fig. 15 The ΔG_{ads} with respect to the electrode potential φ_{SHE} for Au_3M alloys, where M = Cr, Mn, Fe, Co and Ni. Red lines (or first two lines from the right) are for Au(111) for different borohydride coverage. Arrows indicate shifts towards more positive potentials across metals. The opposite is indicated for the coverages. (Reprinted from R. L. Arevalo, M. C. S. Escaño and H. Kasai, Mechanistic Insight into the Au-3d Metal Alloy-Catalyzed Borohydride Electro-Oxidation: From Electronic Properties to Thermodynamics, *ACS Catal.*, 2013, **3**(12), 3031–3040. Copyright 2013, with permission from American Chemical Society.)

5 Metal alloy surfaces – solvated

The effect of water on the borohydride electro-oxidation has been investigated on $Au_3M(111)$ alloys, where M = Cr, Mn, Fe, Co and Ni.[26,27] Using the vacuum slab model, the ΔG_{ads} as a function of electrode potential φ(NHE) is shown in Fig. 15. It is observed that the initial oxidative adsorption of borohydride in $Au_3M(111)$ is favourable at a potential range of −0.60 to −0.32 V *vs.* NHE. This is 0.17 to 0.45 V lower in potential compared to Au(111). The shift towards more positive potentials follows this order: $Au_3Co < Au_3Fe < Au_3Ni < Au_3Mn < Au_3Cr < Au$. For pure Au(111), increase in the coverage of borohydride leads to more positive potential shifts. Comparing the order of potential shifts for the Au_3M with that of the order of adsorption energies on the same systems without solvent effects as shown in Table 3 shows that the trends are not necessarily preserved especially for Au_3Ni, Au_3Mn and Au_3Cr. This may be due to the interaction of water with M.

6 Metal nanoparticles

A Size-dependence

Nanoparticles are different from the surfaces because they contain low coordinated sites such as edges and vertices. The dependence of borohydride adsorption on the size of metal nanoparticle has been studied using Os.[14] In this study, the size ranges from 1.0 nm to 2.0 nm (see Fig. 16). The borohydride prefers the *edge-vertex sites combination* in the adsorption on all the nanoparticle sizes. When the size increases, the adsorption also increases. This is basically due to the increasing Os–Os lateral distance as the nanoparticle size is increased. Larger interatomic

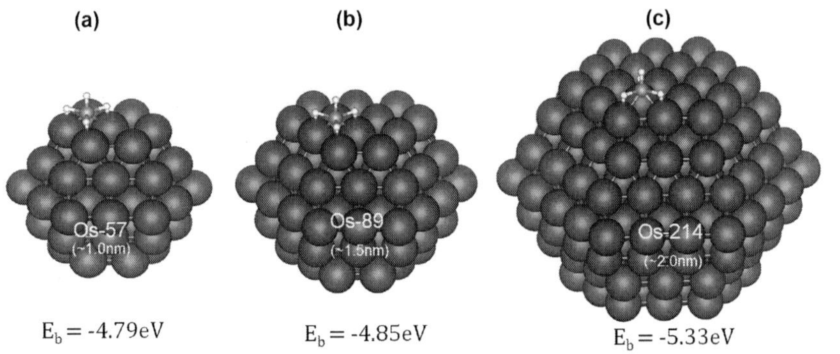

Fig. 16 BH_4^- adsorption structures and energies on (0001) facet of (1) Os-57; (b) Os-89 and (c) Os-214 nanoparticles, which corresponds to ~1 nm, ~1.5 nm and ~2.0 nm nanoparticle sizes, respectively. (Reproduced from ref. 14 with permission from The Royal Society of Chemistry.)

distances suggest less overlap of the d-states of the metal and therefore a higher entire d-band and more enhanced reactivity.

B Facet-dependence

A single nanoparticle can contain several crystal planes or facets. For instance, in the Os-214 in Fig. 16, the (0001), $(10\bar{1}0)$ and $(1\bar{1}00)$ facets can be found. The dependence of adsorption of borohydride on the different facets of Os-214 is also reported in ref. 14. Please note that the facets in the metal surface and in the nanoparticles are quite different, in that the latter contains low coordinated sites. Thus, the configurations of borohydride can vary even on a single facet. For instance, in Fig. 17, three binding sites for (0001) and $(10\bar{1}0)$ and two binding sites for $(1\bar{1}00)$ are identified. The adsorption structure of borohydride is depicted in Fig. 17 and the adsorption energies are given in Table 7. For (0001), all the structures are molecular, BH_4* and the strongest adsorption can be found at site (b), which is a combination of edge-vertex Os atoms. In other metal nanoparticles, such reactivity of "combined" low-coordinated atoms can be expected. For the $(10\bar{1}0)$, the three different binding sites also produce three types of borohydride structures: (d) $BH_2* + 2H*$, (e) BH_4* and (f) $BH_3* + H*$. The strongest adsorption is in site (d), where the boron situates at the bridge site formed by two edge-atoms. For the $(1\bar{1}00)$ facet, the two binding sites also produces two types of adsorption structures, namely: (g) $BH* + 3H*$ and (h) $BH_2* + 2H*$. In this case there is no molecular adsorption. The strongest adsorption is noted in (h), where the B atoms sits in the hollow site. In the Os nanoparticle, the strongest adsorption for all facets is in $(10\bar{1}0)$ which is not the same as that of the Os metal surface. First, it can be noted that the adsorbate structure on the $(1\bar{1}00)$ facet of Os nanoparticle is $BH_2* + 2H*$, which is different from the counterpart facet in the metal surface, $BH* + 3H*$. The breaking of one H from B atom is restricted by the larger distance between the boron and the next neighboring bridge site where H can settle most favorably (marked by X in Fig. 17). In fact, on the nanoparticle, two hydrogen atoms sit instead

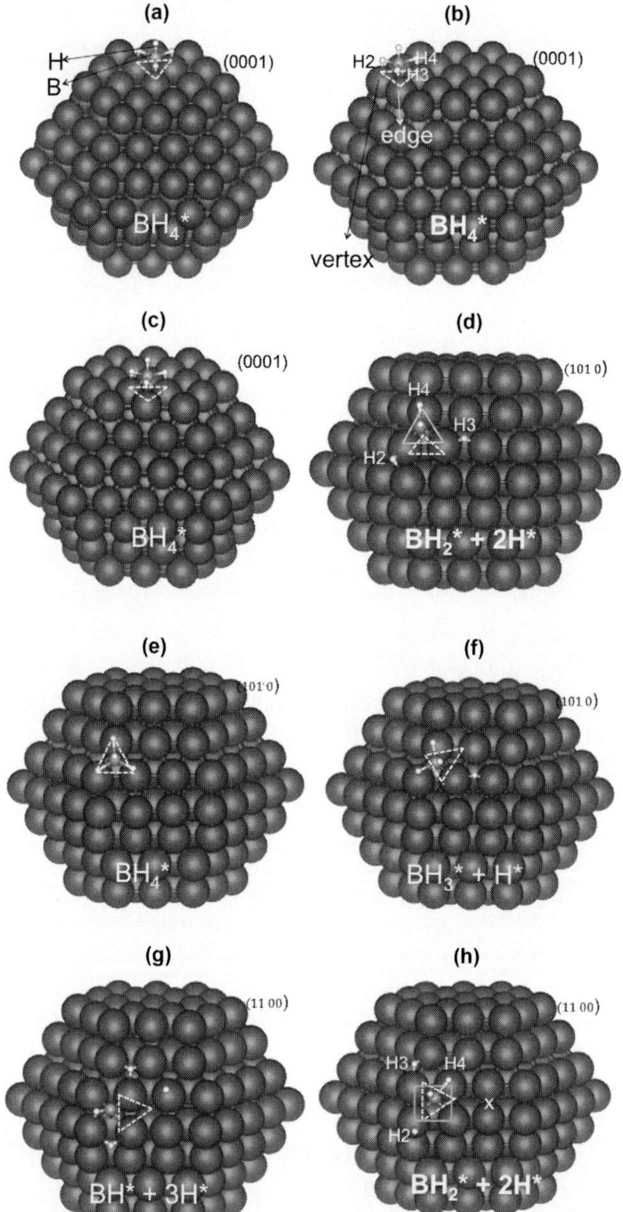

Fig. 17 BH_4^- adsorption structures on different facets of Os-214: (a)–(c) (0001), (d)–(f) (10$\bar{1}$0) and (g)–(h) (1$\bar{1}$00). The final structures of the borohydride is depicted by the smaller atoms. The initial configuration is indicated by a triangle where the vertices represent the triangular comformation of hydrogen with the plane and the center describes the position of B atom, which are: (a) center; (b) edge-vertex site; (c) edge site; (d) bridge site 1; (e) hollow site; (f) bridge site 2; (g) bridge and (h) hollow site. In (d) and (h), the triangle and square in solid line show the conformation of B with the Os-214. (Reproduced from ref. 14 with permission from The Royal Society of Chemistry.)

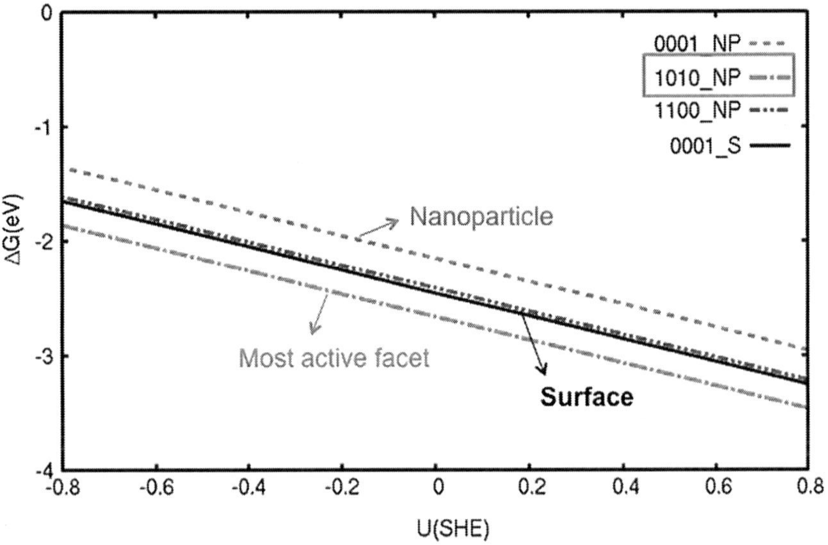

Fig. 18 The ΔG_{ads} with respect to the electrode potential U_{SHE} for (0001), (10$\bar{1}$0) and (1$\bar{1}$00) facets of Os-214 nanoparticle and (0001) facet of Os surface. The (0001) facet of the nanoparticle and the (0001) facet of the Os surface are labelled accordingly.

on the top sites resultingin the less stable final $BH_2^* + 2H^*$ structure. Thus, the accessibility of the favorable hydrogen adsorption site contributes to the difference in the geometry of the adsorbate between the nanoparticle and the surface.

C Facet-dependence-solvated

Using vacuum slab method, the initial oxidative adsorption of borohydride is evaluated for all facets of the Os nanoparticle. Figure 18 shows the comparison of the ΔG_{ads} with respect to the electrode potential U_{SHE} among (0001), (10$\bar{1}$0) and (1$\bar{1}$00) facets of Os-214 nanoparticle and (0001) facet of Os surface.

It can be observed that in the nanoparticle, the shifts towards more positive potentials are in this order: (10$\bar{1}$0) < (1$\bar{1}$00) < (0001), making the (10$\bar{1}$0) most active. To compare the (0001) surface in the Os surface and the same facet in the Os nanoparticle, the latter seemed more active. This is depicted in Fig. 18. The surface almost has the same ΔG_{ads} *vs.* U relationship as that of the (1$\bar{1}$00) facet of the nanoparticle. Moreover, it can be noted that the initial borohydride oxidation is favorable across potentials of interest for all facets in the Os nanoparticles just as in the case of the Os surface.

7 Summary

Theoretically, the electro-oxidation of borohydride on metal anode catalysts is investigated using density functional theory. Most of the study focuses on the initial oxidative adsorption of borohydride on many catalysts structures and compositions. The adsorption structures and energies predict catalysts' selectivity. The agreement of these DFT studies with experiment have been discussed previously,[8] however, comparison of these adsorption properties across many metal catalysts as a function of facets, sizes, alloy composition and solvent effects are still lacking. This review consolidates the studies on such properties and presents comparison among many metal structures and compositions with and without solvent effects. It has been noted that the adsorption energies of borohydride increases as one goes from left to right within the same period of the periodic table, that is, if the same structure (facet) of the metal surface is considered. However, the adsorption structures do not correlate well with adsorption energies. The strongest adsorption in 5d metals is found in Os however, the adsorption structure is still molecular (no dissociated H atoms). 4d transition metal presents the same scenario. For the 3d metals, the strongest adsorption is observed on Cr, with only one H atom dissociated. On the other hand, when different facets are compared within the same metal, the typical correlation between the adsorption energy and the adsorption structure can be observed. For instance, the more open the facet becomes (*i.e.* $(0001)<(10\bar{1}0)<(1\bar{1}00)$), the stronger is the adsorption and the more B–H bonds are dissociated leaving more H atoms on the surface. When the pure 5d metal is alloyed with 3d metal M (= Cr, Mn, Fe, Co and Ni), the adsorption energy increases in the following order: $Au<Au_3Ni<Au_3Mn<Au_3Cr<Au_3Fe<Au_3Co$. The borohydride remains molecular and prefers to adsorb on the M site. When the composition is changed, for instance, in Pd–Ir case, the greater is the composition of the alloying metal, Ir, the less is the adsorption energy. It can be expected that if this alloying metal is a 3d transition metal, M, the opposite trend can be noted.

The effect of the solvent has also been studied, especially on Au, Pt and Os. For the first two, the double reference method is used. It was observed that the water changes the orientation of borohydride on Au but not on Pt. The free energies of the electro-oxidation of borohydride, however is more affected in Pt than in Au, which is attributed to polarizability effect. For the alloyed Au, the inclusion of solvent effects *via* vacuum slab method indicates shifts to more positive potentials in the following order: $Au<Au_3Ni<Au_3Mn<Au_3Cr<Au_3Fe<Au_3Co$. This trend differs slightly from that of the catalysts without the solvent effects. For Os, the effect of surface facets has been studied. It was found that water binds on the surface *via* the edge site of $(1\bar{1}00)$, however, the structure remains molecular for all the facets studied ((0001), $(10\bar{1}0)$ and $(1\bar{1}00)$). This suggests that interaction adsorbed borohydride with water molecules is highly likely in borohydride electro-oxidation. On the nanoparticle, when the size is increased, its interaction with the borohydride also increases. Since, a nanoparticle also contains several facets, the interaction of borohydride with the planes of Os nanoparticle is also studied. It was noted that the most active facet is $(10\bar{1}0)$, which is different from that of Os metal surface interaction. Nanoparticle presents more constricted adsorption sites than on the surface. The accessibility

of hydrogen adsorption site is less favourable in nanoparticle and this contributes to the difference between the nanoparticle and the surface.

Acknowledgements

M. C. S. Escaño extends gratitude to Ministry of Education, Culture, Sports, Science and Technology, Japan (MEXT)/Japan Science and Technology Agency (JST)/Tenure Track Program for Innovative Research and Japan Society for Promotion of Science (JSPS) Grant-in-aid for Young Scientist B – Grant Number 15K21028 for research funds. The calculations are done using the ACCMS, Kyoto University – access funded by CII, University of Fukui and the High-Performance Computing Cluster Fukui (HPCCF) of Escaño Research Group, University of Fukui.

References

1 V. Lam and E. Gyenge, *J. Electrochem. Soc.*, 2008, **155**, B1155.
2 J. Ma, N. A. Choudhury and Y. Sahai, *Renewable Sustainanble Energy Rev.*, 2010, **14**, 183.
3 M. Jimenez, C. P. de León, A. A. Shah and F. C. Walsh, *J. Power Sources*, 2012, **219**, 339.
4 V. W. S. Lam, D. C. W. Kannangara, A. Alfantazi and E. L. Gyenge, *J. Power Sources*, 2012, **212**, 57.
5 A. Aziznia, C. Oloman and E. L. Gyenge, *ChemSusChem*, 2013, **6**, 847.
6 A. Ignaszak and E. L. Gyenge, *Electrochim. Acta*, 2013, **95**, 268.
7 C. Grimmer, R. Zacharias, M. Grandi, B. Cermenek, A. Schenk, S. Weinberger, F. A. Mautner, B. Bitschnau and V. Hacker, *J. Phys. Chem. C*, 2015, **119**, 23839.
8 M. C. S. Escaño, R. L. Arevalo, E. L. Gyenge and H. Kasai, *J. Phys.: Condens. Matter*, 2014, **26**, 353001.
9 M. R. Hartman, J. R. Rush, T. J. Udovic, R. C. Brownman and S. J. Hwang, *J. Solid State Chem.*, 2007, **180**, 1298.
10 M. C. S. Escaño, R. L. Arevalo, E. L. Gyenge and H. Kasai, *J. Phys. Chem. C*, 2011, **115**, 19883.
11 R. L. Arevalo, M. C. S. Escaño, E. L. Gyenge and H. Kasai, *Surf. Sci.*, 2012, **606**, 1954.
12 G. Rostamikia and M. J. Janik, *Electrochim. Acta*, 2010, **55**, 1175.
13 G. Rostamikia and M. J. Janik, *J. Electrochem. Soc.*, 2009, **156**, B86.
14 M. C. S. Escaño, R. L. Arevalo, E. L. Gyenge and H. Kasai, *Catal. Sci. Technol.*, 2014, **4**, 1301.
15 R. L. Arevalo, M. C. S. Escaño, A. Y. Wang and H. Kasai, *Dalton Trans.*, 2013, **42**, 770.
16 I. M. Jimenez, M. J. Janik, C. P. De Leon and F. C. Walsh, *J. Power Sources*, 2014, **269**, 498.
17 J. K. Nørskov, J. Rossmeisl, A. Logadottir and L. Lindqvist, *J. Phys. Chem. B*, 2004, **108**, 17886.
18 M. C. S. Escaño, *Nano Res.*, 2015, **8**, 1689.
19 C. D. Taylor, S. A. Wasileski, J. S. Filhol and M. Neurock, *Phys. Rev. B*, 2006, **73**, 165402.
20 J. S. Filhol and M. Neurock, *Angew. Chem., Int. Ed.*, 2006, **45**, 402.
21 G. Henkelman and H. Jónsson, *J. Chem. Phys.*, 2000, **113**, 9978.

22 S. Meng, E. G. Wang and S. Gao, *Phys. Rev. B*, 2004, **69**, 1954041.
23 A. Michaelides, V. A. Ranea, P. L. de Andres and D. A. King, *Phys. Rev. Lett.*, 2003, **90**, 2161021.
24 A. Michaelides, A. Alavi and D. A. King, *J. Am. Chem. Soc.*, 2003, **125**, 2746.
25 S. Schnur and A. Gross, *New J. Phys.*, 2009, **11**, 1250031.
26 R. L. Arevalo, M. C. S. Escaño and H. Kasai, *ACS Catal.*, 2013, **3**, 3031.
27 R. L. Arevalo, M. C. S. Escaño and H. Kasai, *J. Phys. Chem. C*, 2013, **117**, 3818.

Recent progress in the development of anion exchange membranes for electrochemical devices

Zhongyang Wang,[a,†] Lihui Wang,[b,†] Javier Parrondo[a] and Vijay Ramani*[a]

DOI: 10.1039/9781782622727-00023

1 Types of polymer backbones

We classify the polymer backbones employed in the synthesis of AEMs into five categories: (1) aromatic backbones including polysulfone (PSF),[1] poly(ether ketone),[2] poly(ether imide),[3] poly(ether oxadiazole),[4] poly(phenylene oxide) (PPO),[5–7] polyphenylene,[8,9] polybenzimidazoles[10,11] polystyrene,[12–15] poly(vinylbenzyl chloride);[16,17] (2) aliphatic backbones comprising polynorbornene,[18] polypropylene,[19] unsaturated polyethylene,[20] polyepichlorohydrin (PECH),[21] polyphosphazene,[22] and polyvinyl alcohol (PVA);[23,24] (3) block copolymers such as polystyrene-*b*-poly(ethylene-*co*-butylene)-*b*-polystyrene (SEBS)[25,26] and PPO block copolymers;[27] (4) partially perfluorinated backbones, such as partially fluorinated copoly(arylene ether);[28] and (5) fully perfluorinated backbones.[24,29] The structure of all these polymer backbones can be seen in Table 1.

The most commonly used polymer backbone for the synthesis of AEMs for fuel cells, water electrolyzers, and flow batteries are **aromatic backbones** (*e.g.*, PSF, Cardo-PEEK, PPO, or polyphenylene). PSF has been a very attractive choice due to its excellent oxidative stability, easy functionalization, outstanding film formation properties, good solubility in common organic and aqueous solvents for AEM electrode binders, low cost and widespread availability.[30–32] Cardo-PEEK is also an excellent polymer choice, with strong mechanical properties and good thermal and chemical stability.[33,34] Zeng and co-workers[35,36] have prepared quaternized Cardo-PEEK(QCardo-PEEK) based AEMs and used as electrolytes in direct methanol fuel cells. The QCardo-PEEK AEM had a methanol permeability of less than 10^{-9} mol cm^{-2} min^{-1}, suggesting that the QCardo-PEEK membrane has reasonable methanol resistance for use in methanol fuel cells. PPO-based AEMs have been touted as promising candidates for alkaline fuel cells for more than one decade.[37] PPO is a commercially available and cheap polymer and its backbone hydrophobicity can be tuned by the addition of bromine (increasing the microphase separation and yielding large hydroxide ion conductivities).

[a]*Department of Energy, Environmental and Chemical Engineering, Washington University in St. Louis, USA. E-mail: ramani@wustl.edu*
[b]*Department of Chemical and Biological Engineering, Illinois Institute of Technology, Chicago, USA*
[†]Both authors contributed equally to this book chapter.

Table 1 Summary of common backbones and their attributes.

Chemical structure		Attribute	Ref.
Aromatic backbone	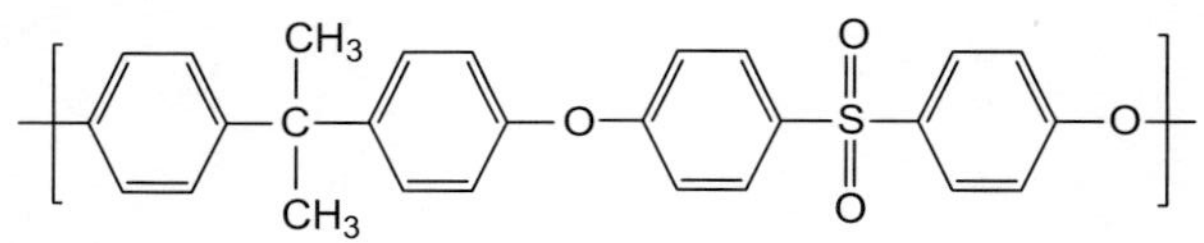 Polysulfone	Widely available; High molecular weight; Chemical stable; Extensively studied; Ease of derivatization; Low cost; Excellent mechanical properties	29, 31, 32
	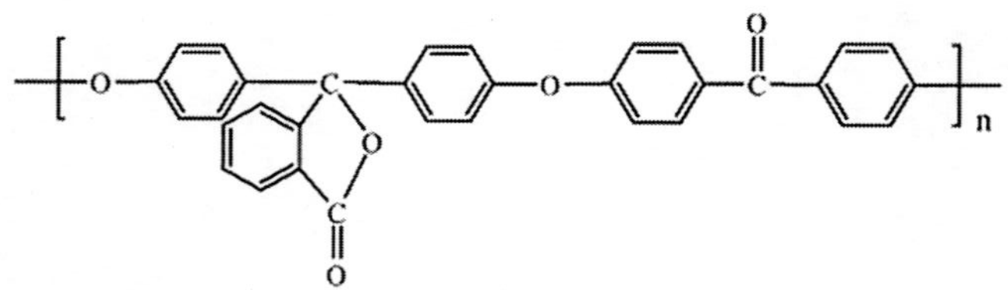 Cardo-poly-(etherketone)	Strong chemical stability; Strong thermal stability; Excellent methanol permeability; Commercially available	2, 34, 36
	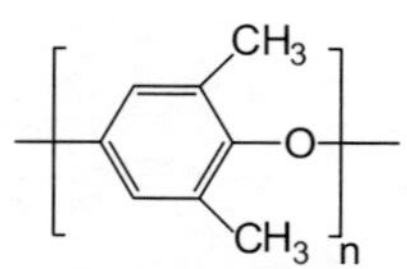 poly-(phenylene oxide)	Widely available; High molecular weight; Chemical stable; Extensively studied; Easy to be functionalized; Low cost	5, 37, 129, 130
	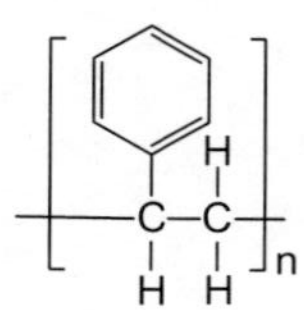 Polystyrene	Strong plastic; Low water uptake; Extensively studied; Low cost	131

	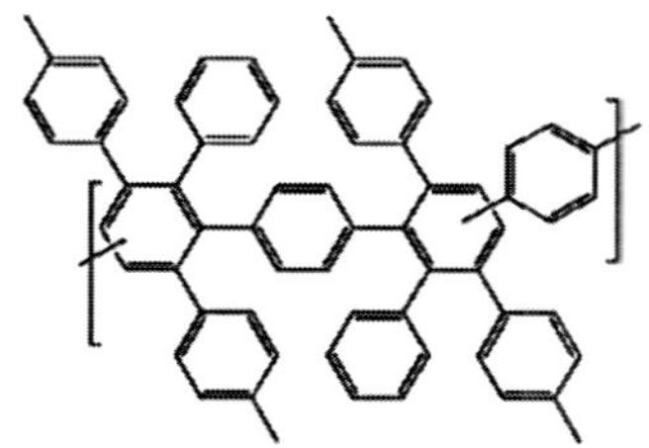 Polyphenylene	Low water uptake; Chemical stable; High molecular weight	8
Aliphatic backbone	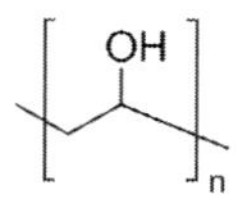 Poly-(vinyl alcohol)	Easy to get high IEC; Good mobility; Excellent thermal and mechanical properties	23
	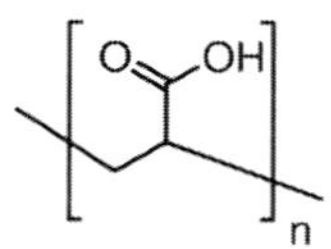 Poly-(acrylic acid)	Easy to get high IEC; Good mobility; Excellent thermal and mechanical properties	23
	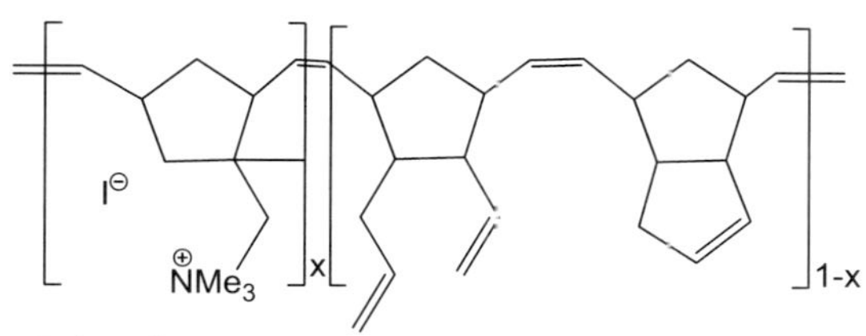 Polynorbornene	Good chemical stability	18

Table 1 *(Continued)*

Chemical structure		Attribute	Ref.
Block copolymer	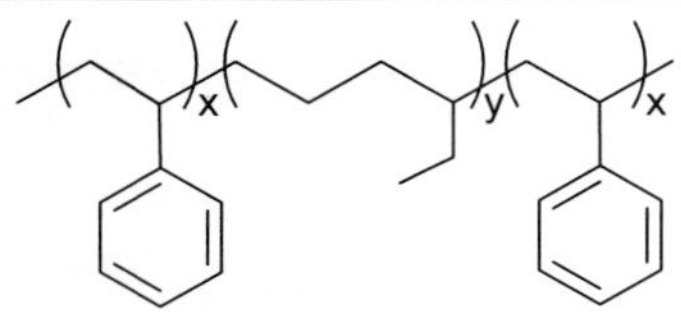 Polystyrene-*b*-poly(ethylene-*co*-butylene)-*b*-polystyrene	Widely available; Excellent morphological control; High molecular weight	25
	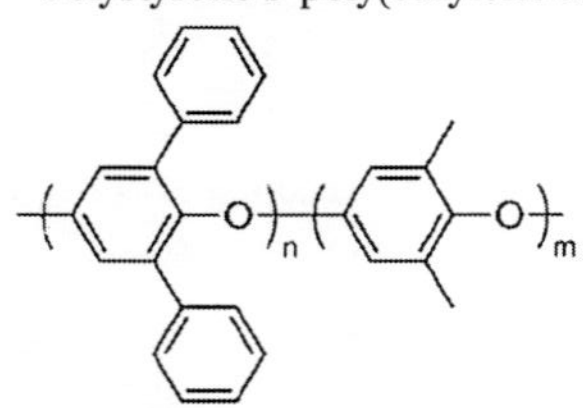 PPO block copolymer	Excellent morphological control; High molecular weight	27
Partially fluorinated backbone	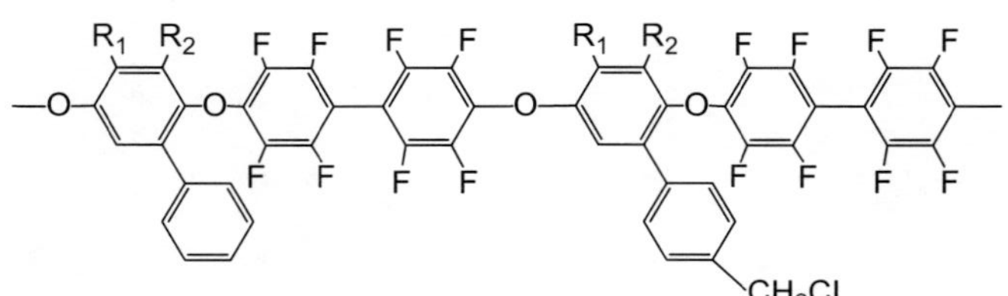 Partially fluorinated copoly(arylene ether)	Hydrophobic nature; Excellent mechanical strength; Good thermal stability; Chemical stability	28

Fluorinated backbone	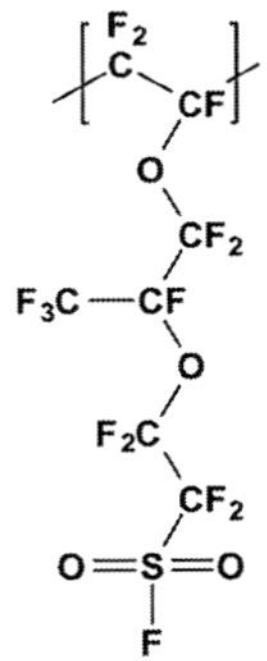	Fluorinated backbones; Chemical stable; Better ionic transport channel; Good permselectivity. Excellent mechanical properties	132

The second group of backbones studied are **aliphatic backbones**, for example, poly (vinyl alcohol) (PVA),[23] polyacrylic acid (PAA),[23] and polynorbornenes.[18] PVA/PAA composite membranes exhibit excellent thermal and mechanical properties and displayed good mechanical strength and ductility. Clark and coworkers synthesized mechanically strong tetraalkylammonium-functionalized norbornene AEMs with an OH^- conductivity of 18 $mS\,cm^{-1}$ (at 20 °C) by employing a facile ring-opening olefin metathesis route.[18]

Recently several researchers have focused on AEMs with backbones based on **tri-block copolymers** (*e.g.*, styrene-ethylene-butylene-styrene triblock copolymer, SEBS). Those backbones promise as highly stable, durable, conductive, and scalable separators for alkaline membrane fuel cells, solid-state alkaline water electrolyzers and redox flow batteries. SEBS-based AEMs have high conductivity, ready availability in industrial quantities, relatively low cost, and good mechanical properties (very similar to Nafion®).[12,13] Knauss and coworkers synthesized PPO-block copolymers for potential use in alkaline exchange membrane applications by oxidative polymerization of 2,6-dimethylphenol and 2,6-diphenylphenol.[27] The copolymers were then converted to benzylammonium functionalized materials by bromination followed by quaternization. The hydroxide ionic conductivity at 80 °C was 84 $mS\,cm^{-1}$.

The **perfluorinated backbone** is also an important backbone in AEM studies. Perfluorosulfonic acid membranes have dominated PEM fuel cell research due to their large proton conductivities and excellent chemical and mechanical stabilities.[24] Nafion® PEM, commercialized by Dupont since the 70s, has a high conductivity and excellent stability, and these facts triggered attempts to synthesize perfluorinated AEMs based on similar chemical structures. However, synthesizing perfluorinated AEMs faces functionalization difficulties, and the conductivity and chemical stabilities in some works remain controversial. Matsui and coworkers reported the synthesis of perfluorinated AEMs starting with Nafion® membranes *via* multiple reaction steps, including conversion to a carboxylic acid type membrane, methylation, conversion to amide-type membrane, reduction, and quaternization.[38] The resultant membrane showed good durability in aqueous saturated chlorine solution at 60 °C for 1000 h. However, the complicated steps and the difficulty to complete each reaction have limited the applications of the work.[39] Fluorinated backbones include: (i) partially fluorinated films such as poly(vinylidene fluoride) (PVDF)[30,40–43] and poly(ethylene-*co*-tetrafluoro ethylene) (ETFE);[44–46] and (ii) fully fluorinated films, like (poly(fluorinated ethylene propylene) (FEP)[42]). The PVDF-based membranes[30,40–43] showed poor mechanical properties and low IECs (0.7×10^{-3} equiv. g^{-1}), which limited their use in electrochemical devices.

2 Types of cations and their alkaline stability

The chemistries of cation head-groups that have been affixed to polymer backbones include N-based, P-based, S-based, and metal-based groups. The alkaline stability of each cationic group depends on different factors affecting the susceptibility of the cation to be attached by strong nucleophiles such as hydroxide. When the head-group is attached to different polymer backbones, the alkaline stability could also differ due to electron donation or withdrawal effects. Aromatic polymer backbones are less electron donating than aliphatic polymer backbones, favoring attacks by hydroxide ions.[26] This reasoning is supported both by theory (DFT studies by Wang and co-workers)[47] and experimental results.

2.1 Nitrogen-based cationic head groups

Ammoniun cations are the most commonly used group for making AEMs due to their low molecular weights and relatively good stability. Benzyl-trimethylammonium[48] has been used as benchmark to evaluate the alkaline stability of cationic head-groups. Its chemical structure can be seen in Fig. 1A. Parrondo and co-workers attached the benzyl-trimethylammonium (TMA^+) onto PPO to make an AEM (PPO-TMA). PPO-TMA was then immersed in 1 M KOH at 60 °C for 30 days. This *ex situ* alkaline stability test condition was stated to be a good mimic for real fuel cell operation environment by Arges and co-workers.[49] After *ex situ* alkaline stability test, there was 18% degradation of the AEM's initial IEC.[50] Long-alkyl-chain quaternary ammonium is another N-based

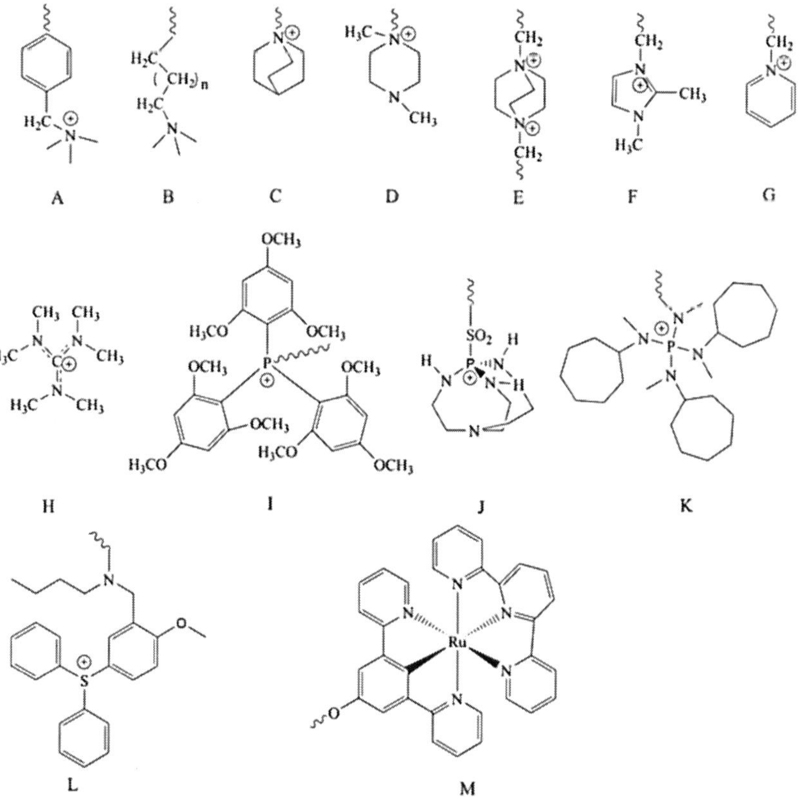

Fig. 1 Cationic head-group chemistries used in the literature to make AEMs.

cationic head-group (see Fig. 1B). Hibbs proved that this kind of ammonium was more stable in alkaline condition than the benchmark cationic chemistry. He synthesized AEMs by attaching long-alkyl chain quaternary ammonium onto poly phenylene (PP). He found that when long-alkyl-chain quaternary ammonium based AEM was immersed into 4 M KOH at 60 °C for 14 days, there only 5% of its initial IEC was lost, whereas for an benzyl-trimethylammonium-based AEM suffered 21% loss of its initial IEC under the same condition.[51] Arges and co-workers attached quaternary 1-azaoniumbicyclo [2.2.2] octane ($ABCO^+$) (see Fig. 1C) onto the PPO backbone to make PPO-ABCO, which they then immersed in 1 M KOH at 60 °C for 30 days. They found there was *ca.* 72% loss of its initial IEC after this alkaline stability test.[52] Arges and co-workers also investigated the alkaline stability of the 1,4-dimethylpiperazinium (DMP^+) (see Fig. 1D) cationic head-group. They attached DMP^+ onto polysulfone (PSF) to make a PSF-DMP AEM. PSF-DMP was immersed into 1 M KOH at 60 °C for 30 days. They found that 35% cation sites of PSF-DMP suffered IEC lost. PSF-DMP was less alkaline stable than benchmark chemistry (PSF-TMA).[53] Morandi and co-workers attached 1,4-diazabicyclo[2.2.2]octane (DABCO) (see Fig. 1E) to PPO to make a PPO-DABCO AEM. They also attached *N*,*N*,*N*′,*N*′-tetramethylethylenediamine (TMEDA) to PPO to make a PPO-TMEDA (which has alkaline stability similar to the benchmark head-group chemistry) as reference. The membrane samples were immersed into 1 M KOH at 90 °C for 10 days. PPO-DABCO lost *ca.* 28% of its initial IEC and PPO-TMEDA lost *ca.* 18% of its initial IEC.[54] The DABCO cationic head-group is not as stable as the benchmark head-group. Wang and co-workers synthesized a series of imidazolium-based AEMs. For one, they attach 1,2-dimethylimidzole (1,2DMI) (see Fig. 1F) to PPO to make PPO-1,2DMI, which they immersed in 1 M KOH at 60 °C for 13.5 hours. PPO-1,2DMI remained 75% cationic sites after alkaline treatment.[55] Although imidazolium cation initially attracted researchers' attention mainly for their high hydroxide ion conductivity,[56] their applications in alkaline fuel cells and other alkaline electrochemical devices have been limited by their low alkaline stability. It is well known that pyridinium (see Fig. 1G) exhibits low stability in alkaline media, due to hydroxide attack in the *ortho* position followed by irreversible oxidation to pyridone and consequent loss of positive charges, which are needed for the transport of hydroxide anions.[57,58] Kim and co-workers synthesized pentamethylguanidinium functionalized, perfluorinated hydroxide conducting ionomers (see Fig. 1H). The alkaline stability of the ionomers largely depended on adjacent groups that connected the cation. Sulfone guanidinium functionalized ionomer degraded almost 100% after immersion into 0.5 M NaOH at 80 °C for 24 hours, while phenylguanidinium did not degrade under the same conditions for 72 hours.[59]

2.2 Phosphorous-based cationic head groups

Gu and co-workers functionalized polysulfone with the benzyl-tris(2,4,6-trimethoxyphenyl) phosphonium ($TTMP^+$) cation (see Fig. 1I) to make AEMs. They clearly showed that $TTMP^+$ cations result in highly alkaline-stable AEMs that did not experience any loss of ionic conductivity after immersion in 1 M KOH at room temperature for 30 days.[60–62] $TTMP^+$ as P-based cationic head-group was more alkaline-stable than the benchmark chemistry. $TTMP^+$ is a bulky cation that can prevent hydroxide ion attacks on phosphorus atoms. If we can increase the IEC and ion conductivity of TTMP-based AEM, $TTMP^+$ cations can be good candidates to make alkaline-stable AEMs. Kong and co-workers chemically attached proazaphosphatranium (see Fig. 1J) cations under microwave treatment to the sulfonic groups of Nafion-F to make a novel AEM. This P–N cationic head-group was confirmed by solid-state nuclear magnetic resonance techniques.[63] The alkaline stability of this novel P–N system was not evaluated. Noonan and co-workers attached tetrakis (dialkylamino) phosphonium group (see Fig. 1K) onto a polyethylene backbone. The AEM was exposed to 1 M KOH at 80 °C. A small initial loss of hydroxide ion conductivity (from 22 to 18 mS cm^{-1}) was observed after 3 days, but no further loss in conductivity was evident up to 22 days. They suggested that the alkaline stability of the phosphonium-functionalized AEM in 1 M KOH at 80 °C makes it an excellent candidate for higher-temperature alkaline membrane fuel cell devices.[64] The IEC of Noonan's AEM was 0.67 mmol OH^-/g Cl^-. Although some phosphonium-based AEMs have shown great alkaline stability, they are limited by the requirement of high ion conductivity and IEC.

2.3 Sulfur-based cationic head groups

Zhang and co-workers attached a sulfonium cation (see Fig. 1L) onto polysulfone. The AEM remained intact after an immersion in 1 M KOH at 60 °C for 10 days, confirmed by ^1H NMR spectra. They claimed that there was no obvious loss of conductivity for this AEM after immersion in 1 M KOH for 30 days at room temperature, indicating good long-term stability.[65] The IEC for this sulfonium-based AEM was 0.69, with a hydroxide ion conductivity of 15.4 mS cm^{-1} at 20 °C. Higher IEC values and greater hydroxide ion conductivity are desirable for sulfonium-based AEMs. The AEM showed great alkaline stability at room temperature for 30 days, and it may be a good candidate to be used as alkaline fuel cell at room temperature, but at higher operation temperatures such as 60 °C, further studies need to be done.

2.4 Metal-based cationic head groups

Disabb-Miller and co-workers introduced bis(terpyridine)Ru(II)-based AEM (see Fig. 1M), using ring-opening metathesis polymerization techniques with dicyclopentadiene cross-linking moieties. The use of the metal Ru(II) complex in an AEM was targeted toward increased anion conductivity by allowing for two counter ions to be involved in ion transport, instead of the single counter ion typically associated with tethered cations such as benzyl-trimethylammonium. The highest AEM conductivity they obtained was 7.9 mS cm^{-1} for the 1.6 mequiv. g^{-1} IEC in chloride form at 30 °C.[66] The alkaline stability of the metal(Ru)-based AEM was not evaluated by the researchers.

2.5 Methods to improve cation stability

To mitigate cation degradation, it is necessary to avoid or in the worst scenario, to minimize Hoffman Elimination and direct nucleophilic (S_N2) reactions occurring at the cation sites. Liu and co-workers substituted all the hydrogens on an imidazolium ring with methyl groups, which improved the imidazolium-based AEM's stability (no degradation in 1 M KOH at 80 °C for 24 hours) compared with an unsubstituted imidazolium-based AEM (24% degradation of IEC in 1 M KOH at 80 °C for 24 hours).[67] The absence of α-hydrogens prevented hydroxide ion attack at the imidazolium ring. Similarly, Gu and co-workers synthesized a quaternary bulky phosphonium cation-based AEM that was stable in 1 M KOH at room temperature for 30 days.[68] Here, too, the steric hindrance effect was shown to protect the bulky cation ($TTMPP^+$) from hydroxide ion attack. Another approach to improve AEM chemical stability was to affix Gemini cations to the polymer backbone. Steric hindrance (the effect of side chains), could decrease the chance of hydroxide ions attacking benzyl-connected quaternary ammonium groups.[69]

Zhuang and co-workers grafted an alkyl long chain onto the polymer backbone and affixed a quaternary ammonium group at the end of the chain.[70,71] An approximate 10% loss in IEC was observed after exposure to 1 M KOH at 60 °C for 500 hours. Due to hydrophilic/hydrophobic microphase separation, enhanced by the separation between the backbone and the polar groups, the polymer backbone was well protected by the additional hydrophobic structure. Hibbs and co-workers grafted a long alkyl chain between the quaternary ammonium and polyphenylene backbone, and they found no IEC drop after immersion into 4 M KOH at 90 °C for 14 days. They suggested that benzyl quaternary ammonium separated from the polymer backbone by an alkyl chain (with carbon atom number >3) was less susceptible to S_N2 and β-hydrogen attacks than benzyltrimethylammonium cations.

3 Degradation mechanisms

The mechanisms of AEM degradation under alkaline conditions are related to the different degradation modes of the tethered cation groups. For example, quaternary ammonium can degrade through the following known mechanisms: (1) Hoffman elimination, where the quaternary cation is cleaved, leaving tertiary amine and resulting in the formation of an alkene at the carbon in the polymer backbone where the ammonium was bonded; (2) a direct nucleophilic (S_N2) reaction where the cation is completely cleaved too, resulting in the formation of a tertiary amine and an alcohol at the carbon where the ammonium was bonded; and (3) an additional S_N2 reaction where the organic moiety adjacent to the inorganic atom (usually a methyl or alkyl group) is cleaved, resulting in the formation of a tertiary amine and an alcohol.[72–75] There are less frequently observed degradation pathways that involve the formation of ylide intermediates (commonly encountered in phosphonium and sulfonium cations), known as Sommelet–Hauser rearrangement and Stevens rearrangement.[76,77] All these mechanisms involve the presence of a

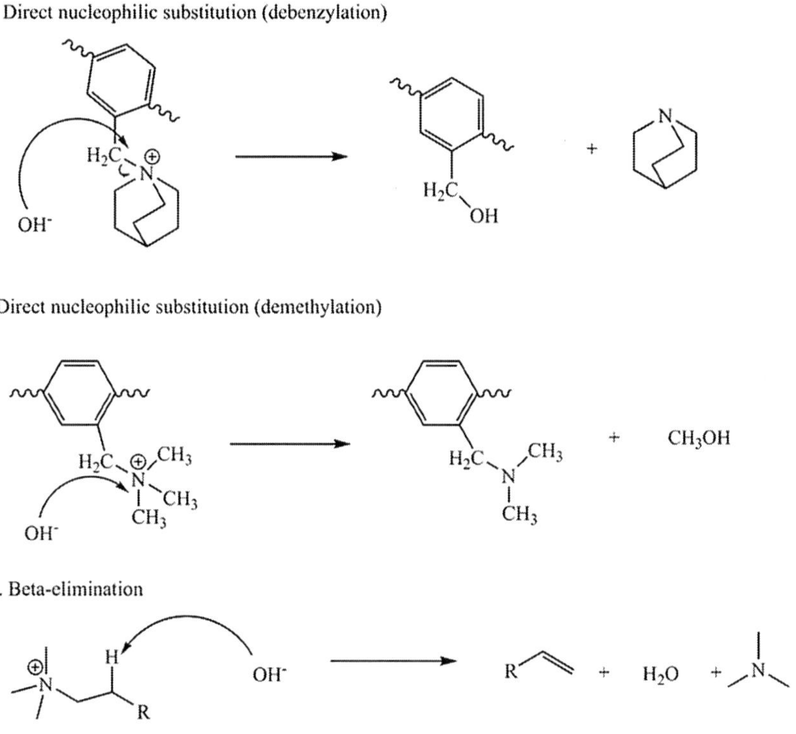

Fig. 2 Degradation mechanism (Direct nucleophilic attach and β-elimination) for quaternary ammonium-based AEMs. Reproduced from C. G. Arges, L. Wang, J. Parrondo and V. Ramani, Best Practices for Investigating Anion Exchange Membrane Suitability for Alkaline Electrochemical Devices: Case Study Using Quaternary Ammonium Poly(2,6-dimethyl 1,4-phenylene)oxide Anion Exchange Membranes, *J. Electrochem. Soc.*, 2013, **160**, F1258–F1274.

strong base and cause the formation of tertiary amines (or other groups, depending on the cation used) with the subsequent loss of ion exchange capacity and ionic conductivity.[52,53,78,79] In summary, the primary reasons for degradation are nucleophilic substitutions[52,80] and β-elimination[81] through reaction with hydroxide ions. The degradation mechanisms are shown in Fig. 2. Below we will discuss studies on the degradation of AEMs that our group and others have performed using 1-D ^{1}H-NMR and 2-D NMR spectroscopy. NMR spectroscopy is the preferred technique to quantify and investigate AEM degradation mechanisms.

3.1 1-D NMR as tool to detect the degradation of AEM

Wang and coworkers synthesized a series of imidazolium-based AEMs and used ^{1}H NMR spectroscopy to investigate the degradation mechanisms.[55] Figure 3 shows the ^{1}H NMR spectra of an imidazolium-based AEM before and after exposure to 1 M KOH (at 60 °C for 6 hours). After immersing the membrane in alkali media, we observed the appearance of a new peak [6′*] at a chemical shift of 3.93 ppm (see red rectangular box

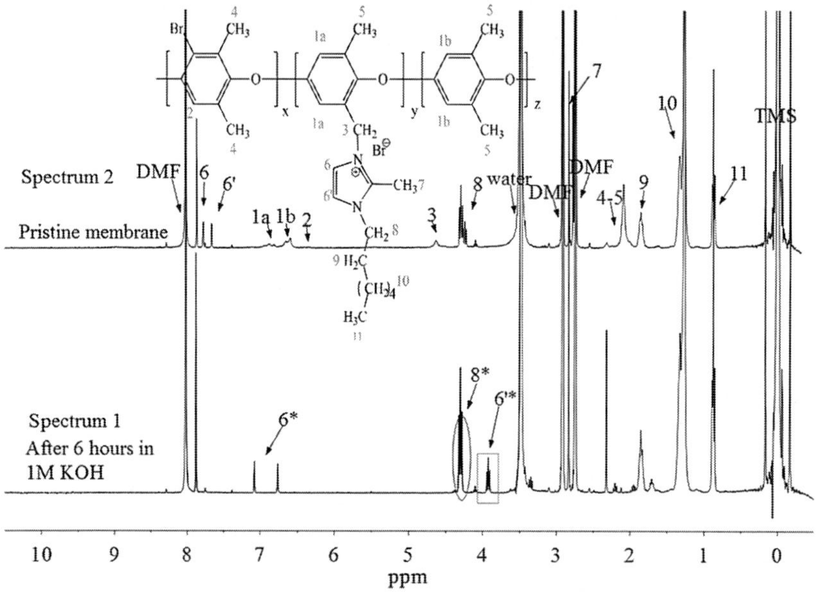

Fig. 3 ^{1}H NMR spectra of imidazolium-based AEM before (1) and after immersion in 1 M KOH at 60 °C for 6 hours (2). Reproduced from Z. Wang, J. Parrondo and V. Ramani, *J. Electrochem. Soc.*, 2016, **163**, F824–F831. Copyright The Authors 2016. Published by ECS under a Creative Commons Attribution 4.0 License (CC BY, http://creativecommons.org/licenses/by/4.0/).

in Spectrum 1 in Fig. 3). Ye and coworkers[82] observed that imidazolium-based anion exchange ionic liquids treated with alkali experienced hydroxide ion attack at the C-2 carbon, resulting in a ring-opening reaction. The protons labeled as 6 and 6′ (Spectrum 2 in Fig. 3), corresponding to the aromatic protons in the imidazolium ring disappeared in the case of a ring opening reaction. The protons were in an aliphatic chain in the final product [6* and 6′*] and had considerably lower chemical shifts (*ca.* 4 ppm). We also observed all these changes in our spectra, confirming that the ring opening reaction occurred upon exposure to alkali. We also observed that the peak area for the protons in the methylene bridge (peak 8, linking the backbone to the base) decreased after exposure to alkali. This observation also confirmed the loss of cationic sites and confirmed the ring opening reaction in the imidazolium group. The degradation mechanism (Fig. 4) was postulated according to conclusions derived by analyzing the 1-D NMR spectrum. Figure 4 shows the proposed degradation mechanism based on the ^{1}H-NMR spectrum (Fig. 3). The degradation process is triggered by the nucleophilic attack of hydroxide ion at the C-2 position of the cation. A covalent bond is formed between the hydroxide ion and the C-2 carbon (the alpha carbon with respect to both nitrogen atoms). Subsequently, there is a ring-opening reaction that results in the loss of cation sites. The covalent bond between C-2 and N-1 or N-3 is then broken, and, after a rearrangement reaction, formyl groups (–CHO) are formed (steps 2 and 3).

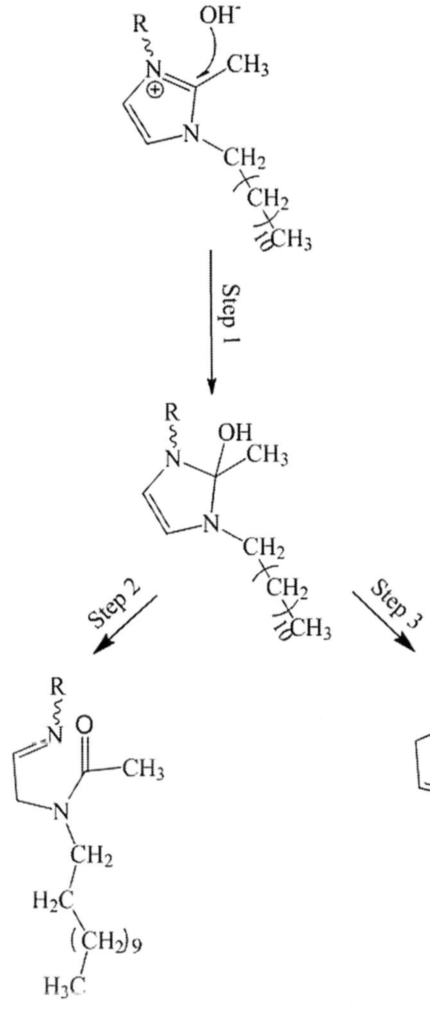

Fig. 4 Mechanism proposed for the alkaline degradation AEMs containing imidazolium cations (R stands for the polymer backbone, with or without the alkyl spacer). The mechanism is based on the work of Ye and coworkers.[82] Reproduced from Z. Wang, J. Parrondo and V. Ramani, *J. Electrochem. Soc.*, 2016, **163**, F824–F831.

3.2 2-D NMR as tool to investigate AEM degradation

Arges and co-workers first used 2-D NMR spectroscopy to deduce the degradation pathways in PSF-TMA AEMs.[49] The Heteronuclear Multiple Quantum Coherence (HMQC) NMR spectrum for PSF–TMA (Fig. 5) shows the formation and growth of two different signals at the same ^{1}H chemical shift and slightly different ^{13}C chemical shifts after exposure to 1 M and 6 M KOH solutions at 60 °C. Specifically, Fig. 5B (zoomed-in HMQC spectrum) shows the formation of a degradation product "A", which was assigned to phenylpropane alcohol (^{1}H at 1.2 ppm and ^{13}C at 30 ppm). The stability experiment was performed as before by immersion in 1 M KOH.

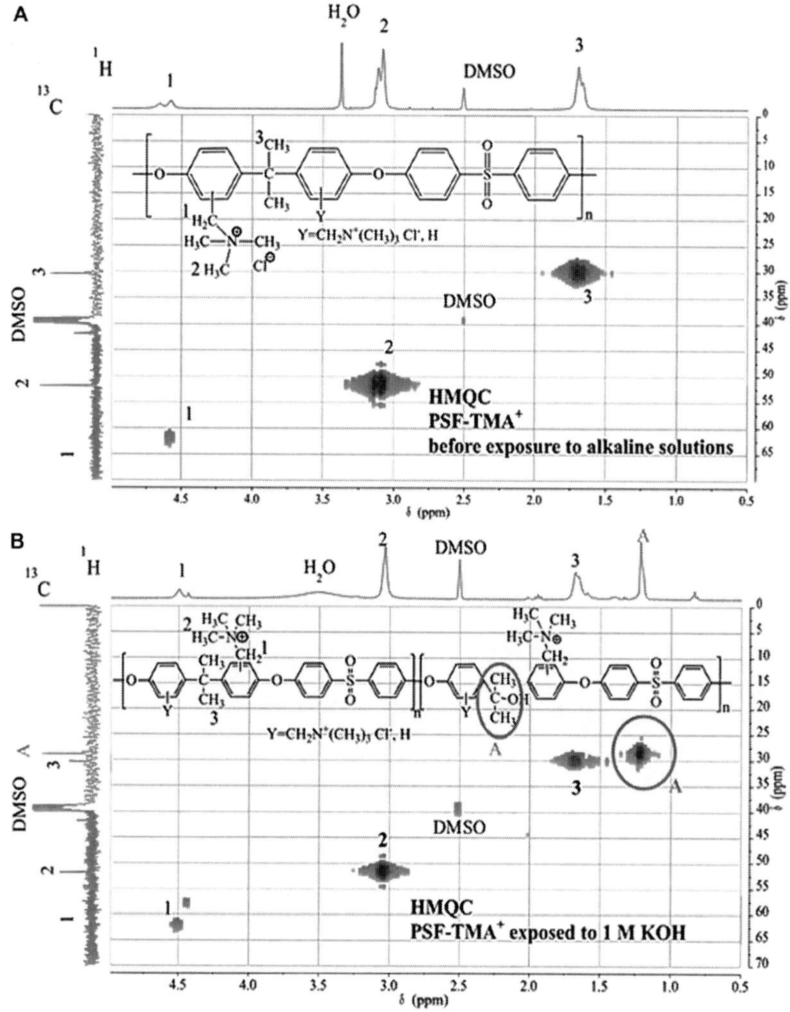

Fig. 5 2-D HMQC NMR spectra of PSF-TMA (A) before and (B) after exposure to 1 M KOH. Group A is the dimethyl group from the hydrolyzed diphenylpropane moiety in PSF-TMA. The formation of the phenylprpane alcohol group is a result of quaternary carbon hydrolysis of PSF-TMA. Reproduced from C. G. Arges and V. Ramani, Two-dimensional NMR spectroscopy reveals cation-triggered backbone degradation in polysulfone-based anion exchange membranes, *Proc. Natl. Acad. Sci.*, 2013, **110**(7), 2490–2495.

Figure 6 depicts the zoomed-in homonuclear correlation (COSY) spectrum for PSF–TMA before and after exposure to 6 M KOH. The COSY spectrum of the PSF–TMA AEM after degradation shows cross-couplings between the ^{1}H signals at 6 ppm and 7.2 ppm (Fig. 6B). The cross-couplings indicate that the protons in the alcohol group and in the phenyl ring were in adjacent carbons (group B) in PSF (Fig. 6B), confirming ether hydrolysis in the backbone. In conclusion, the pioneering use of 2D NMR spectroscopy to study AEM degradation can be employed to elucidate the mechanism of backbone hydrolysis in a PSF-based AEM exposed to alkali. Such 2D NMR spectra offer a more precise information than 1D NMR.

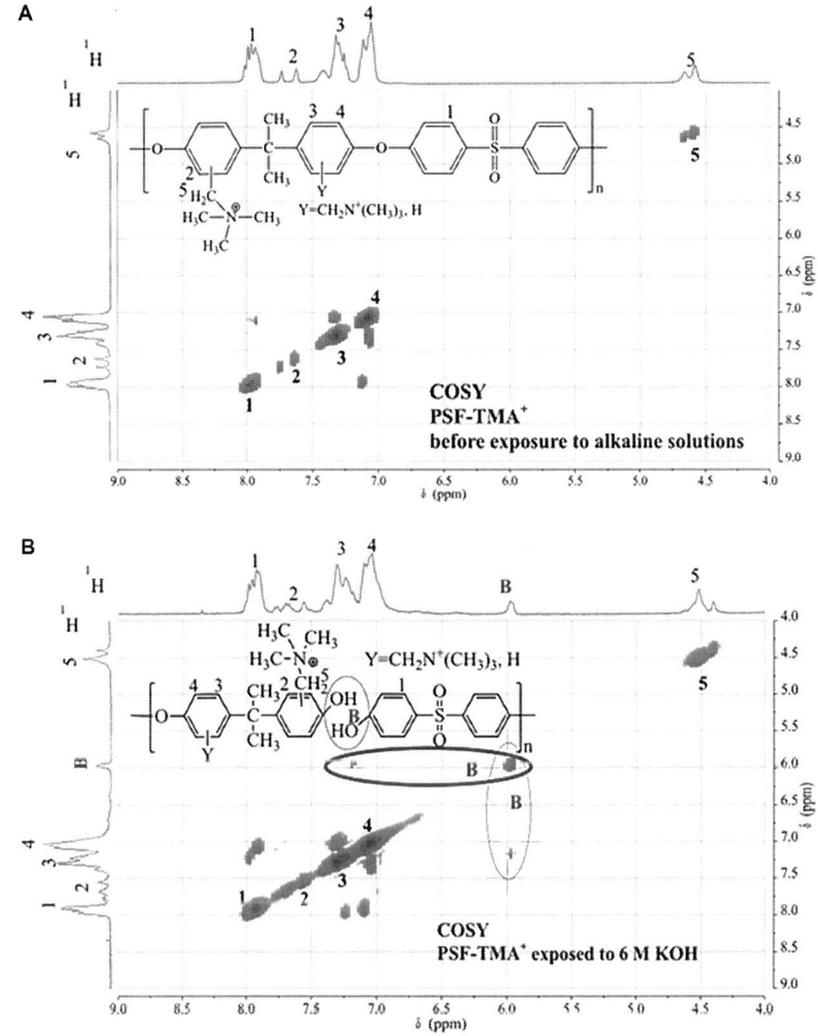

Fig. 6 Zoomed-in 2D COSY NMR spectra of PSF–TMA (A) before and (B) after exposure to 6 M KOH. Group B is the alcohol coupled to protons in the phenyl ring. The formation of phenyl alcohol is a result of ether hydrolysis of PSF–TMA. Reproduced from C. G. Arges and V. Ramani, Two-dimensional NMR spectroscopy reveals cation-triggered backbone degradation in polysulfone-based anion exchange membranes, *Proc. Natl. Acad. Sci.*, 2013, **110**(7), 2490–2495.

4 AEM characterization methods

4.1 IEC and ionic conductivity

The IECs of AEMs in chloride form can be determined by using the Volhard titration method.[52,83] A vacuum dried SEBS-based AEM in chloride form was weighed (W_{dry}) and immersed in 20 mL of 1 M sodium nitrate for 48 h. Later, 5.0 mL of 0.1 M silver nitrate was added to precipitate all the chloride ions. The remaining silver in the solution was titrated with 0.1 M potassium thiocyanate after adding 2–3 drops of 11 wt% iron (III) nitrate as an indicator. A control sample containing

20 mL of 1 M $NaNO_3$ and 5.0 mL of 0.1 M $AgNO_3$ was also titrated. The difference in the volume used to titrate the control (V_C) and the sample (V_S) was used for calculating the IEC:

$$\mathrm{IEC} = (V_C - V_S) \times 100/\mathrm{W_{dry}} \tag{1}$$

In-plane ionic conductivity measurements can be performed in a 4-point conductivity cell using electrochemical impedance spectroscopy (EIS) to measure its resistance. Typically, a 1 cm×3 cm membrane is placed in the conductivity cell in contact with the four platinum electrodes immersed in a temperature controlled water bath. A potentiostat is employed to measure the impedance in the frequency range 100 kHz to 0.1 Hz. The high frequency resistance (R) is estimated from the Bode plots:

$$\sigma = L/(R \times t \times w) \tag{2}$$

where σ is the in-plane membrane conductivity (mS cm^{-1}); R is the in-plane membrane resistance (Ohms); t is the membrane thickness (cm); w is the membrane width (cm) and L is the distance between the two inner electrodes (cm).

4.2 Tensile testing – evaluation of mechanical properties

A differential mechanical analyzer equipped with a humidity chamber is recommended for obtaining adequate results when performing tensile testing. AEMs contain polar cations and tend to uptake relatively large amounts of water, significantly affecting the mechanical properties and necessitating accurate control of humidity. In a typical experiment, a membrane sample is fixed in a film tension clamp using a torque betwen 3–9 $\mathrm{lb_F} \times \mathrm{in}$. The membrane is then stretched at 0.5 MPa min^{-1} until rupture. The Young's modulus, ultimate tensile strength, and elongation at the break point can be used to characterize the membranes, compare with other polymers, and also investigate their chemical and mechanical degradation during use in the electrochemical cell.

4.3 AEM permselectivity and transport numbers

The AEM permselectivity and transport numbers can be determined by using the membrane potential method.[84] The AEM is clamped between the two well-stirred compartments of a diffusion cell, containing different concentrations of the same salt (e.g., 0.1 M and 0.5 M KCl). Two identical Calomel reference electrodes are used to measure the potential difference (E_m (mV)) between the two solutions, arising from the different mobilities of chloride and potassium ions through the membrane. The membrane permselectivity (selectivity of the anion exchange membrane towards anions) and the transport numbers (in this case for chloride and potassium) are calculated by using the following equation:

$$E_m = (2t_- - 1)\left(\frac{RT}{F}\right)\ln\left(\frac{a_A}{a_B}\right) = (t_- - t_+)\left(\frac{RT}{F}\right)\ln\left(\frac{a_A}{a_B}\right) \tag{3}$$

Here, t_+ and t_- are respectively the transport numbers for the cation (K^+) and the anion (Cl^-), a_A and a_B are the activities of the electrolyte (KCl) in the concentrated and diluted compartments, T is the absolute temperature, R is the gas constant, and F is the Faraday constant. The coefficient $(2t_- - 1)$, commonly referred as the membrane permselectivity, represents the difference between the transport numbers for anions and cations, and can also be interpreted as the selectivity of the membrane to anions.[85]

4.4 Water uptake and swelling ratio

Vacuum dried AEM samples (<1 in Hg at 60 °C for at least 12 hours), typically in the chloride form, are weighed (approx. weight of 0.1 g), then immersed in DI water and kept in an oven at 30 °C for at least 24 hours. The samples are quickly swabbed to remove surface water and then immediately placed into previously tared Ziploc® bags to measure their weight. Water uptake is determined as follows:

$$WU = \frac{Wt_{hydrated} - Wt_{dry}}{Wt_{dry}} \cdot 100\% \tag{4}$$

Here, WU is the water uptake (%), $Wt_{hydrated}$ is the weight of the fully-hydrated membrane (g), and Wt_{dry} is the weight of the dry membrane (g). The dimensions (thickness, length and width) of the dry and fully hydrated membrane are also measured and used to estimate the swelling ratio. The swelling ratio is defined as the percentage of volume increase from the completely dry membrane to the fully hydrated membrane (at the specified temperature).

4.5 Nuclear Magnetic Resonance (NMR) and Fourier transform infrared (FTIR) spectroscopies

The chemical structure and composition (and variations therein with time of exposure to external stresses) of AEM polymers can be monitored at the molecular level by NMR and FTIR spectroscopies. The presence of reactive functional groups can be qualitatively confirmed by FTIR spectroscopy. All samples should be dried in a vacuum oven before FTIR measurements to minimize the presence of water and avoid a large water peak.

NMR (^1H, ^{13}C, ^{31}P, and ^{19}F) spectroscopies are employed to quantify the extent of derivatization of each polymer sample, to assess the formation of the AEM material, and to quantify and investigate degradation pathways. The following NMR experiments are very adequate for the characterization of the AEMs and identification of any degradation products: (1) ^1H NMR (spectra collected at 360 MHz), (2) ^{13}C NMR (spectra collected at 90 MHz - proton decoupled), (3) ^1H–^{13}C heteronuclear multiple-quantum correlation spectroscopy (HMQC), and (4) ^1H–^1H correlation spectroscopy (COSY). The 1-D and 2-D NMR techniques are very useful to quantify the functionalization degree (during the steps required to synthesize the final product) and to investigate any degradation products. The NMR samples are prepared by dissolving approximately 0.1 g of AEM material in 1 mL of deuterated solvent (DMSO-d_6, $CHCl_3$-d_6, or $(CD_3)_2NCOD$, depending on each particular case). The selection of the solvents is based on the AEM solubility. Finally, 35 μL of tetramethylsilane are added as internal standard to calibrate the chemical shifts ($\delta = 0$ ppm for ^1H and ^{13}C). For the analysis of the degradation products, 1D and 2D NMR techniques can be employed together to determine the chemical structure changes.

4.6 Thermogravimetric analysis (TGA) and differential scanning calorimetry (DSC)

TGA provides an estimate of the decomposition onset temperature, and DSC can be employed to obtain the glass transition temperature. The experiments are usually performed while purging the furnace with nitrogen gas (inert gas). Approximately 2 mg of the AEM are used in a typical DSC experiment. Samples are first heated to 190 °C at 10 °C min^{-1}, maintained at that temperature for 5 minutes, and then quenched to 0 °C. After, the samples being heated to final temperature (to the onset decomposition temperature determined through TGA minus 20 degrees). The heating temperature rate is typically in the range 2–5 °C min^{-1}. To help increase the resolution and allow identification of the different phase transitions, the experiment can also be performed using modulated temperature, mostly available in newer instruments. In a typical experiment, the modulation amplitude is 1 °C and the period is 60 s. Modulated DSC experiments allow measuring the total heat flow and sample heat capacity in a single experiment.

4.7 Alkaline stability testing

For all *ex situ* alkaline stability tests, dry AEM samples of uniform thickness are cut into several pieces of an appropriate size for each individual test (water uptake, ion exchange capacity, ionic conductivity, NMR analysis, membrane permeability, *etc.*), weighed (to establish a baseline for post-experiment re-weighing), and placed in N_2-degassed KOH solution (1 M or 6 M KOH concentration) in properly sealed polypropylene containers. The samples are maintained at constant temperature in an oven (30 or 60 °C). AEM samples are removed from the solution at regular time intervals (for example, every 1, 2, 3, 7, 14, 30, 60, 90, and 180 days), washed with abundant DI water to remove the KOH, and dried in an oven at 60 °C overnight prior to analysis. The protocols for measuring NMR, FTIR, water uptake, membrane conductivity, TGA, DSC, DMA, tensile test and transport numbers were described in previous sections.

5 Anion exchange membrane fuel cells: performance and durability

To achieve better fuel cell performance with an anion exchange membrane, we need not only a highly conductive membrane with good mechanical and thermal properties but also a very durable AEM while operating. The benchmark we are using now is Tokuyama A201®. Ren and *et al.* have reported that under 50 °C the H_2/air fuel cell performance of Tokuyama A201® is relatively high, the power density is 148 mW cm^{-2} (see Table 2 for the most recent and relevant fuel cell data).[86] Arges and *et al.* reported that with PPO-TMA$^+$ AEMs, the peak power density for a single cell using PPO-TMA$^+$ as the AEM and an electrode binder was 294 mW cm^{-2} (hydrogen as the fuel and oxygen as the oxidant).[5] A general procedure for making MEA and performing the fuel cell testing is given below.[5,7,33]

Preparation of the membrane electrode assembly (MEA) – PPO AEM catalyst ink was painted on a PPO AEM in bromide form. The PPO AEM in bromide form did not dissolve or swell excessively when the PPO AEM catalyst ink was applied. The ink was sprayed onto the PPO AEM (thickness of about 30 to 40 μm – active area was 5 cm^2) using an air-brush to give electrodes with 0.4 mg$_{Pt}$ cm^{-2} for the platinum-ebased electrodes or 0.6 mg$_{Ag}$ cm^{-2} for the silver-based electrodes. The concentration of ionomer was 30 wt% (dissolved PPO AEM). The catalyst loading was measured by the weight difference before and after painting with the ink. The MEA was carefully exchanged to the hydroxide form (by immersion in 1 M KOH overnight) before evaluating its fuel cell performance.

Fuel cell testing – The MEA was assembled into the fuel cell hardware (Fuel Cell Technologies, Inc.) using gaskets that resulted in 1–2 mil pinch, and applying a torque of 25 lbF×in. The fuel cell hardware containing the MEA was connected to a fuel cell test station (Scribner Associates, Inc.), and heated to 70 °C while nitrogen was passed along both sides (0.2 L min^{-1}, 98% relative humidity). After reaching the test temperature, the gases were switched to hydrogen and oxygen (both 0.2 L min^{-1}, 98% relative humidity) and a constant voltage of 0.55 V was applied for 15 minutes to condition the MEA (this step drives out the residual bicarbonate/carbonate ions formed from exposing the MEA to air). The fuel cell polarization curve was collected with hydrogen (H_2) and the following oxidants: oxygen (O_2), CO_2-free air (air with carbon dioxide removed), helox (same concentration of oxygen as in air but balanced with helium instead of nitrogen), and 4 vol% oxygen balanced nitrogen. The polarization curves were collected by scanning the current (15 points per decade) from 0.02 A to 15 A. The acquisition was stopped when the voltage dropped below 0.1 V. Stoichiometric flow rates (2 for hydrogen and 3 for oxidants) with minimum flows of 0.2 L min^{-1} were used. Gases were humidified by keeping the humidifier at 69 °C (resulting in a relative humidity around 98% at 70 °C).

It is necessary to establish a standard procedure for testing fuel cell performance since the results varied due to several parameters, such as catalyst loading, the MEA fabrication techniques and testing conditions. In Table 1, Arges and *et al.* summarize the top SAFC performances reported in the peer-reviewed and non-peer-reviewed literature that exceed power densities of 200 mW cm^{-2} before 2014.[5]

Table 2 Summary of literature values for SAFC performance with oxygen or CO_2-free air that exceeds peak power densities of 200 mW cm^{-2}.[a,b,c,d,e,f,g,h]

Oxidant	AEM	Binder	Catalyst loading	MEA fabrication techniques	Testing conditions	Peak power density	Ref.
Oxygen	PPO-TMA^+	PPO-TMA^+ 30 wt% ionomer in electrode	0.5 mg_{Pt} cm^{-2} (46% Pt/C) (both electrodes)	Painted electrode directly to membrane	70 °C (cell temp.); atmospheric pressure; 98% RH (both H_2/O_2); flow rates 0.2 L min^{-1} (both H_2/O_2)	294 mW cm^{-2}	5
Oxygen	A201	AS-4	1.5 mg_{Pt} cm^{-2} (Pt/C) for anode 0.75 mg_{Pt} cm^{-2} (Pt/C) for cathode 30 wt% ionomer in electrode	Painted electrodes onto GDL	60 °C (cell temp.); atmospheric pressure; 62% RH for anode and 100% for cathode; flow rates 0.25 L min^{-1} for anode and 0.5 L min^{-1} for cathode	387 mW cm^{-2}	133, 134
Oxygen	A201	AS-4	1.5 mg_{Pt} cm^{-2} (Pt/C) for anode 0.75 mg cm^{-2} (CoO/rGO(N)) for cathode 30 wt% ionomer in electrode	Painted electrodes onto GDL	60 °C (cell temp.); atmospheric pressure; 62% RH for anode and 100% for cathode; flow rates 0.25 L min^{-1} for anode and 0.5 L min^{-1} for cathode	248 mW cm^{-2}	133
Oxygen	PPO-$TTMPP^+$	PPO-$TTMPP^+$	0.2 mg_{Pt} cm^{-2} (Pt black) (both electrodes) 20 wt% ionomer in electrode	Painted electrode on GDL	70 °C (cell temp.); 250 kPa pressure on both anode and cathode; humidifier temp. for anode 70 °C and cathode 80 °C; flow rates 0.2 L min^{-1} (both H_2/O_2)	242 mW cm^{-2}	135
Oxygen	Fuma-Tech FAA	PSF-$TTMPP^+$	0.5 mg_{Pt} cm^{-2} (Pt black) for anode 0.5 mg_{Pt} cm^{-2} (Ag nanoparticles) 20 wt% ionomer in electrode	Painted electrode on GDL	70 °C (cell temp.); 250 kPa pressure on both anode and cathode; humidifier temp. for anode 70 °C and cathode 80 °C; flow rates 0.2 L min^{-1} (both H_2/O_2)	208 mW cm^{-2}	136

Oxygen	Poly(tetrafluoroethylene-*co*-hexafluoropropylene) with quaternary benzyl ammonium	Poly(vinylbenzyltrimethylammonium)	0.5 $mg_{Pt}\,cm^{-2}$ (40% Pt/C) (both electrodes)	Painted electrode on GDL	50 °C (cell temp.); atmospheric pressure; 100% RH (both H_2/O_2); flow rates 2 $L\,min^{-1}$ (both H_2/O_2)	≈270 $mW\,cm^{-2}$	137
Oxygen	Polyphenylene – TMA^+ (to the benzyl position)	Perfluorinated ionomer with pentamethylguanidium	3 $mg_{Pt}\,cm^{-2}$ (Pt black)	Not disclosed	80 °C (cell temp.); no other info disclosed	577 $mW\,cm^{-2}$	138
Oxygen	Not disclosed	Not disclosed	Not disclosed	*Unknown (but catalyzed coated membrane prepared)	80 °C (cell temp.); pressure of 2 bar on anode and cathode	≈730 $mW\,cm^{-2}$	139
Oxygen	A901	AS-X 30 wt% ionomer in electrode	0.5 $mg_{Pt}\,cm^{-2}$ 46% Pt/C) (both electrodes)	Screen printed electrode directly to membrane	50 °C (cell temp.); atmospheric pressure 95% RH (both H_2/O_2); flow rates: 1 $L\,min^{-1}$ (both H_2/O_2)	450 $mW\,cm^{-2}$	140
Oxygen	C18BQAPPO-3	Not disclosed	0.4 $mg_{Pt}\,cm^{-2}$ both cathode and anode	*Unknown	60 °C (cell temp.); back-pressure of 0.1 MPa; 98% RH (both H_2/O_2); flow rates 0.2 $L\,min^{-1}$ (both H_2/O_2)	369.3 $mW\,cm^{-2}$	141
CO_2-free air	PPO-TMA^+	PPO-TMA^+ 30 wt% ionomer in electrode	0.5 $mg_{Pt}\,cm^{-2}$ (46% Pt/C) (both electrodes)	Painted electrode directly to membrane	70 °C (cell temp.); atmospheric pressure; 98% RH (both H_2/O_2); flow rates 0.2 $L\,min^{-1}$ (both H_2/O_2)	246 $mW\,cm^{-2}$	5
CO_2-free air	Polyphenylene – TMA^+ (to the benzyl position)	Perfluorinated ionomer with pentamethylguanidium	3 $mg_{Pt}\,cm^{-2}$ (Pt black)	Not disclosed	80 °C (cell temp.); no other info disclosed	466 $mW\,cm^{-2}$	138

Table 2 *(Continued)*

Oxidant	AEM	Binder	Catalyst loading	MEA fabrication techniques	Testing conditions	Peak power density	Ref.
CO_2-free air	Not disclosed	Not disclosed	Not disclosed	*Unknown (but catalyzed coated membrane prepared)	80 °C (cell temp.); pressure of 2 bar on anode and cathode	≈550 mW cm^{-2}	139
CO_2-free air	A901	AS-X 30 wt% ionomer in electrode	0.5 mg_{Pt} cm^{-2} (46% Pt/C) (both electrodes)	Screen printed electrode directly to membrane	50 °C (cell temp.); atmospheric pressure 95% RH (both H_2/O_2); flow rates: 1 L min^{-1} (both H_2/O_2)	340 mW cm^{-2}	140
Air	A201	The electrodes were multi-walled carbon nanotubes doped with KOH containing pyridine polybenzimidazole (MWNT/KOH-PyPBI) with platinum deposited on it	0.45 mg_{Pt} cm^{-2} (Pt deposited on to MWNT/KOH-PyPBI)	Painted electrode on GDL	50 °C (cell temp.); atmospheric pressure 95% RH (both H_2/O_2); flow rates: 1 L min^{-1} (H_2) and 2 L min^{-1} (O_2)	256 mW cm^{-2}	142

[a] $TTMPP^+$ = tris(2,4,6-trimethoxyphenyl) phosphonium cation.
[b] A901/A201 – polystyrene based AEMs with divinylbenzene crosslinks that contains quaternary ammonium groups (manufactured by Tokuyama Corporation).
[c] Pt/C = platinum on high surface area carbon.
[d] GDL = gas diffusion layer.
[e] Temp. = temperature.
[f] RH = relative humidity.
[g] AS-X/AS-4 = a soluble polystyrene based anion exchange polymer electrolyte used as a binder in the electrode (manufactured by Tokuyama corporation).
[h] CoO/rGO(N) – cobalt oxide/nitrogen-doped reduced graphene oxide.

6 Other applications

6.1 Separators in redox flow batteries

There has been a resurgence of interest in redox flow batteries (RFBs) for large-scale storage applications, due to the rapid development of renewable electricity plants, such as wind and solar farms. Since wind and solar are intermittent energy sources, they need appropriate storage to buffer and modulate the spikes and slumps in the energy production and demand. RFBs are modular and can be easily scaled-up by increasing the area of the electrodes and the volume of the electrolyte solutions. Additionally, power and energy are decoupled in an RFB, which facilitates the design and operation of plants for large-scale electric energy storage. Examples of RFB for large-scale energy storage include solar energy storage systems[87–89] and MW-scale smart grids.[90] In an RFB, the energy is stored in reversible redox couples that are recirculated through the electrodes of the battery during the charging/discharging process.[91] Redox species are stored in external tanks, allowing the independent management of capacity, energy, and power densities. Figure 7 shows the basic concept of a redox flow battery, where the negative and positive compartments are connected to reservoir tanks and pumps. The positive and negative electrolytes are kept separated inside the electrochemical conversion cell by a membrane that acts both as separator and as an ionic conductor. In practical systems, stacks of 10–200 single cells are employed.[91] There are many different types of RFB systems, which can be classified based on the redox couples they employ to store the electrical energy: all-vanadium,[87,92–94] bromide-vanadium,[95] iron-vanadium,[96–98] iron-chromium,[99] polysulfide-bromine,[100] zinc-nickel,[101] and zinc-cerium.[102–104] All-vanadium, polysulfide-bromide, and zinc-bromide RFBs have reached high levels of technological readiness.[88] For several decades, now different redox couples, electrodes, membrane separators,

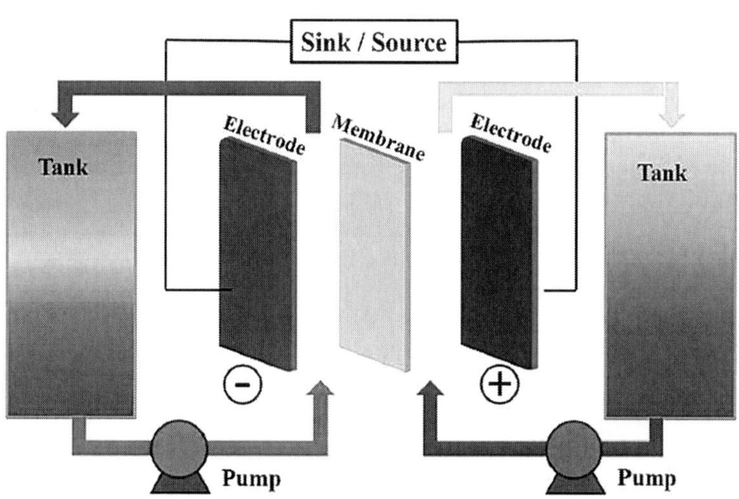

Fig. 7 Schematic of redox flow battery.

and operating systems have been actively studied. The details for each RFB system will be discussed in the following sections. In our group, AEM separator for all-vanadium redox flow batteries (VRFB) and vanadium/cerium batteries were extensively investigated.

All-vanadium redox flow batteries. VRFBs were first proposed by Skyllas-Kazacos and co-workers at the University of New South Wales. Commercial VRFB systems have been manufactured and deployed by a number of companies, including UniEnergy Technologies and Ashlawn Energy in the United States, and Prudent Energy and Rongke Power in China.[89,94,105,106] The main advantage and characteristic of VRFBs is the use of the same active element (vanadium) in both electrodes. Hence, the intermixing of positive and negative electrode solutions (the main cause of degradation in other flow batteries) does not cause irreversible damage to the battery. The electrolytes can be regenerated electrochemically by completely mixing all the electrolytes and recharging again. This procedure facilitates the reuse of the redox-active element.[92,107] Rahman and coworkers reported that the solubility of $VOSO_4$ decreased monotonically with an increase in H_2SO_4 concentration and the decrease in temperature.[108] Commonly, there is not active cooling or heating to simplify the operation and reduce operational costs, and the temperature is just slightly above room temperature due to irreversible heat loss during operation. The operational temperature should be limited to 10–40 °C to avoid V^{5+} in the positive electrolyte solution and V^{2+} in the negative electrolyte solutions, which would form V_2O_5 at high temperatures and VSO_4 precipitates at low temperatures.[96] The reactions of the VO^{2+}/VO^{2+} and V^{2+}/V^{3+} redox couples at the positive and negative electrodes are the following:

$$\text{Positive side: } VO^{2+} + H_2O \leftrightarrow VO^{2+} + 2H^+ + e^- \ (\Delta E^\circ = 1.00 \text{ V}) \quad (5)$$

$$\text{Negative side: } V^{3+} + e^- \leftrightarrow V^{2+} \ (\Delta E^\circ = -0.25 \text{ V}) \quad (6)$$

The use of carbon materials (carbon felts or carbon paper) as electrodes in VRFBs has economic advantages over other electrochemical systems (*e.g.*, fuel cells) that require precious materials (rare earths or platinum group elements) as catalysts.[109] The electrochemical reversibility (activity towards the electrochemical reactions) of vanadium redox couples strongly depends on the surface treatment of the carbon electrodes. Skyllas-Kazacos and co-workers modified the surface of graphite electrode materials by thermal or acid treatments, and showed that optimum performance of the battery was achieved when the graphite electrodes were heated in air in an oven at 400 °C for 30 h or immersed in 98% H_2SO_4 for 5 h.[110,111] Han and coworkers investigated the thermally treated materials and found that they had a relatively high active surface area and a large number of active functional groups (hydroxyl and carboxylate).[112]

AEM based separators for VRFB. Capacity decay during the charge-discharge cycles is one of the major problems encountered during long term operation of VRFBs, resulting in substantial capacity and energy density losses. Luo and coworkers demonstrated that capacity faded due to both imbalanced vanadium active species and the asymmetrical valence of vanadium ions in the positive and negative electrolytes.[113] Capacity decay is limited by the negative electrolyte, since V^{2+} and V^{3+} diffuse faster than VO^{2+} and VO^{2+}.[107] AEMs have suitable characteristics to prevent capacity decay since AEMs exclude cations (Donnan exclusion)[114] and selectively allow the passage of anions, enabling the passage of current during charging/discharging. This ability permits RFB operation with very low vanadium (or any other cations in the case of other chemistries) crossover, which is a great advantage for RFBs. Despite the promising characteristics of anion exchange membranes, there is a relatively small number of works related to the use of AEM in vanadium redox flow batteries.[92,115–117] Several polymers, such as polysulfone,[118] poly(fluorenyl ether),[116] poly(arylene ether),[119] poly(phthalazinone ether sulfone ketone),[120] poly(ether sulfone)-cardo[121] and poly(ether ether ketone)[122] have been used for the synthesis of AEMs. In our group, we developed a novel AEM based on a tri-block copolymer (Polystyrene-*block*-poly(ethylene-*ran*-butylene)-*block*-polystyrene; SEBS). The synthesis scheme is shown in Fig. 8. We have evaluated this AEM in a redox flow battery (RFB) operated with 1.5 M V^{2+}/V^{3+} (anolyte) and 1.5 M V^{5+}/V^{4+} (catholyte) (both in 3 M H_2SO_4). The RFB was run continuously with the same AEM separator for 60 charge-discharge cycles. After 20 charge-discharge circles (see Figs. 9 and 10) the voltage efficiency remained constant (*ca.* 98%), the energy efficiency dropped from 90% to 82%, and the coulombic efficiency decreased from 92% to 84%. The efficiency dropped not only from membrane degradation, but also from the loss of activity of the carbon felt electrodes. After 60 charge-discharge cycles (350 hours), the battery was disassembled and the AEM membrane was characterized to evaluate its degradation. The IEC and the ionic conductivity of the SEBS30-TMA had dropped by 18% and 20%, respectively. A tensile test to assess mechanical degradation did not show any decrease in the ultimate tensile strength or elongation at break. We also operated an "electrode-decoupled" RFB with 0.5 M V^{2+}/V^{3+} anolyte and 0.5 M Ce^{4+}/Ce^{3+} catholyte (both in 1 M H_2SO_4). The RFB was run continuously with an SEBS30-TMA AEM for 20 charge-discharge cycles. Figure 11 shows the normalized capacity (the capacity after each complete cycle of charge-discharge divided by the initial capacity), which dropped by less than 10% for the SEBS-based AEM (in contrast to a 40% drop for Nafion212®). The AEM's stability was also evaluated *ex situ* by immersing the membranes in 1.5 M $VO_2^+ + 3$ M H_2SO_4 for up to 500 hours. The IEC, ionic conductivity and permselectivity were measured after 0, 220, and 500 hours (see Fig. 12). The IEC decreased modestly from 1.35 mmol g^{-1} to 1.2 mmol g^{-1} after 500 hours in the highly oxidizing solution. The membrane permselectivity also decreased from 0.73 to 0.66 (the chloride transport number declined from 0.87 to 0.83). No significant changes in the ionic conductivity

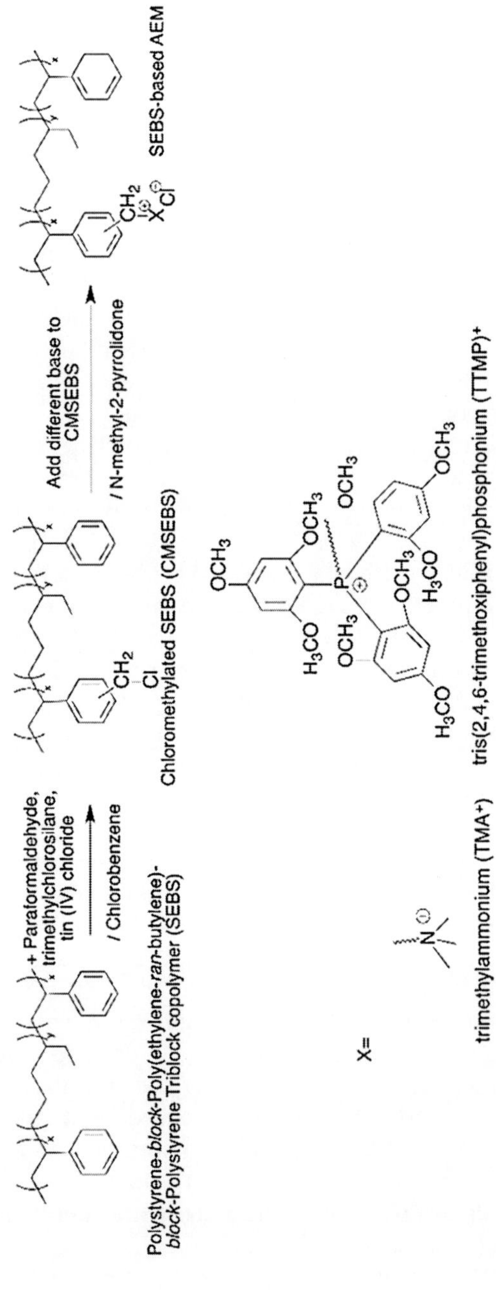

Fig. 8 The synthesis scheme of SEBS-based AEMs.

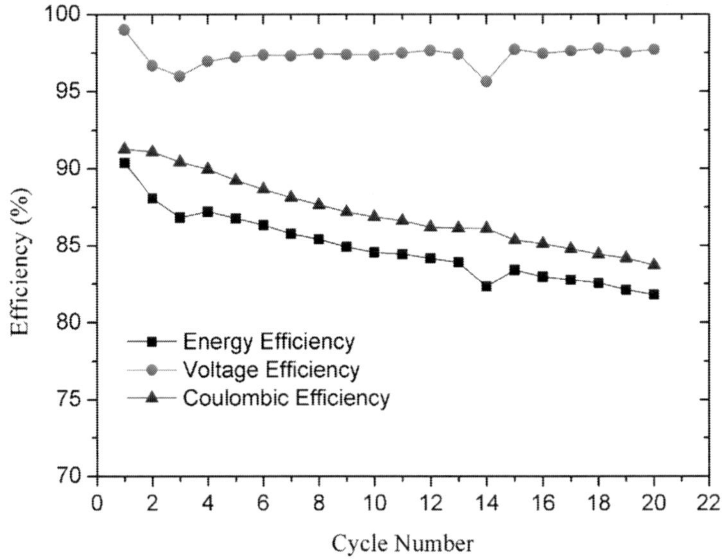

Fig. 9 The battery efficiency obtained using SEBS-TMA as separator to run an all vanadium redox flow battery.

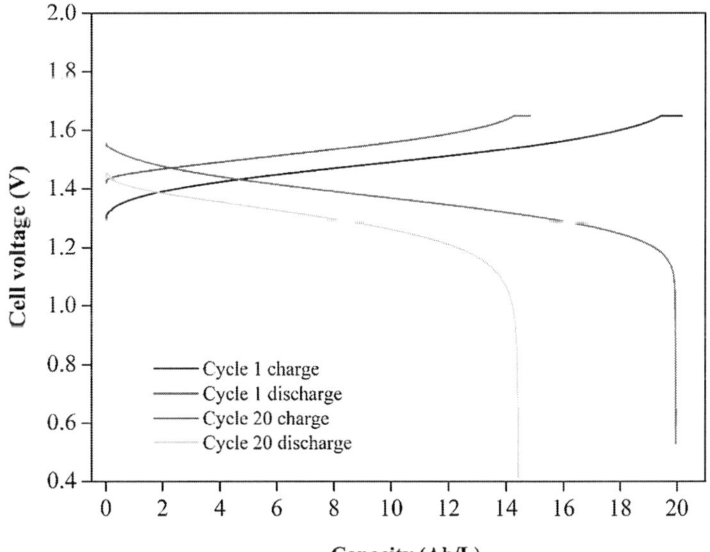

Fig. 10 All vanadium RFB charge/discharge curves using SEBS-TMA AEM as separator.

were found. Consequently, these SEBS-based AEMs revealed good stability in acidic electrochemical devices.

6.2 AEMs as separators for solid-state alkaline membrane electrolyzers

H_2 can be produced using chemical, electrochemical, catalytic, thermal, or biological processes.[123] H_2 production draw researchers' attention

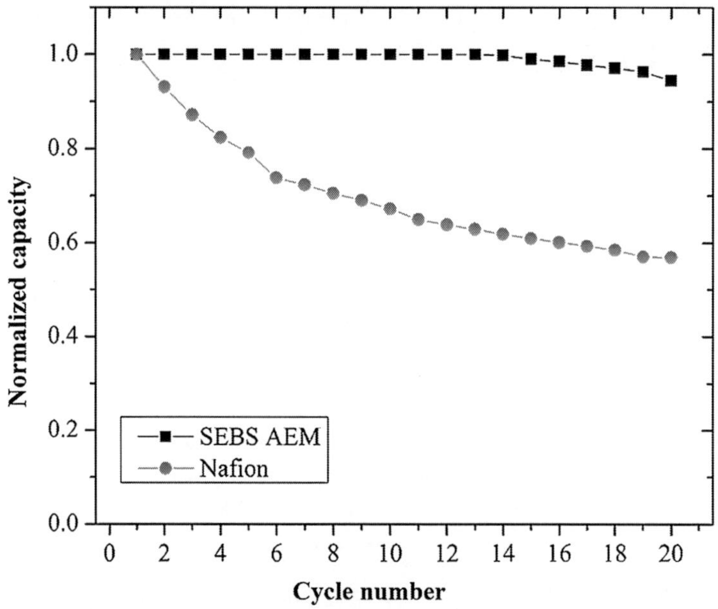

Fig. 11 SEBS-TMA and Nafion212® were used as separators to run an electrode-decoupled vanadium/cerium redox flow battery for 20 charge-discharge cycles continuously. The normalized capacities are plotted here.

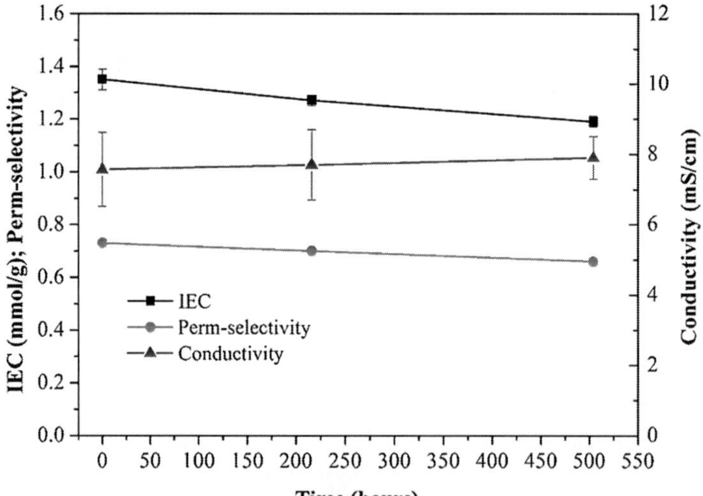

Fig. 12 Chemical stability of SEBS-TMA AEMs in 1.5 M VO_2^+ dissolved in 3 M H_2SO_4.

because of its potential use as a fuel produced from renewable and sustainable energies.[124] Currently, 95% of the H_2 used worldwide is produced by catalytic steam reforming of methane. The remaining commercial production of H_2 is mainly *via* electrolysis of water, a convenient and simple route, which produces very high purity H_2 for use in several applications: semiconductor manufacture, hydrogenation of food products, and the production and refining of high purity metals.[48] Most commercial electrolyzers are based on alkaline electrolysis, operate at current densities of 1000–3000 A m^{-2}, and contain 30% wt KOH aqueous electrolyte (ionic conductivity of 1.5 S cm^{-1} at 80 °C).[125] In an alkaline electrolyzer, H_2 is generated at the cathode (hydrogen evolution reaction: HER) and O_2 at the anode (oxygen evolution reaction; OER), through the following half-cell reactions:

$$\text{Cathode (HER): } 4H_2O + 4e^- \rightarrow 2H_2 + 4OH^-; \; E = -0.83\text{V } vs. \text{ SHE} \quad (7)$$

$$\text{Anode (OER): } 4OH^- \rightarrow O_2 + 2H_2O + 4e^-; \; E = 0.40 \; vs. \text{ SHE} \quad (8)$$

In the electrolyzer cell, a porous separator keeps the H_2 isolated from the O_2 gas (to avoid formation of a potentially explosive mixture and increase the purity of the product). As with the alkaline polymer electrolyte fuel cell (APEFC), an alkaline electrolyte enables the use of a low cost non-platinum group metal, such as Ni, which helps to keep capital cost of the cells relatively low.

A traditional alkaline electrolyzer uses a porous diaphragm to prevent intermixing of the O_2 and H_2 gases and obtain high gas purities and high current efficiencies.[48] The diaphragm ideally needs to prevent the formation of a gas bubble "curtain" at the surface of the electrodes (when pressing the electrodes onto the elastic diaphragm) to guarantee low ohmic and contact resistances. The separator (thickness of 1–2 mm) can be nickel oxide, asbestos, or a polymer. Diaphragm materials have also been made from polyantimonic acid and polysulfone/zirconium oxide (Zirfon®) composites.[48]

Replacing the separator/diaphragms mentioned above with ion exchange membranes offers advantages such as reduced gas crossover and area resistance (especially when using thin membranes), and the possibility of using differential pressure operation to minimize hydrogen compression and storage costs. The use of AEMs in electrolyzers provides the opportunity to consolidate the advantages of both types of traditional electrolyzers: alkaline and PEM electrolyzers. The use of electrode materials and catalysts with lower prices than those employed in PEMFCs, and the absence of cationic species present in liquid alkaline electrolyzers (normally leading to precipitates), benefit solid-state alkaline membrane water electrolyzers. The presence of cations (as impurities in the feed water) is also a major issue with PEM electrolyzers. A major reason for gradual deterioration in their performances related to the cations binding to the proton conducting (exchange) sites of the PEM.[126] In the case of solid-state alkaline membrane water electrolyzers, the AEM serves as the ion conducting electrolyte and also as the separator for H_2 and O_2 gases. The research challenges for AEMs used in alkaline electrolyzers are to obtain AEMs with high conductivity, good alkaline stability, low gas crossover, and durable mechanical properties.

In our group, we extensively investigated the AEM alkaline stability of the electrolyzer. Parrondo and coworkers investigated the degradation of AEMs and of solubilized AEM electrode binders during operation of a solid-state alkaline membrane water electrolyzer.[50,123] The performance and degradation of polyphenylene oxide (PPO) based AEMs in an operating solid-alkaline water electrolyzer at 50 °C was investigated by using electrochemical testing and 1-D and 2-D NMR spectroscopy. The PPO AEMs were derivatized with trimethylamine (TMA) and quinuclidine (ABCO) to yield AEMs with TMA^+ and $ABCO^+$ cations. The AEMs underwent chemical degradation during electrolysis. The cell voltage at 0.2 A cm^{-2} for BrPPO-$ABCO^+$ and BrPPO-TMA^+ increased by 0.4 V and 0.2 V respectively after 5 hours of operation at 0.1 A cm^{-2}. Postmortem analysis of the membrane and AEM binder recovered from the electrodes, using 1D and 2D NMR spectroscopy, revealed degradation *via* the backbone hydrolysis mechanism. The degradation occurred preferentially in the vicinity of the oxygen evolution electrode. Backbone hydrolysis resulted in a loss of AEM's mechanical integrity as well as solubilization and loss of binder in the electrodes. Impedance spectroscopy revealed an increase in both the high frequency resistance (from 0.41 to 0.53 Ohm cm^{-2} for BrPPO-TMA^+, and from 0.83 to 1.86 Ohm cm^{-2} for BrPPO-$ABCO^+$) and in the charge transfer resistance (from 0.26 to 1.47 Ohm cm^{-2} for BrPPO-TMA^+, and from 0.28 to 6.77 Ohm cm^{-2} for BrPPO-$ABCO^+$) over this timeframe, corroborating degradation of the membrane separator and the electrode binder.

Figure 13 shows the initial electrolysis polarization curves in our experiments, which were obtained on MEAs prepared with BrPPO-TMA^+ OH^- and BrPPO-$ABCO^+$ OH^-. The performance of the solid alkaline

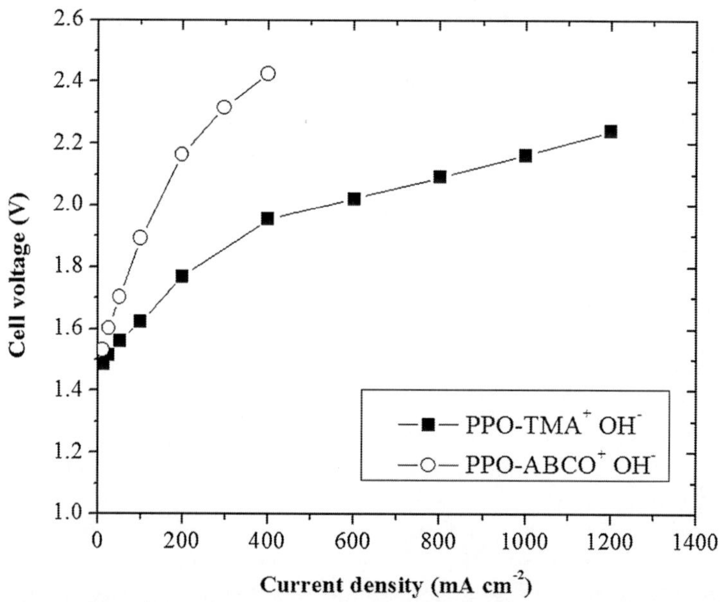

Fig. 13 Initial polarization curves for alkaline water electrolysis at 50 °C using BrPPO-TMA^+OH^- and BrPPO-$ABCO^+OH^-$ AEMs. Reproduced from J. Parrondo and V. Ramani, Stability of Poly(2,6-dimethyl 1,4-phenylene)Oxide-Based Anion Exchange Membrane Separator and Solubilized Electrode Binder in Solid-State Alkaline Water Electrolyzers, *J. Electrochem. Soc.*, 2014, **161**(10), F1015–F1020. Copyright the Authors 2014. Published by ECS under the terms of the Creative Commons Attribution 4.0 License (CC BY http://creativecommons.org/licenses/by/4.0/).

water electrolyzer using the BrPPO-TMA$^+$OH$^-$ membrane and binder was slightly lower than the performances reported by Hickner and co-workers[127] and Xiao and coworkers[128] for polysulfone based MEAs, where current densities of 400 mA cm^{-2} at 1.8 V were achieved. The lower performance attributed to the lower ionic conductivities of BrPPO-TMA$^+$ and BrPPO-ABCO$^+$ when compared to the corresponding polysulfone-based AEMs. Chloride ion conductivity at 50 °C for trimethylammonium functionalized polysulfone was 35 mS cm^{-1}, compared to 20 mS cm^{-1} for trimethylammonium functionalized PPO (this comparison was made for AEMs having approximately the same ion exchange capacity of 2 mmol g^{-1}).[52,53] A second factor responsible for the lower performance was the use of non-optimized GDEs instead of catalyzed membranes.[127] The lower performance of the electrolyzer prepared with the BrPPO-ABCO$^+$ AEM compared with that prepared with the BrPPO-TMA$^+$ AEM was attributed to the lower ion conductivity of BrPPO-ABCO$^+$.

To study AEM and solubilized AEM binder degradation during the operation of the electrolyzer, the MEA was held at a current density of 0.1 A cm^{-2} for 5 hours. Samples of the membrane separator and the AEM binder were recovered from the tested MEAs and used for postmortem NMR spectroscopy analysis. Figure 14 shows the post-mortem ^{1}H NMR spectra of the BrPPO-TMA$^+$ membrane separator, and of the cathode and

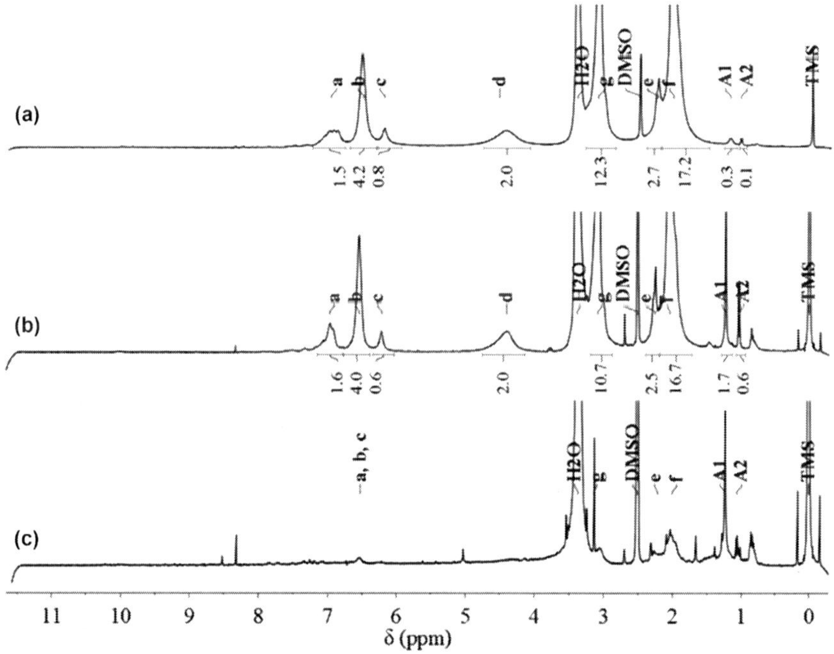

Fig. 14 ^{1}H NMR spectra of (a) BrPPO-TMA$^+$ membrane, (b) cathode AEM binder and (c) anode AEM binder after electrolyzer operation at 0.1 Acm^{-2} for 5 hours. Reproduced from J. Parrondo and V. Ramani, Stability of Poly(2,6-dimethyl 1,4-phenylene)Oxide-Based Anion Exchange Membrane Separator and Solubilized Electrode Binder in Solid-State Alkaline Water Electrolyzers, *J. Electrochem. Soc.*, 2014, **161**(10), F1015–F1020.

anode binders. We first probed the degradation of the quaternary ammonium groups in BrPPO-TMA$^+$ by evaluating the ratio between the areas under the peaks corresponding to the protons "g" and "d". The absence of degradation of the cation groups involved in ionic conduction gave raise to added questions regarding the reasons for the observed loss in the performance. Physical observation of the GDEs after the current hold experiment showed significant deterioration of the catalyst layer, with most of the catalyst (especially at the anode side) lost.

The ^{1}H NMR spectra of the AEM membrane and binders after testing in the electrolyzer (Fig. 14) for 5 hours (at 0.1 A cm^{-2}) displayed two new peaks (A1 and A2) at approximately 1.2 and 0.9 ppm, which increased in height for samples exhibiting larger degradation extents. The peaks A1 and A2 were larger for the binder in the anode than for the binder in the cathode. In our previous studies of AEM alkaline stability (in KOH solutions), we assigned these peaks to backbone degradation products.[52,78,79] Due to their upfield shift, the two new peaks were believed to correspond to protons in alkyl chains far from carbonyls, aromatic rings, or any other electron withdrawing groups. Both PSF and PPO AEMs yielded similar peaks after alkaline stability tests.[52,78] The most probable mechanism for the backbone degradation reaction was ether hydrolysis (leading to chain scission) followed by a ring opening reaction.[52,78] The backbone degradation mechanism was further supported by the brittleness of the membranes in stability experiments, the reduction in their tensile strength,[52] and by the poor adherence of the catalyst to the GDE that we observed during our experiments.

7 Summary and concluding remarks

There are ongoing efforts for the development of chemically and mechanically stable anion exchange membranes for application as electrolytes and separators in electrochemical devices like alkaline polymer electrolyte fuel cells, solid-state alkaline membrane water electrolysers, and redox flow batteries. The conductivities of AEMs in OH^- form can be as high as the H^+ conductivity in proton exchange membranes such as Nafion®. The main challenge is chemical stability under alkaline environments at the high temperature commonly required for cell operation (80 °C). The cationic ammonium, phosphonium, and metal cation groups are prone to OH^- nucleophilic attachment, especially at high temperatures, leading to a loss of membrane conductivity. Even when the polymer backbone employed to functionalize the cationic groups is stable in alkali conditions, after functionalization with cations, if may lose stability and experience chain scission, leading to mechanical failure. To make stable AEM, research should be focused not only on the cation alkaline stability but on the effects and interactions of the cation with the polymer backbone, and on methods to minimize the interactions (*e.g.*, spacer chains).

For application as separators in redox flow battery, the AEMs need to be stable under strong acids and in the presence of redox species (*e.g.*, vanadium species). The idea behind the use of AEMs is that the membrane permselectivity can permit enough rejection of the active cationic species (*i.e.* vanadium cations) to minimize active species crossover, maintain high energy efficiency, and exhibit low capacity fade over repeated cycling.

Acknowledgements

We would like to thank the Office of Naval Research (Grant number N00014-14-1-0705) for funding the work performed in our laboratory summarized in this book chapter.

References

1 C. G. Arges and V. Ramani, *J. Electrochem. Soc.*, 2013, **160**, F1006–F1021.
2 S. Yun, J. Parrondo and V. Ramani, *J. Mater. Chem. A*, 2014, **2**, 6605–6615.
3 G. Wang, Y. Weng, D. Chu, D. Xie and R. Chen, *J. Membr. Sci.*, 2009, **326**, 4–8.
4 Q. Hu, Y. Shang, Y. Wang, M. Xu, S. Wang, X. Xie, Y. Liu, H. Zhang, J. Wang and Z. Mao, *Int. J. Hydrogen Energy*, 2012, **37**, 12659–12665.
5 C. G. Arges, L. Wang, J. Parrondo and V. Ramani, *J. Electrochem. Soc.*, 2013, **160**, F1258–F1274.
6 C. G. Arges, L. Wang, M.-S. Jung and V. Ramani, *J. Electrochem. Soc.*, 2015, **162**, F686–F693.
7 J. Parrondo, M.-S. J. Jung, Z. Wang, C. G. Arges and V. Ramani, *J. Electrochem. Soc.*, 2015, **162**, F1236–F1242.
8 M. R. Hibbs, *J. Polym. Sci., Part B: Polym. Phys.*, 2013, **51**, 1736–1742.
9 C. Fujimoto, D.-S. Kim, M. Hibbs, D. Wrobleski and Y. S. Kim, *J. Membr. Sci.*, 2012, **423–424**, 438–449.
10 A. G. Wright and S. Holdcroft, *ACS Macro Lett.*, 2014, **3**, 444–447.
11 O. D. Thomas, K. J. W. Y. Soo, T. J. Peckham, M. P. Kulkarni and S. Holdcroft, *J. Am. Chem. Soc.*, 2012, **134**, 10753–10756.
12 R. Vinodh, A. Ilakkiya, S. Elamathi and D. Sangeetha, *Mater. Sci. Eng. B*, 2010, **167**, 43–50.
13 L. Sun, J. Guo, J. Zhou, Q. Xu, D. Chu and R. Chen, *J. Power Sources*, 2012, **202**, 70–77.
14 G. Sudre, S. Inceoglu, P. Cotanda and N. P. Balsara, *Macromolecules*, 2013, **46**, 1519–1527.
15 T.-H. Tsai, S. P. Ertem, A. M. Maes, S. Seifert, A. M. Herring and E. B. Coughlin, *Macromolecules*, 2015, **48**, 655–662.
16 H. Herman, R. C. T. Slade and J. R. Varcoe, *J. Membr. Sci.*, 2003, **218**, 147–163.
17 Y.-C. Cao, X. Wang, M. Mamlouk and K. Scott, *J. Mater. Chem.*, 2011, **21**, 12910–12916.
18 T. J. Clark, N. J. Robertson, H. A. Kostalik Iv, E. B. Lobkovsky, P. F. Mutolo, H. D. Abruña and G. W. Coates, *J. Am. Chem. Soc.*, 2009, **131**, 12888–12889.
19 M. Zhang, J. Liu, Y. Wang, L. An, M. D. Guiver and N. Li, *J. Mater. Chem. A*, 2015, **3**, 12284–12296.
20 H. A. Kostalik, T. J. Clark, N. J. Robertson, P. F. Mutolo, J. M. Longo, H. D. Abruña and G. W. Coates, *Macromolecules*, 2010, **43**, 7147–7150.
21 J. Hu, D. Wan, W. Zhu, L. Huang, S. Tan, X. Cai and X. Zhang, *ACS Appl. Mater. Interfaces*, 2014, **6**, 4720–4730.

22 C. G. Arges, S. Kulkarni, A. Baranek, K.-J. Pan, M.-S. Jung, D. Patton, K. A. Mauritz and K. Ramani Vijay, *Prepr. Symp. – Am. Chem. Soc., Div. Fuel Chem.*, 2010, **55**, 286–287.
23 G. M. Wu, S. J. Lin and C. C. Yang, *J. Membr. Sci.*, 2006, **275**, 127–133.
24 M. A. Vandiver, J. L. Horan, Y. Yang, E. T. Tansey, S. Seifert, M. W. Liberatore and A. M. Herring, *J. Polym. Sci., Part B: Polym. Phys.*, 2013, **51**, 1761–1769.
25 A. D. Mohanty, C. Y. Ryu, Y. S. Kim and C. Bae, *Macromolecules*, 2015, **48**, 7085–7095.
26 A. D. Mohanty, S. E. Tignor, J. A. Krause, Y.-K. Choe and C. Bae, *Macromolecules*, 2016, **49**, 3361–3372.
27 N. T. Rebeck, Y. Li and D. M. Knauss, *J. Polym. Sci., Part B: Polym. Phys.*, 2013, **51**, 1770–1778.
28 J. Zhou, M. Ünlü, I. Anestis-Richard, H. Kim and P. A. Kohl, *J. Power Sources*, 2011, **196**, 7924–7930.
29 C. G. Arges, M.-S. Jung, G. Johnson, J. Parrondo and V. Ramani, *ECS Trans.*, 2011, **41**, 1795–1816.
30 G. Merle, M. Wessling and K. Nijmeijer, *J. Membr. Sci.*, 2011, **377**, 1–35.
31 C. G. Arges and V. Ramani, *Proc. Natl. Acad. Sci. U. S. A.*, 2013, **110**, 2490–2495, S2490/2491–S2490/2419.
32 C. G. Arges, J. Parrondo, G. Johnson, A. Nadhan and V. Ramani, *J. Mater. Chem.*, 2012, **22**, 3733–3744.
33 S. Yun, J. Parrondo and V. Ramani, *Int. J. Hydrogen Energy*, 2016, **41**, 10766–10775.
34 Y. Xiong, Q. L. Liu and Q. H. Zeng, *J. Power Sources*, 2009, **193**, 541–546.
35 O. I. Deavin, S. Murphy, A. L. Ong, S. D. Poynton, R. Zeng, H. Herman and J. R. Varcoe, *Energy Environ. Sci.*, 2012, **5**, 8584–8597.
36 R. Zeng and J. R. Varcoe, *Recent Pat. Chem. Eng.*, 2011, **4**, 93–115.
37 T. Xu and W. Yang, *J. Membr. Sci.*, 2001, **190**, 159–166.
38 K. Matsui, E. Tobita, K. Sugimoto, K. Kondo, T. Seita and A. Akimoto, *J. Appl. Polym. Sci.*, 1986, **32**, 4137–4143.
39 S. Banerjee and D. E. Curtin, *J. Fluorine Chem.*, 2004, **125**, 1211–1216.
40 T. N. Danks, R. C. T. Slade and J. R. Varcoe, *J. Mater. Chem.*, 2002, **12**, 3371–3373.
41 T. N. Danks, R. C. T. Slade and J. R. Varcoe, *J. Mater. Chem.*, 2003, **13**, 712–721.
42 J. R. Varcoe, R. C. T. Slade and E. Lam How Yee, *Chem. Commun.*, 2006, 1428–1429.
43 N. Tzanetakis, J. R. Varcoe, R. C. T. Slade and K. Scott, *Desalination*, 2005, **174**, 257–265.
44 S. D. Poynton, J. P. Kizewski, R. C. T. Slade and J. R. Varcoe, *Solid State Ionics*, 2010, **181**, 219–222.
45 J. R. Varcoe and R. C. T. Slade, *Electrochem. Commun.*, 2006, **8**, 839–843.
46 J. R. Varcoe, R. C. T. Slade, E. Lam How Yee, S. D. Poynton, D. J. Driscoll and D. C. Apperley, *Chem. Mater.*, 2007, **19**, 2686–2693.
47 W. Wang, S. Wang, X. Xie, Y. lv and V. Ramani, *Int. J. Hydrogen Energy*, 2014, **39**, 14355–14361.
48 J. R. Varcoe, P. Atanassov, D. R. Dekel, A. M. Herring, M. A. Hickner, P. A. Kohl, A. R. Kucernak, W. E. Mustain, K. Nijmeijer, K. Scott, T. Xu and L. Zhuang, *Energy Environ. Sci.*, 2014, 7, 3135–3191.
49 C. G. Arges and V. Ramani, *Proc. Natl. Acad. Sci.*, 2013, **110**, 2490–2495.

50 J. Parrondo, M.-S. J. Jung, Z. Wang, C. G. Arges and V. Ramani, *J. Electrochem. Soc.*, 2015, **162**, F1236–F1242.
51 M. R. Hibbs, *J. Polym. Sci., Part B: Polym. Phys.*, 2013, **51**, 1736–1742.
52 C. G. Arges, L. Wang, J. Parrondo and V. Ramani, *J. Electrochem. Soc.*, 2013, **160**, F1258–F1274.
53 C. G. Arges, J. Parrondo, G. Johnson, A. Nadhan and V. Ramani, *J. Mater. Chem.*, 2012, **22**, 3733–3744.
54 C. G. Morandi, R. Peach, H. M. Krieg and J. Kerres, *J. Mater. Chem. A*, 2015, **3**, 1110–1120.
55 Z. Wang, J. Parrondo and V. Ramani, *J. Electrochem. Soc.*, 2016, **163**, F824–F831.
56 Y. Ye, S. Sharick, E. M. Davis, K. I. Winey and Y. A. Elabd, *ACS Macro Lett.*, 2013, **2**, 575–580.
57 M. Sudoh, S. Niimi, N. Takaoka and M. Watanabe, *ECS Trans.*, 2010, **25**, 61–70.
58 A. Voge, V. Deimede and J. K. Kallitsis, *RSC Adv.*, 2014, **4**, 45040–45049.
59 D. S. Kim, C. H. Fujimoto, M. R. Hibbs, A. Labouriau, Y.-K. Choe and Y. S. Kim, *Macromolecules*, 2013, **46**, 7826–7833.
60 S. Gu, R. Cai, T. Luo, Z. Chen, M. Sun, Y. Liu, G. He and Y. Yan, *Angew. Chem., Int. Ed.*, 2009, **48**, 6499–6502.
61 S. Gu, R. Cai, T. Luo, K. Jensen, C. Contreras and Y. Yan, *ChemSusChem*, 2010, **3**, 555–558.
62 S. Gu, R. Cai and Y. Yan, *Chem. Commun.*, 2011, **47**, 2856–2858.
63 X. Kong, K. Wadhwa, J. G. Verkade and K. Schmidt-Rohr, *Macromolecules*, 2009, **42**, 1659–1664.
64 K. J. T. Noonan, K. M. Hugar, H. A. Kostalik, E. B. Lobkovsky, H. D. Abruña and G. W. Coates, *J. Am. Chem. Soc.*, 2012, **134**, 18161–18164.
65 B. Zhang, S. Gu, J. Wang, Y. Liu, A. M. Herring and Y. Yan, *RSC Adv.*, 2012, **2**, 12683–12685.
66 M. L. Disabb-Miller, Y. Zha, A. J. DeCarlo, M. Pawar, G. N. Tew and M. A. Hickner, *Macromolecules*, 2013, **46**, 9279–9287.
67 Y. Liu, J. Wang, Y. Yang, T. M. Brenner, S. Seifert, Y. Yan, M. W. Liberatore and A. M. Herring, *J. Phys. Chem. C*, 2014, **118**, 15136–15145.
68 S. Gu, R. Cai, T. Luo, Z. Chen, M. Sun, Y. Liu, G. He and Y. Yan, *Angew. Chem., Int. Ed.*, 2009, **48**, 6499–6502, S6499/6491–S6499/6494.
69 Y. He, J. Pan, L. Wu, Y. Zhu, X. Ge, J. Ran, Z. Yang and T. Xu, *Sci. Rep.*, 2015, **5**, 13417.
70 J. Pan, C. Chen, Y. Li, L. Wang, L. Tan, G. Li, X. Tang, L. Xiao, J. Lu and L. Zhuang, *Energy Environ. Sci.*, 2014, **7**, 354–360.
71 L. Zhu, J. Pan, Y. Wang, J. Han, L. Zhuang and M. A. Hickner, *Macromolecules*, 2016, **49**, 815–824.
72 G. Merle, M. Wessling and K. Nijmeijer, *J. Membr. Sci.*, 2011, **377**, 1.
73 C. G. Arges, V. Ramani and P. N. Pintauro, *Electrochem. Soc. Interface*, 2010, **19**, 31.
74 J. R. Varcoe and R. C. T. Slade, *Fuel Cells*, 2005, **187**, 5.
75 S. Chempath, B. R. Einsla, L. R. Pratt, C. S. Macomber, J. M. Boncella, J. A. Rau and B. S. Pivovar, *J. Phys. Chem. C*, 2010, **114**, 11977.
76 W. R. Brasen and C. R. Hauser, *Org. Synth.*, 1963, **4**, 585.
77 S. H. Pine, *The Base-Promoted Rearrangements of Quaternary Ammonium Salts*, New York, 2011.
78 C. G. Arges and V. Ramani, *Proc. Natl. Acad. Sci. U. S. A.*, 2013, **110**, 2490–2495, S2490/2491–S2490/2419.

79 C. G. Arges and V. Ramani, *J. Electrochem. Soc.*, 2013, **160**, F1006–F1021.
80 S. Chempath, B. R. Einsla, L. R. Pratt, C. S. Macomber, J. M. Boncella, J. A. Rau and B. S. Pivovar, *J. Phys. Chem. C*, 2008, **112**, 3179–3182.
81 M. G. Marino and K. D. Kreuer, *ChemSusChem*, 2015, **8**, 513–523.
82 Y. Ye and Y. A. Elabd, *Macromolecules*, 2011, **44**, 8494–8503.
83 C. G. Arges, J. Parrondo, G. Johnson, A. Nadhan and V. Ramani, *J. Mater. Chem.*, 2012, **22**, 3733.
84 F. Helfferich, *Ion Exchange*, Dover Publications, 1995.
85 N. Lakshminarayaniah, *J. Membr. Biol.*, 1975, **21**, 175–189.
86 X. Ren, S. C. Price, A. C. Jackson, N. Pomerantz and F. L. Beyer, *ACS Appl. Mater. Interfaces*, 2014, **6**, 13330–13333.
87 V. Mazinani, S. Tabaian, M. Rezaei, M. Mallahi, M. Mohammadijoo and H. Omidvar, *J. Fuel Cell Sci. Technol.*, 2014, **11**, 031004.
88 M. Schreiber, M. Harrer, A. Whitehead, H. Bucsich, M. Dragschitz, E. Seifert and P. Tymciw, *J. Power Sources*, 2012, **206**, 483–489.
89 M. Skyllas-Kazacos, M. H. Chakrabarti, S. A. Hajimolana, F. S. Mjalli and M. Saleem, *J. Electrochem. Soc.*, 2011, **158**, R55–R79.
90 G. Kear, A. A. Shah and F. C. Walsh, *Int. J. Energy Res.*, 2012, **36**, 1105–1120.
91 C. Ponce de León, A. Frías-Ferrer, J. González-García, D. A. Szánto and F. C. Walsh, *J. Power Sources*, 2006, **160**, 716–732.
92 B. Schwenzer, J. Zhang, S. Kim, L. Li, J. Liu and Z. Yang, *ChemSusChem*, 2011, **4**, 1388–1406.
93 B. Dunn, H. Kamath and J.-M. Tarascon, *Science*, 2011, **334**, 928–935.
94 M. Skyllas-Kazacos, D. Kasherman, D. R. Hong and M. Kazacos, *J. Power Sources*, 1991, **35**, 399–404.
95 A. Z. Weber, M. M. Mench, J. P. Meyers, P. N. Ross, J. T. Gostick and Q. Liu, *J. Appl. Electrochem.*, 2011, **41**, 1137–1164.
96 W. Wang, S. Kim, B. Chen, Z. Nie, J. Zhang, G.-G. Xia, L. Li and Z. Yang, *Energy Environ. Sci.*, 2011, **4**, 4068–4073.
97 M. Skyllas-Kazacos and Y. Limantari, *J. Appl. Electrochem.*, 2004, **34**, 681–685.
98 W. Wang, Z. Nie, B. Chen, F. Chen, Q. Luo, X. Wei, G.-G. Xia, M. Skyllas-Kazacos, L. Li and Z. Yang, *Adv. Energy Mater.*, 2012, **2**, 487–493.
99 Y. K. Zeng, T. S. Zhao, L. An, X. L. Zhou and L. Wei, *J. Power Sources*, 2015, **300**, 438–443.
100 D. P. Scamman, G. W. Reade and E. P. L. Roberts, *J. Power Sources*, 2009, **189**, 1220–1230.
101 J. Cheng, L. Zhang, Y.-S. Yang, Y.-H. Wen, G.-P. Cao and X.-D. Wang, *Electrochem. Commun.*, 2007, **9**, 2639–2642.
102 P. K. Leung, C. Ponce de León and F. C. Walsh, *Electrochem. Commun.*, 2011, **13**, 770–773.
103 P. K. Leung, C. Ponce-de-León, C. T. J. Low, A. A. Shah and F. C. Walsh, *J. Power Sources*, 2011, **196**, 5174–5185.
104 Z. Xie, D. Zhou, F. Xiong, S. Zhang and K. Huang, *J. Rare Earths*, 2011, **29**, 567–573.
105 M. Skyllas-Kazacos, M. Rychcik, R. G. Robins, A. G. Fane and M. A. Green, *J. Electrochem. Soc.*, 1986, **133**, 1057–1058.
106 M. Rychcik and M. Skyllas-Kazacos, *J. Power Sources*, 1988, **22**, 59–67.
107 S. Kim, J. Yan, B. Schwenzer, J. Zhang, L. Li, J. Liu, Z. Yang and M. A. Hickner, *Electrochem. Commun.*, 2010, **12**, 1650–1653.
108 F. Rahman and M. Skyllas-Kazacos, *J. Power Sources*, 1998, **72**, 105–110.

109 C. Ding, H. Zhang, X. Li, T. Liu and F. Xing, *J. Phys. Chem. Lett.*, 2013, **4**, 1281–1294.
110 B. Sun and M. Skyllas-Kazacos, *Electrochim. Acta*, 1992, **37**, 1253–1260.
111 B. Sun and M. Skyllas-Kazacos, *Electrochim. Acta*, 1992, **37**, 2459–2465.
112 P. Han, H. Wang, Z. Liu, X. Chen, W. Ma, J. Yao, Y. Zhu and G. Cui, *Carbon*, 2011, **49**, 693–700.
113 Q. Luo, L. Li, W. Wang, Z. Nie, X. Wei, B. Li, B. Chen, Z. Yang and V. Sprenkle, *ChemSusChem*, 2013, **6**, 268–274.
114 T. Mohammadi and M. Skyllas-Kazacos, *J. Membr. Sci.*, 1995, **98**, 77–87.
115 H. Prifti, A. Parasuraman, S. Winardi, T. M. Lim and M. Skyllas-Kazacos, *Membranes*, 2012, **2**, 275.
116 D. Chen, M. A. Hickner, E. Agar and E. C. Kumbur, *Electrochem. Commun.*, 2013, **26**, 37–40.
117 B. Zhang, S. Zhang, D. Xing, R. Han, C. Yin and X. Jian, *J. Power Sources*, 2012, **217**, 296–302.
118 M.-S. J. Jung, J. Parrondo, C. G. Arges and V. Ramani, *J. Mater. Chem. A*, 2013, **1**, 10458–10464.
119 D. Chen, S. Kim, L. Li, G. Yang and M. A. Hickner, *RSC Adv.*, 2012, **2**, 8087–8094.
120 J. Fang and P. K. Shen, *J. Membr. Sci.*, 2006, **285**, 317–322.
121 S. Yun, J. Parrondo and V. Ramani, *ChemPlusChem*, 2015, **80**, 412–421.
122 L. Li and Y. Wang, *J. Membr. Sci.*, 2005, **262**, 1–4.
123 J. Parrondo, C. G. Arges, M. Niedzwiecki, E. B. Anderson, K. E. Ayers and V. Ramani, *Rsc Adv.*, 2014, **4**, 9875–9879.
124 W. Kreuter and H. Hofmann, *Int. J. Hydrogen Energy*, 1998, **23**, 661–666.
125 H. Wendt, *Electrochemical Hydrogen Technologies: Electrochemical Production and Combustion of Hydrogen*, Elsevier, Distributors for the U.S. and Canada, Elsevier Science Pub. Co., Amsterdam, New York, NY, U.S.A., 1990.
126 J. Xu, G.-P. Sheng, H.-W. Luo, W.-W. Li, L.-F. Wang and H.-Q. Yu, *Water Res.*, 2012, **46**, 1817–1824.
127 Y. Leng, G. Chen, A. J. Mendoza, T. B. Tighe, M. A. Hickner and C. Wang, *J. Am. Chem. Soc.*, 2012, **134**, 9054–9057.
128 L. Xiao, S. Zhang, J. Pan, C. Yang, M. He, L. Zhuang and J. Lu, *Energy Environ. Sci.*, 2012, **5**, 7869.
129 Y. Li, T. Xu and M. Gong, *J. Membr. Sci.*, 2006, **279**, 200–208.
130 S. Zhang, C. Wu, T. Xu, M. Gong and X. Xu, *J. Solid State Chem.*, 2005, **178**, 2292–2300.
131 G. Couture, A. Alaaeddine, F. Boschet and B. Ameduri, *Prog. Polym. Sci.*, 2011, **36**, 1521–1557.
132 B. Pivovar, *DOE Hydrogen and Fuel Cells Program Review: Advanced Ionomers & MEAs for Alkaline Membrane Fuel Cells*, 2016, https://www.hydrogen.energy.gov/pdfs/review16/fc147_pivovar_2016_o.pdf.
133 Q. L. He, S. Khene, X. Ren, F. E. López-Suárez, D. Lozano-Castelló, A. Bueno-López and G. Wu, *J. Phys. Chem. C*, 2013, **117**, 8697–8707.
134 R. P. Pandey and V. K. Shahi, *J. Mater. Chem. A*, 2013, **1**, 14375–14383.
135 S. Gu, J. Skovgard and Y. S. Yan, *ChemSusChem*, 2012, **5**, 843–848.
136 S. Gu, W. Sheng, R. Cai, S. M. Alia, S. Song, K. O. Jensen and Y. Yan, *Chem. Commun.*, 2013, **49**, 131–133.
137 R. Zeng, J. Handsel, S. D. Poynton, A. J. Roberts, R. C. T. Slade, H. Herman, D. C. Apperley and J. R. Varcoe, *Energy Environ. Sci.*, 2011, **4**, 4925–4928.

138 Y. S. Kim, *Resonance-Stabilized Anion Exchange Polymer Electrolytes, DOE: Annual Merit Review*, 2012, http://www.hydrogen.energy.gov/pdfs/review12/fc043_kim_2012_p.pdf.
139 S. Gottesfeld, *CellEra Perspective on AMFCs*, 2011, http://www1.eere.energy.gov/hydrogenandfuelcells/wkshp_alkaline_membrane.html.
140 K. Fukuta, Electrolyte Materials for AMFCs Electrolyte Materials for AMFCs and AMFC Performance and AMFC Performance.
141 Y. He, J. Si, L. Wu, S. Chen, Y. Zhu, J. Pan, X. Ge, Z. Yang and T. Xu, *J. Membr. Sci.*, 2016, **515**, 189–195.
142 K. Matsumoto, T. Fujigaya, H. Yanagi and N. Nakashima, *Adv. Funct. Mater.*, 2011, **21**, 1089–1094.

Anodic materials for electrooxidation of alcohols in alkaline media

Sadia Kabir and Alexey Serov*
DOI: 10.1039/9781782622727-00061

1 Introduction

The tremendous amount of effort that was invested towards developing sustainable alternatives to conventional hydrocarbon-based energy technologies are finally beginning to pay off. The growing distribution of fuel cell vehicles (FCVs) in the public markets of Japan, USA and Europe pioneered by some of the world leaders in automobile manufacturing such as Toyota and Hyundai have demonstrated that fuel cells are no longer limited to niche markets. For example, fuel cell powered busses and other mobile vehicles (forklifts *etc.*) have been widely deployed in many cities around the globe by various municipal authorities as well as governmental entities. The increasing demand of reliable backup systems for critical civilian infrastructures such as hospitals, cell phones towers *etc.* have also resulted in an aggressive penetration of the existing market by fuel cell manufacturing companies such as Ballard and Intelligent Energy. Countries with a decentralized heating supply to the residential properties are now looking forward to integrating Combined Heat and Power (CHP) devices in newly constructed buildings. These CHP devices can inherently produce both electricity and heat for everyday needs of tenants from a single fuel source.[1,2]

The first practical fuel cell resulted from the work begun in England in 1932 by F. T. Bacon resulted in design of 1.5 kW Apollo alkaline fuel ce11[3] operating with a KOH–H_2O electrolyte solution. However, the technology that revolutionized the current state-of-the-art fuel cells resulted from the Gemini earth-orbiting space mission (1962 to 1965) that utilized a solid polymer electrolyte – called an ion-exchange membrane (IEM) at that time – consisting of a cationic membrane of sulfonated polystyrene resin. Since then, different types of fuel cells developed over the years – primarily classified according to (i) the type of fuel they use, (ii) the temperatures they operate at and (iii) the electrolyte or membrane utilized for transferring the ions generated during the oxidation/reduction reactions. The membranes are a part of fuel cell Membrane Electrode Assembly (MEA) that consist of a separator (membrane) and a proton conductive binder (ionomer). In general, fuel cells with a proton exchange Nafion® membrane – comprising of a polytetrafluoroethylene (PTFE) backbone with sulfonate heads – is

Department of Chemical & Biological Engineering and Center for Micro-Engineered Materials (CMEM), The University of New Mexico, Advanced Materials Laboratory, 1001 University Blvd. SE, Suite 103, Albuquerque, New Mexico, USA.
E-mail: serov@unm.edu

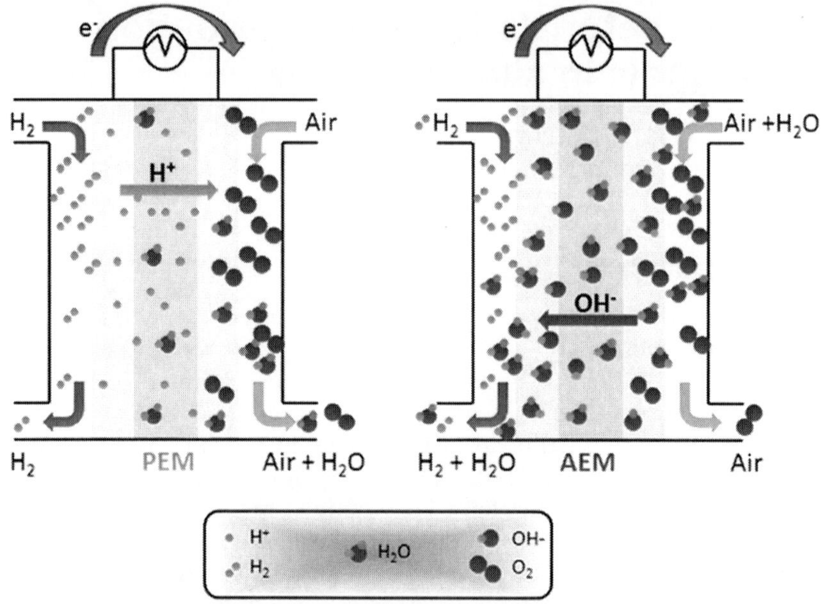

Fig. 1 Schematic comparison of a proton exchange membrane fuel cell (PEMFC, left) and an anion exchange membrane fuel cell (AEMFC, right) that are supplied with H_2 and air. Reproduced from ref. 4 with permission from the Royal Society of Chemistry.

widely used for facilitating the transfer of cations (H^+) between anode and cathode catalyst layers (Fig. 1).[4] The operation of fuel cells is based on the oxidation of fuels in the anode compartment and electro-reduction of oxygen from the air in the cathode side of MEA. Electro-catalysts such as platinum (Pt) are routinely utilized to facilitate the kinetics of the redox reactions by decreasing the activation barrier. The electricity is then generated by harvesting the flow of electrons from the anode to the cathode compartment *via* an external circuit, while protons from the oxidation reactions are transported from the anode to the cathode side of MEA.

Since the 1960s, proton exchange membranes such as Nafion® by Dupont have been widely deployed due to several important factors: (i) the membranes and ionomers have high proton conductivity, (ii) the materials are thermally stable up to 100 °C, (iii) the membranes possess the appropriate mechanical properties and (iv) the overall performance and durability of PEMFCs comply with the design points required for different applications. However, one of the major disadvantages of existing proton exchange membrane fuel cells (PEMFCs) is the limitation on the type of fuels which can be electro-oxidized on anode. Due to the slower kinetics of carbonaceous fuels oxidation, and low tolerance of platinum to the intermediate byproducts generated from the oxidation of organic fuels, the majority of low temperatures PEMFCs have continued to operate with hydrogen. However, safety issues with high pressure H_2 storing tanks (700 atm) on board cars and buses have been a matter of additional concern. The limitations associated with hydrogen production, purification, storage and transportation logistics are posing a considerable barrier to commercialization of the hydrogen-based fuel cells. Moreover, the efficient operation of PEM fuel cells require a high concentration of H^+ ions, which results in extremely low pH (<1). Such harsh and corrosive conditions substantially limit the range under which the electrocatalysts can perform with stability.

1.1 Anion exchange membranes and their advantages

There is an alternative to the proton exchange membrane which utilizes the separator and ionomer made from polymeric materials. The alternative membranes are called Anion Exchange Membranes (AEM) or Alkaline Electrolyte Membranes due to their ability to conduct hydroxyl ions (OH^-) (see Fig. 1). AEMs are solid polymer electrolyte membranes that contain positive ionic groups (usually quaternary ammonium (QA) functional groups such as poly-NMe_3^+) and mobile negatively charged anions. The replacement of the traditional caustic aqueous potassium hydroxide [KOH(aq)] electrolyte with suitable commercial-available such as those from Tokuyama Co (Japan), Fumatech (Germany) or other AEM suppliers have shown great potential to substitute PEMs due to their enhanced catalyst electrokinetic performances, described below.[5,6] The fabrication of membrane electrode assemblies (MEAs) and their corresponding performance/durability in AEM fuel cells were recently reviewed[7] and their advantages summarized as follows:

(i) Reaction kinetics are comparatively faster under alkaline conditions. For example, experimental data obtained from Fourier-Transform Infrared (FTIR) spectroscopic measurements showed considerably higher anodic performances in carbonate/bicarbonate solutions (alkaline conditions) compared to sulfuric acid (acid conditions).[8]

(ii) Due to the fact that corrosion effect is minimized in alkaline environments, the use of AEMs also eliminates the need for precious metal catalysts such as Pt, allowing other types of Pt-free electrocatalysts to be utilized for fuel oxidation on the anode as well as oxygen reduction on cathode,[9] substantially decreasing the total price of final fuel cell systems and making them more affordable for customers.

(iii) The issues associated with fuel crossover from the anode to the cathode compartment are also minimized.[10,11]

(iv) Substitution of the conventional liquid electrolytes with AEMs – where cationic moieties are fixed to polymer chains (and are not freely mobile as in a liquid electrolyte) – can minimize the formation and precipitation of carbonates (CO_3^{2-}/HCO_3^-) as there is no liquid-phase dissociated cation which can react with carbonate anions to form insoluble species). Hence solid AEMs can enhance the durability of MEAs under operating conditions.[5,12]

(v) One of the other opportunities provided by AEM based fuel cells is the ability to utilize logistically favourable fuels and oxidants without being limited by hydrogen.

With improvements made in AEM technology in terms of membrane conductivity and stability, a great deal of attention has gone into developing low-temperature solid anion conducting membranes direct liquid fuel cell (DLFC) devices that can directly electro-oxidize (i) inorganic fuels such as ammonia, hydrazine hydrate, ammonium borane, sodium borohydride *etc.*[13–15] and (ii) small organic carbonaceous molecules (for example: alcohols, aldehydes, salts of organic acids[16,17]) for providing an alternative path to power (Table 1, Fig. 2).[18,19]

1.2 AEM based Direct Alcohol Fuel Cells (DAFC)

Although the oxidation of small organic oxygenated molecules such as alcohols: ethanol, methanol, ethylene glycol and glycerol in acidic media[21] has been extensively investigated since the early 1970's as potential fuels, the developments in AEMs have recaptured the attention of researchers due to increased oxidation kinetics (particularly for organic fuels) from lower overpotentials and reduced fuel crossover in alkaline media.[20,22–26] For instance, the oxidation of methanol, ethanol and ethylene glycol and glycerol has shown much higher current densities in alkaline media compared to the acidic media. Furthermore, the C–C bonds present in alcohols with higher molecular weights are more easily broken in an alkaline environment when compared to that in an acidic environment. In electrolytes with a high pH, the de-protonation of alcohols with higher molecular weight is also facilitated by the high concentration of OH^- in the electrolyte and adsorbed hydroxyl on the surface of catalysts, significantly lowering the energy barrier of alcohol oxidation reaction.

For example, the rate constant for methanol adsorption in 1 M NaOH was found to be about 20-fold of that in 0.5 M H_2SO_4, and the removal of adsorbed poisoning species such as CO was much faster in alkaline media due to the availability of oxygen-containing species.[28] Moreover, the usage of liquid fuels such as alcohols in AEM fuel cells compared to hydrogen-powered PEM fuel cells are considered to be very promising due to their higher gravimetric and volumetric energy densities, as shown in Table 2.

One of the other advantages of using alcohols as fuels is that they are liquid under ambient temperature and pressure, which eliminates the need for compression, making them easier to transport and viable for portable electronic devices and other commercial applications. Alcohol based fuels also have an already existing infrastructure from gasoline distribution in place, making them convenient to store and transport, in comparison to compressed hydrogen. Furthermore, these alcohols can be directly obtained from low cost bio-mass fermentation or as by-products from other industrial processes and used as a fuel without any prior reforming, whereas pure hydrogen must be produced through complex processes involving steam reforming of hydrocarbons[30] or other methods such as electrochemical water-splitting. Moreover, the oxidation of larger alcohols such as glycerol co-generates electricity as well as other value added chemicals such as formic acid, which can be utilized towards other applications, such as Direct Formic Acid Fuel cells.[31–34]

Although the electrooxidation of alcohol fuels in alkaline media has been the subject of increasing interest, the kinetics of alcohol oxidation

Table 1 Summary of fuels reported for direct oxidation alkaline fuel cells applications and their theoretical cell potentials. Reproduced from ref. 27 with permission from the Royal Society of Chemistry.

Fuel	Anode reactions	E°_a / V	Cathode reactions	E°_c / V	E°_{cell} / V	Energy density/W h kg^{-1}
Methanol CH_3OH	$CH_3OH + 6OH^- \rightarrow CO_2 + 5H_2O + 6e^-$	−0.81	$\frac{1}{2}O_2 + 3H_2O + 6e^- \rightarrow 6OH^-$	0.40	1.21	6100
Ethanol C_2H_5OH	$CH_3CH_2OH + 2OH^- \rightarrow CH_3CHO + 2H_2O + 2e^-$ $CH_3CH_2OH + 4OH^- \rightarrow CH_3COOH + 3H_2O + 4e^-$ $CH_3CH_2OH + 12OH^- \rightarrow 2CO_2 + 9H_2O + 12e^-$	−0.77	$3O_2 + 6H_2O + 12e^- \rightarrow 12OH^-$	0.40	1.17	8030
Iso-propanol $CH_3CHOHCH_3$	$CH_3CHOHCH_3 + 2OH^- \rightarrow CH_3COCH_3 + 2H_2O + 2e^-$ $CH_3COCH_3 + 16OH^- \rightarrow 3CO_2 + 11H_2O + 16e^-$	−0.67	$\frac{9}{2}O_2 + 9H_2O + 18e^- \rightarrow 18OH^-$	0.40	1.07	8600
Ethylene glycol $(CH_2OH)_2$	$(CH_2OH)_2 + 14OH^- \rightarrow 2CO_3^{2-} + 10H_2O + 10e^-$ or $(CH_2OH)_2 + 10OH^- \rightarrow (CO^2)_2^{2-} + 8H_2O + 6e^-$	−0.72	$\frac{1}{2}O_2 + H_2O + 2e^- \rightarrow 2OH^-$	0.40	1.12	5200
Glycerol $HOCH_2CHOHCH_2OH$	$HOCH_2CHOHCH_2OH + 20OH^- \rightarrow 3CO_3^{2-} + 14H_2O + 14e^-$ or $HOCH_2CHOHCH_2OH + 12OH^- \rightarrow COO^-\text{–}COH\text{–}COO^- + 10H_2O + 10e^-$	−0.69	$\frac{1}{2}O_2 + H_2O + 2e^- \rightarrow 2OH^-$	0.40	1.09	5000
Urea $CO(NH_2)_2$	$CO(NH_2)_2 + 6OH^- \rightarrow N_2 + CO_2 + 5H_2O + 6e^-$	−0.75	$\frac{1}{2}O_2 + H_2O + 2e^- \rightarrow 2OH^-$	0.40	1.15	2927
Glucose $C_6H_{12}O_6$	$C_6H_{12}O_6 + 2OH^- \rightarrow C_6H_{12}O_7 + 2e^-$	−0.85	$\frac{1}{2}O_2 + H_2O + 2e^- \rightarrow 2OH^-$	0.40	1.25	4430
Dimethyl ether CH_3OCH_3	$CH_3OCH_3 + 12OH^- \rightarrow 9CO_2 + 2H_2O + 12e^-$	−0.80	$3O_2 + 6H_2O + 12e^- \rightarrow 12OH^-$	0.40	1.20	7100
Cellulose $(C_6H_{10}O_5)_n$	Converted to different cellulose derivatives	—	$\frac{1}{2}O_2 + H_2O + 2e^- \rightarrow 2OH^-$	0.40	—	
Ammonia NH_3	$2NH_3 + 6OH^- \rightarrow N_2 + 3H_2O + 6e^-$	−0.77	$\frac{3}{2}O_2 + 3H_2O + 6e^- \rightarrow 6OH^-$	0.40	1.17	3300
Hydrogen H_2	$2H_2 + 4OH^- \rightarrow 4H_2O + 4e^-$	−0.83	$\frac{1}{2}O_2 + H_2O + 2e^- \rightarrow 2OH^-$	0.40	1.23	32 160
Hydrazine N_2H_4	$N_2H_4 + 4OH^- \rightarrow 4H_2O + N_2 + 4e^-$	−1.17	$\frac{1}{2}O_2 + H_2O + 2e^- \rightarrow 2OH^-$ $H_2O_2 + 2e \rightarrow 2OH^-$	O_2 : 0.40 H_2O_2 : 0.88	1.57 2.05	5260
Sodium borohydride $NaBH_4$	$NaBH_4 + 8OH^- \rightarrow NaBO_2 + 6H_2O + 8e^-$	−1.24	$\frac{1}{2}O_2 + H_2O + 2e^- \rightarrow 2OH^-$ $H_2O_2 + 2e^- \rightarrow 2OH^-$	O_2 : 0.40 H_2O_2 : 0.88	1.64 2.12	9000

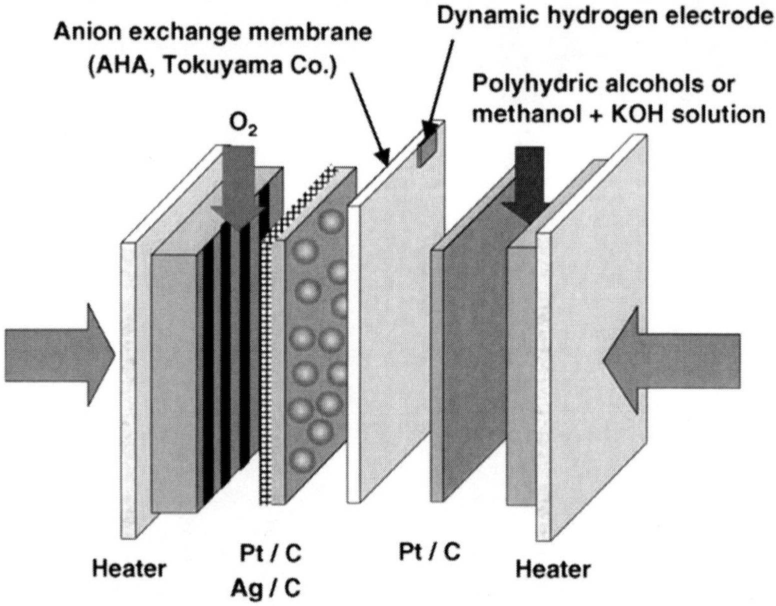

Fig. 2 Schematic representation of a direct alkaline fuel cell assembled using an anion exchange membrane. Reproduced from K. Matsuoka, Y. Iriyama, T. Abe, M. Matsuoka and Z. Ogumi, Alkaline direct alcohol fuel cells using an anion exchange membrane, *J. Power Sources*, **150**, 27–31. Copyright 2005, with permission from Elsevier.

Table 2 The comparison of energetic properties of alcohols with hydrogen as fuels. Reproduced from ref. 29 with permission from Elsevier.

Property	Hydrogen	Methanol	Ethanol	Ethylene glycol
Formula	H_2	CH_3OH	C_2H_5OH	$(CH_2OH)_2$
$-\Delta G^{\circ a}$ (kJ mol^{-1})	237	702	1325	1180
$-\Delta H^b$ (kJ mol^{-1})	286	725	1367	1192
Energy density, LHV (kW h kg^{-1})	33	6.09	8.0	5.29
Energy density, LHV (kW h L^{-1})	2.96×10^{-3}	4.80	6.32	5.80
$E^{\circ}_{cell}{}^{c}$ (V)	1.23	1.21	1.14	1.22
$\eta_{theo}{}^{d}$	2	6	12	10

^{a} $-\Delta G^{\circ}$: Gibbs free energy change of the reaction (at standard temperature and pressure).
^{b} $-\Delta H$: to the enthalpy change.
^{c} E°_{cell}: the overall efficiency of the cell.
^{d} η_{theo}: theoretical number of electrons involved in the complete fuel oxidation reaction.

on known catalysts are still considerably slower than hydrogen oxidation reaction. The slower kinetics corresponds to one of the significant challenges in the development of direct alcohol fuel cell (DAFC) technologies. Previous reviews on the same subject have shown that the overall efficiency and performance of DAFCs is primarily influenced by the nature of the electrocatalyst.[26,35–37] It has therefore become imperative to design and fabricate highly active electrocatalysts that can activate the C–C bond cleavage mechanism involved in the oxidation of alcohols[38] for yielding the desired efficiencies and achieve maximum power densities. In view of that, this chapter will summarize and provide a comprehensive overview of:

(i) The fundamental and mechanistic understanding of alcohols electrooxidation pathways on the surface of catalysts – in particular ethanol, methanol, ethylene glycol and glycerol – in alkaline media.

(ii) Current state-of-the-art mono-, bi- and multi-metallic of Pt and Pd based alloys that have been recently developed for promoting the electrooxidation of alcohols.

(iii) A molecular level understanding of the interaction between the alcohols and the catalytic active sites from Density Functional Theory (DFT) calculations.

(iv) The electrochemical activity, selectivity towards CO_2 formation, yield of value added products, as well as tolerance towards CO and adsorbed intermediates poisoning will be discussed in detail.

2 Mechanisms of alcohol oxidation reactions in alkaline media

Harvesting the maximum energy of an alcohol molecule requires its complete conversion to CO_2 and H_2O, as it releases the theoretically possible maximum number of electrons. Since the alcohol radical group contains only one oxygen atom (R-OH), the extra oxygen atom must be provided from chemisorption of water molecules, or water adsorbed residues (*e.g.* adsorbed OH^-) produced by the oxygen reduction reaction (ORR) in the cathode compartment according to the following equation:

$$O_2 + 2H_2O + 4e^- \rightarrow 4OH^- \tag{1}$$

However, the reaction mechanism of alcohol oxidation is generally complex in nature, typically involving various successive and parallel reaction pathways, encompassing several adsorbed species. The pathway also almost always involves the formation of numerous intermediates and by products, including poisonous species such as adsorbed CO on the surface of the catalyst. A significant number of fundamental studies on alcohol oxidation mechanisms have been performed on state-of-the-art Pt based catalysts using various analytical techniques, such as Fourier transform infrared (FTIR) spectroscopy,[39] and differential electrochemical mass spectrometry (DEMS).[53] These techniques are routinely used to analyze the reaction intermediates adsorbed or formed on catalyst surfaces and/or in liquid electrolyte during methanol, ethanol, ethylene glycol or glycerol electrooxidation. The identification of reaction intermediates and poisoning species on different catalysts provides valuable information for developing efficient anode electrocatalysts for DAFCs. In the following sub section, the mechanisms of alcohols electrooxidation on the surface of platinum electrocatalysts (as a model well-studied systems) are summarized.

2.1 Methanol Oxidation Reaction (MOR)

The mechanism of methanol oxidation reaction is generally considered to be a complex process, which depends on the structure and composition of the catalysts, reaction conditions, such as pH of the electrolyte, temperature as well as fuel and electrolyte concentration. The oxidation of methanol in alkaline media usually takes place in a series of reactions involving successive electron transfers (oxidation) to adsorbed species which react with adsorbed OH^- to potentially form CO_2 and other products. One of the proposed MOR mechanism proceeds as follows:[40]

$$CH_3OH + 4OH^- \rightarrow CO_{ads} + 4H_2O + 4e^- \quad (2)$$

$$OH^- \rightarrow OH_{ads} + e^- \quad (3)$$

$$CO_{ads} + 2OH_{ads} + 2OH^- \rightarrow CO_3{}^{2-} + 2H_2O \quad (4)$$

The formation of other intermediates are also possible:

$$CH_3OH + 3OH^- \rightarrow COH_{ads} + 3H_2O + 3e^- \quad (5)$$

$$CH_3OH + OH^- \rightarrow CH_3O_{ads} + H_2O + e^- \quad (6)$$

$$CH_3OH \rightarrow CH_3OH_{ads} \quad (7)$$

The overall reaction of MOR in alkaline media can be simplified as:

$$CH_3OH + 6OH^- \rightarrow CO_2 + 5H_2O + 6e^- \quad (8)$$

$$(E_0 = 0.810 \text{ V } \textit{vs.} \text{ SHE @ pH} = 14)$$

It is widely accepted that the most significant reactions of the overall process are the adsorption of methanol and the oxidation of adsorbed CO. On the other hand, the tightly bound CO can also poison the catalyst's surface.[40] While some studies confirmed that the main product of methanol oxidation reaction is CO_2,[41,42] most other studies have also concluded that the reaction usually proceed according to the multiple step mechanism, where significant amounts of formaldehyde,[43] and formate[44,45] can form due to the variety adsorbed intermediates such as R-CO and OH^- species, which can be further oxidized to CO_2.

2.2 Ethanol Oxidation Reaction (EOR)

Ethanol is considered to be an attractive alternative fuel to methanol due to several reasons: (i) ethanol has a higher energy density (8.1 kWh kg^{-1}) compared to methanol (6.1 kWh kg^{-1}) (ii) ethanol also has lower toxicity; and more importantly, (iii) ethanol can be obtained from renewable resources like corn; all of which can improve the safety and sustainability of direct liquid fed fuel cells.

The mechanism of ethanol electrooxidation and their kinetics have been extensively studied on Pt/C catalysts. In alkaline media, the complete electrooxidation of ethanol releases up to twelve electrons:

$$CH_3CH_2OH + 12OH^- \rightarrow 2CO_2 + 9H_2O + 12e^- \quad (9)$$

But similar to methanol oxidation, the ethanol oxidation reaction is believed to be an equally complex multi-step reaction proceeding through sequential and parallel reaction steps, C1- and C2:[46] involving active intermediate such as CH_3CHO_{ads} and poisoning intermediate CO_{ads} according to two pathways:

C2 pathway:

$$CH_3CH_3OH \rightarrow CH_3CHO_{ads} \rightarrow CH_3COOH/CH_3COO^- \text{ (alkaline)} \quad (10)$$

$$CH_3CH_3OH \rightarrow CH_3CHO_{ads} \rightarrow CH_3CHO \quad (11)$$

C1 pathway:

$$CH_3CH_3OH \rightarrow CH_3CHO_{ads} \rightarrow CO_{ads}, CH_{x,ads} \rightarrow CO_2/CO_3{}^{2-} \text{ (alkaline)} \quad (12)$$

$$CH_3CH_3OH \rightarrow CO_{ads}, CH_{x,ads} \rightarrow CO_2 \text{ or } CO_3{}^{2-} \text{ (alkaline)} \quad (13)$$

The C1 pathway involves the complete oxidation of ethanol to CO_2 or carbonates *via* CO_{ads} intermediate by delivering 12 electrons according to the following equations:

$$CH_3CH_2OH + 3H_2O \rightarrow 2CO_2 + 12H^+ + 12e^- \quad (14)$$

$$CH_3CH_2OH + 5H_2O \rightarrow 2HCO_3{}^- + 14H^+ + 12e^- \quad (15)$$

$$CH_3CH_2OH + 5H_2O \rightarrow 2CO_3{}^{2-} + 16H^+ + 12e^- \quad (16)$$

On the other hand, unlike the complete oxidation of ethanol to CO_2 that yields 12 electrons per ethanol molecule, the C2 pathway involves the partial oxidation of ethanol to acetaldehyde ($CH_3CH{=}O$) or the adsorption of ethanol molecule on the active sites followed by reaction with hydroxyl ions to produce CH_3COOH (acetic acid), which exists in the form of CH_3COO^- ions (acetate) in alkaline solution.[47] This incomplete oxidation of ethanol liberates only up to 2 and 4 electrons respectively:

$$CH_3CH_2OH \rightarrow CH_3CHO + 2H^+ + 2e^- \quad (17)$$

$$CH_3CH_2OH + H_2O \rightarrow CH_3COOH + 4H^+ + 4e^- \quad (18)$$

More specifically, it has been proposed that oxidation of ethanol on the surface of the metallic catalysts (M) involves the formation of M-$(CH_3CO)_{ads}$ upon C–H cleavage and the removal of the adsorbed ethoxy by the adsorbed hydroxyl as the rate-determining step (r.d.s.), according to the following equations:[48,49]

$$M + CH_3CH_2OH \leftrightarrow M\text{-}(CH_3CH_2OH)_{ads} \tag{19}$$

$$M\text{-}(CH_3CH_2OH)_{ads} + 3OH^- \leftrightarrow M\text{-}(CH_3CO)_{ads} + 3H_2O + 3e^- \tag{20}$$

$$M + OH^- \leftrightarrow M\text{-}OH_{ads} + e^- \tag{21}$$

The rate determining step involves the reaction of the M-$(CH_3OH)_{ads}$ with the activated M-OH_{ads}, forming acetic acid species:

$$M\text{-}(CH_3CO)_{ads} + M\text{-}OH_{ads} \rightarrow M\text{-}CH_3COOH + M \tag{22}$$

$$M\text{-}CH_3COOH + OH^- \rightarrow M + CH_3COO^- + H_2O \tag{23}$$

As it can be seen from the equations above, the C2 pathway primarily involves the activation of both the C–H and O–H bonds in ethanol, instead of the C–C bond. That is because the complete oxidation of ethanol to CO_2 *via* C–C bond cleavage is mechanistically or sterically suppressed. Moreover, strongly adsorbed species such as CO_{ads} can potentially poison the catalysts' surface and reduce the rate of charge transfer considerably. As a result of the incomplete oxidation and poisonous intermediates, the Faradic efficiency of the fuel cell diminishes greatly.[36] Hence, the challenge is to completely oxidize ethanol, which releases up to 12 electrons, and eliminates or minimize the generation of aldehydes and acetic acid or acetates.

2.3 Ethylene Glycol Oxidation Reaction (EGOR)

In the last few years, several research groups have focused on development of a relatively new type of fuel cell that uses ethylene glycol (EG) as the liquid fuel.[50] Ethylene glycol can be directly obtained from catalytic conversion of cellulose with high yields, and its relatively good reactivity makes it a feasible candidate for fuel cell applications. Similar to ethanol and methanol, ethylene glycol has as a specific energy density of 8.6 $kWh\,kg^{-1}$.[51] The complete oxidation of ethylene glycol to CO_2 generated up to 10 electrons according to the equation:

$$(CH_2OH)_2 + 10OH^- \rightarrow 2CO_2 + 8H_2O + 10e^- \tag{24}$$

However, further investigations performed by various groups have shown that the majority of the EG is only partially oxidized, as C–C bond is comparatively strong and cannot be easily cleaved. Particularly, ethylene glycol can be oxidized into oxalate species (–OOC=COO–), where 8 electrons are produced for each mole of ethylene glycol, with a Faradic efficiency (FE) of 80%.[52] Similar to ethanol, methanol and glycerol oxidation, a number of reaction intermediates[53] including glycolaldehyde, glyoxal, glycolic acid, glyoxylic acid and oxalic acid are usually generated. For example, in a recent analysis of ethylene glycol oxidation by Pt/C in alkaline media detected two intermediate products of formate and glycolate.[54] Glycolate was produced *via* the 4-electron oxidation (eqn (25)), in which C–C bond remained intact, whereas formate was generated *via* 6-electron oxidation (eqn (26)) *via* the C–C bond cleavage of ethylene glycol:

$$(CH_2OH)_2 + 5OH^- \rightarrow CH_2OHCOO^- + 4H_2O + 4e^- \quad (25)$$

$$(CH_2OH)_2 + 8OH^- \rightarrow 2HCOO^- + 6H_2O + 6e^- \quad (26)$$

The oxidation of ethylene glycol in alkaline media was also investigated using a three-electrode cell with on-line sample collection system.[55] Glycolic acid, oxalic acid and formic acid were sequentially produced from the oxidation of ethylene glycol on Pt/C with increasing linear staircase scan voltammetry.

2.4 Glycerol Oxidation Reaction (GOR)

Recently, the electrochemical conversion and oxidation of glycerol – an abundant and inexpensive poly-alcoholic compound generated as a byproduct in biodiesel and bioethanol production processes – has gained increasing attention for its potential to cogenerate electrical energy and added value chemicals, such as glyceric acid, tartronic acid, mesoxalic acid, glycolic acid, *etc.*[56] In an alkaline exchange membrane fuel cells, the complete oxidation of glycerol involves the generation of 14 electrons per molecule of glycerol:[57]

$$CH_2OH\text{–}CHOH\text{–}CH_2OH + 20OH^- \rightarrow 3CO_3^{2-} + 14H_2O + 14e^- \quad (27)$$

It also should be mentioned that the complete electrooxidation of glycerol requires C–C bond breaking in a complex multi-step mechanism, especially taking into account its larger molecular weight. Similarly, the generation of partially oxidized byproducts through a series of parallel and sequential reactions involves the exchange of less than 14 electrons and loss in total theoretical efficiency.

The viability of utilizing glycerol as a fuel not only depends on the energy yield of the oxidation reaction, but also its ability to generate valuable oxidized products and produce a large number of useful chemicals from the complex oxidation pathways.[58] For example, the formation of specific value added products (VAP) such as formic acid would be of great interest since formic acid can be further utilized towards other electrochemical applications, such as Direct Formic Acid Fuel Cells.[32] For such purposes, glycerol – which is a non-valued residue of biodiesel production – can be considered to be a secondary as well as primary source of alternative energy. Therefore, it is important that the product-current trade-off in the selectivity of glycerol oxidation reaction is precisely controlled towards the formation of desired products as well as CO_2 formation in order to maximize the energy output. The control of activity and selectivity can be achieved by range of applied electrode potential and the utilization of an electrocatalysts with different chemistries.

3 Anode materials for DAFCs

For the alcohol fed AEM fuel cells to be highly efficient, electrocatalysts should be designed to effectively activate the C–C bond cleavage mechanism and completely oxidize the alcohol to CO_2 or desired value added products in a cogeneration system. The analysis of intermediates produced during alcohol oxidation as described in the previous section demonstrated that the degree of complete oxidation of the alcohols as well as selectivity of the catalysts towards CO_2 and product distribution significantly depend on the efficiency of the C–C bond cleavage. Therefore, identifying highly effective anode electrocatalysts that can activate C–C bond cleavage mechanism involved in the oxidation of alcohols has becomes imperative to improve the overall reaction rate of the DAFCs

Currently, platinum and platinum based alloys are considered to the most active electrocatalysts for alcohol oxidation, which is mainly due to the fact that majority of works in this field was performed with Pt and Pt-compounds. The comprehensive study of the electrooxidation of the several C1–C3 alcohols by various research groups have demonstrated that platinum and platinum-based materials can be used for generation of electricity in fuel cells as well production of different value added products (VAPs) as intermediates of incomplete fuel oxidation.[16,17,59] In addition to the extensively studied platinum catalysts, palladium-based systems have also received increased interest, especially due to its increased stability in alkaline media.[60] Understanding the intrinsic properties of Pt and Pd is important for the rational design of catalysts for alcohol electrooxidation in alkaline media.

3.1 Platinum and its alloys for alcohol electrooxidation

Until now, the most widely used catalyst for alcohol oxidation reactions has been platinum because of its ability to adsorb and dissociate of small organic molecules by cleaving the C–H bond cleavage.[22,61,62] However, it is well-known that platinum surface can be easily poisoned during the oxidation of oxygenated organic molecules due to formation of CO and other poisonous intermediates. This limits the application of Pt and Pt supported on carbon in real fuel cell systems. On the other side, the understanding the mechanistic aspects of alcohols electrooxidation reaction requires performing the experiments on model and well-defined mono-metallic platinum surfaces. Several detailed studies on utilization of Pt/C for alcohol electrooxidation were reported by multiple groups of researchers and will be summarized in this section.

Recently, platinum electrocatalysts supported on mono-dispersed spherical carbon support were studied in the reaction of methanol and ethanol electrooxidation (Fig. 3a–b).[63] The support materials were prepared by the composite-molten-salt (CMS) method, which is a modification of the low temperature carbonization of glucose as a carbon source in the melting flux of potassium and lithium nitrates. The mono-dispersity of support was confirmed by scanning electron microscope (SEM) imaging and prepared material was used for deposition of platinum nanoparticles by hydrothermal method. The electrooxidation of two alcohols: methanol

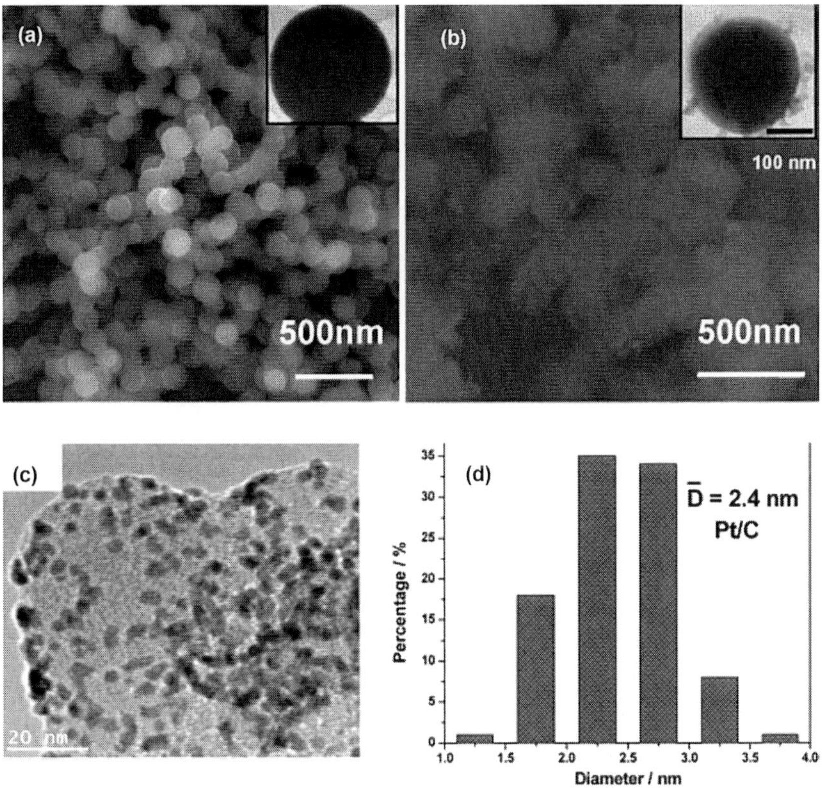

Fig. 3 (a) FESEM image of carbon nanospheres synthesized by the composite-molten-salt (CMS) method. TEM image in inset; (b) FESEM image of Pt nanoparticles supported on CMS carbon nanospheres. TEM image in inset. Reproduced from X. Wang, C. Hu, Y. Xiong, H. Liu, G. Du and X. He, Carbon-nanosphere-supported Pt nanoparticles for methanol and ethanol electro-oxidation in alkaline media, *J. Power Sources*, **196**, 1904–1908. Copyright 2011 with permission from Elsevier. (c) TEM images and (d) particle size distributions of Pt/C catalysts. Reproduced from L. Xin, Z. Zhang, J. Qi, D. Chadderdon and W. Li, Electrocatalytic oxidation of ethylene glycol (EG) on supported Pt and Au catalysts in alkaline media: Reaction pathway investigation in three-electrode cell and fuel cell reactors, *Appl. Catal., B*, **125**, 85–94. Copyright 2012 with permission from Elsevier.

and ethanol was performed in 1 M KOH solution on three Pt/C materials, where the carbons were: spherical mono-dispersed carbon made by CMS method, hydrothermally synthesized carbon and commercial carbon black. The activity for platinum nanoparticles supported on spherical carbon support synthesized by CMS were found to be higher compared to other supports. It was speculated that the increase was due to better mass transfer properties of Pt/C-CMS. However, since cyclic voltammograms usually do not provide detailed information about the mass transfer behaviour of catalysts, the performance of these Pt/C-CMS catalysts can be further investigated using polarization curves in operating DAFCs.

The electrooxidation of ethylene glycol on the platinum nanoparticles supported on carbon was studies with detailed analysis of products formed at different potentials (Fig. 3c–d).[55] It was shown that platinum can electrooxidize ethylene glycol with formation of glycolic, oxalic and formic acids. The formation of glycolic acid (GA) was detected by High

Performance Liquid Chromatography (HPLC), with the maximum concentration at 0.9 V. A decreased amount of GA was also observed at 0.85 V, which was attributed to formation of surface oxides on the platinum resulting in decrease of active sites available for electrooxidation of ethylene glycol. A similar behaviour was also noticed in the case of oxalic acid, where the reactions started at higher potentials (~0.6 V), with the maximum concentrations recorded at 0.9 V, followed by decrease in concentration due to PtO_x coverage. The formation of formic acid was shown to have two distinct peaks, between 0.6 and 0.9 V and in the range of 1.1–1.7 V.

Additional experiments on electrooxidation of glycolic acid in alkaline media by Pt/C catalysts revealed detailed insights into the mechanistic propagation of EG electrooxidation into two possible parallel pathways: quick electrooxidation of EG to glycolic acid, followed by further oxidation to oxalic acid and C–C bond cleavage in EG with formation of formic acid and later carbonate (Fig. 4).

The same group of researchers performed electrochemical studies on co-generation of electricity and value added products by Pt/C in fuel cell configuration using glycerol as a liquid fuel.[64] It was shown that parameters such as the concentration of alkaline electrolyte, fuel, as well as operational potential can have an effect on selectivity of catalysts towards the desirable products. These parameters were optimized mentioned to achieve a remarkable power density of 125 mW cm^{-2}, while selectivity to C3 acids was 91%.

In another study, platinum supported on nitrogen-doped graphene (N-doped graphene) materials were synthesized by low temperature polyol reduction method and tested for the electroreduction of methanol and ethanol.[65] The morphology and phase composition of catalysts were studied by SEM and XRD methods, however, the particle size of the Pt nanoparticles were neither determined by TEM particle size distribution nor from the Scherrer equation. The activity of Pt/N-doped Graphene was also compared with the Pd/N-doped Graphene and PdPt/N-doped Graphene. Platinum was shown to have the lowest value for half wave potentials ($E_{1/2}$), which was selected as an indication of intrinsic catalytic activity. The analysis of half-waves led to the conclusion that platinum was intrinsically more active towards oxidation of light alcohols (methanol and ethanol) compared to palladium. It should be noted that these observations contradict the results obtained in other studies (described below), where palladium catalysts were found to be more active than platinum. Moreover, based on the XRD analysis, the intrinsic activities referred in this study could also be due to the substantially different particles size of platinum and palladium nanoparticles. Since the Pt/N-doped GNS catalyst was reported to have the smallest particles with a highest number of exposed crystallographic facets – the shift of $E_{1/2}$ to the more negative value can be explained by increase of the kinetics of the rate limiting step (which is usually associated with adsorption of alcohol molecule on the surface of catalysts). The shift in half wave could also be due the difference in orientations of the Pt(110)[45] and Pt(111)[44] facets, where previous studies have shown the surface

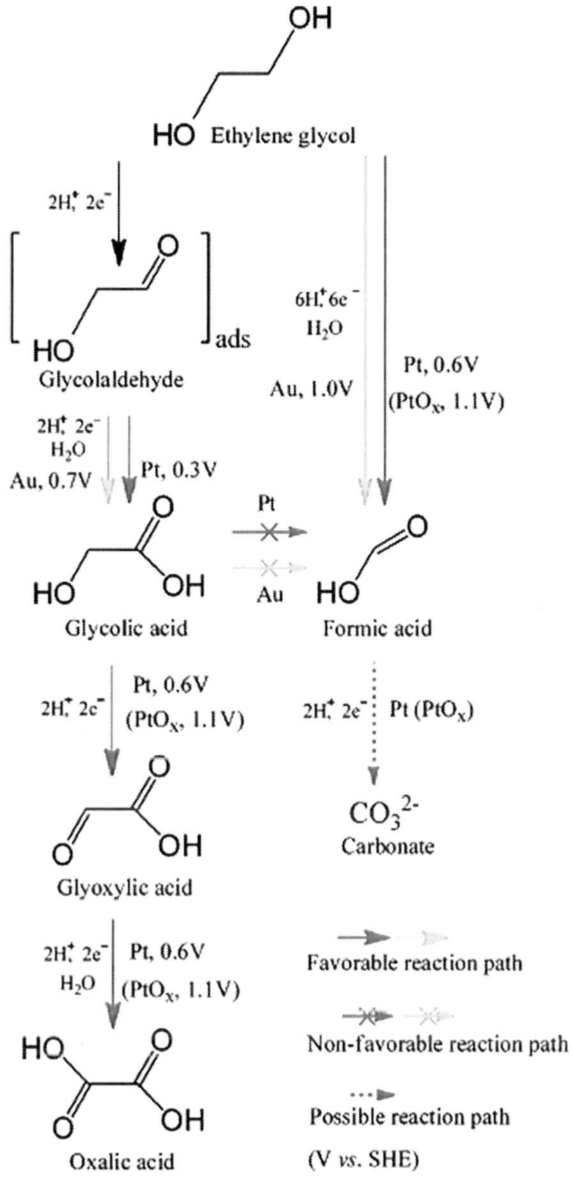

Fig. 4 The proposed pathways for electrocatalytic oxidation of ethylene glycol on Au/C and Pt/C in alkaline media, the starting potentials for observed reaction paths are marked. Reproduced from L. Xin, Z. Zhang, J. Qi, D. Chadderdon and W. Li, Electrocatalytic oxidation of ethylene glycol (EG) on supported Pt and Au catalysts in alkaline media: Reaction pathway investigation in three-electrode cell and fuel cell reactors, *Appl. Catal., B*, **125**, 85–94. Copyright 2012 with permission from Elsevier.

activities were directly correlated to the adsorbed species coverage. The chronoamperometric studies on prepared materials revealed that palladium is more stable towards poisoning effect of CO and intermediates compared to platinum, which is in perfect agreement with previously published data. The effect of poisoning by CO and other products of electrooxidation was more pronounced in the case of ethanol, which can be explained by production of different poisoning species (aldehyde and acetate).

However, platinum has a tendency to be readily poisoned by CO_{ads} and other intermediate species. In order to overcome these limitations, various studies have investigated alloying Pt with other metals such as Ru or Sn. For example, PtSn/C catalysts synthesized by polyol method with varying NaOH concentration during the synthesis protocol.[14] As a result of this synthetic approach, the two groups of catalysts were prepared: PtSn/C and PtSn-SnO_x/C, where the latter has a presence of tin oxide. It was demonstrated that high concentrations of tin oxide in combination with increased amount of platinum lead to an increase in ethanol electrooxidation even at lower overpotentials. Despite the initially promising results, chronoamperometric studies revealed the relatively low stability of Pt-Sn catalysts towards deactivation, with all catalysts losing more than 80% of initial activity after 3500 seconds.

Based on these studies, it seems that although Pt-based catalysts can sometimes exhibit a high intrinsic activity – they however have a tendency to deactivate due to CO poisoning during alcohol electrooxidation, which decreases the number of active sites available for alcohol oxidation and leads to a loss in performance.[66,67] Moreover, the high price of Pt catalysts is a major impediment in the commercialization of fuel cell technology, since it alone accounts for approximately half of the total fuel cell stack cost. This resulted in the increased attention directed towards the utilization of other comparatively abundant and relatively cheaper palladium (Pd) and Pd-based alloys as suitable catalyst for alcohol oxidation has been reported before.[35,37,47] Various studies have reported the higher activity and stability of Pd for ethanol electrooxidation than Pt in alkaline media.

For example, the electrochemical behaviour of commercially available platinum supported on carbon Pt/C and Pd/C catalysts in the reaction of ethanol electrooxidation reaction was studied by potentiodynamic and potentiostatic methods (Fig. 5).[68] Parameters such as concentration of electrolyte (NaOH), ethanol concentration and temperature were varied in this study. Initial CVs of both Pt/C and Pd/C in the electrolyte saturated with argon revealed the features which were attributed to hydroxyl ions adsorption on the surface of electrocatalysts (Fig. 5a–b). Particularly, peaks corresponding to the hydrogen adsorption/desorption between 0.1–0.3 V, and the cathodic peaks between 0.6–0.8 V from reduction of Pt and Pd surface oxides such as Pt/Pd-OH or Pt/Pd-O_x can be observed. It should also be noted that the area under the reduction peak between 0.6–0.8 V corresponding to the charge integration of oxide reduction is significantly greater for Pd/C catalysts compared to Pt/C, which is indicative of a higher electrochemically accessible surface area of the former.[69]

The negative (cathodic) scan resulted in reduction of PtO_x and PdO_x which is typical behaviour of these platinum group metals. The electrooxidation of ethanol was performed in two different concentrations of NaOH (0.1 M and 1 M) as well as four concentrations of ethanol (0.01, 0.1, 1 and 2 M). It was found that in 0.1 M NaOH, platinum and palladium revealed a characteristic peak of ethanol electrooxidation with similar

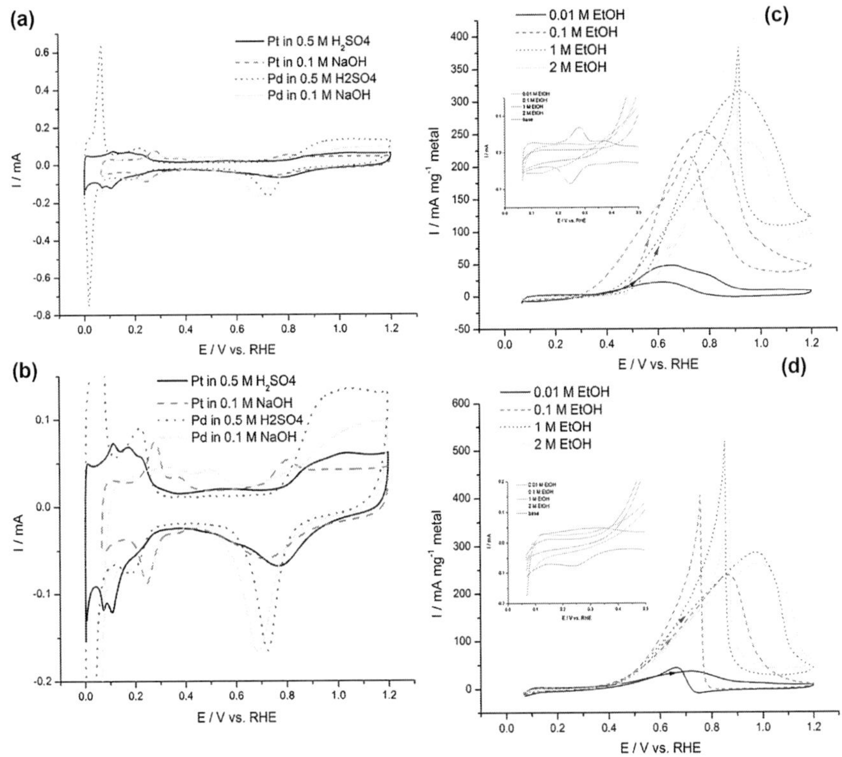

Fig. 5 Cyclic voltammograms (CV) of Pt/C and Pd/C in an Ar-saturated 0.1 M NaOH or 0.5 M H_2SO_4 solution, (a) full CVs, (b) enlarged CVs. CV in a 0.1 M NaOH ethanol solution, (c) Pt/C, and (d) Pd/C. Reproduced from L. Ma, D. Chu and R. Chen, Comparison of ethanol electro-oxidation on Pt/C and Pd/C catalysts in alkaline media, *Int. J. Hydrogen Energy*, **37**, 11185–11194. Copyright 2012 with permission from Hydrogen Energy Publications, LLC and Elsevier.

features of onset potential in the range of 0.35–0.4 V (*vs.* RHE) and highest peak current density with 1 M ethanol. Despite the similarities, the onset potential of oxidation peak was shifted to the positive part of potential range with increase of ethanol concentration. In contrast the shift dependence on fuel concertation in case of palladium was less pronounced. It was hypothesized that such trend was due to the formation of larger amounts of poisonous intermediates on the surface of Pt/C. Second, the Pt/C catalysts had a dual peak characteristic in the range of 0.65–0.8 V (*vs.* RHE) while lowest concentration of fuel was used (0.01 M), which is indicative of two main pathways: CO_2 and acetate formation. Third observation dealt with the fact that hysteresis of forward and backward scans on the platinum catalyst is significantly larger than one in case of palladium. That is another confirmation that intermediate products of ethanol electrooxidation have a higher adsorption affinity to platinum surface. Increase in electrolyte concentration from 0.1 to 1 M resulted in overall increase of peak current densities for platinum and palladium with the maximum at 1 M of ethanol. It was concluded that an increase in NaOH concentration resulted in an increase of OH_{ads}, which facilitated the complete oxidation of ethanol to CO_2 on platinum catalysts. Moreover, the effect of temperature was highly pronounced in the case of platinum with direct indication of higher conversion of ethanol to CO_2 with increase of the temperature. Chronoamperometric studies of ethanol oxidation on Pt/C showed a sharp decrease in current density at lower potentials. The current density also increased with increase of temperature and potential, which can be explained by the promotion of reaction pathways to full fuel oxidation with cleavage of C–C bonds. By comparing two noble metals (Pt/C and Pd/C), it can be concluded that despite the fact that Pd/C has higher current densities in alkaline solutions – the main pathway was primarily related to formation of acetate, which is undesirable for fuel cells applications. On the other side, Pt/C was found to be more prone to poisoning by C1 intermediate species with ability to mitigate it by operating at higher temperature and potentials.

Nevertheless, Pd based catalysts are now being considered to be promising alternatives due to several other reasons, such as their superior performance for the selective oxidation of alcohols to CO_2 or other value added products, and better resistance to CO_{ads} poisoning compared to Pt catalysts.[26,35,37,70–75] For example, a recent study showed that although both Pt and Pd activity decreased during the ethanol oxidation, the extent of Pd deactivation was smaller than that of Pt one.[65] Various groups have studied the alcohol mechanism reaction on Pd based catalysts using cyclic voltammetry, *in situ* Fourier transform infrared (FTIR) reflection spectroscopy, and Differential electrochemical mass spectrometry *etc.*[76,77] For example, ionic chromatography and ^{13}C Nuclear magnetic resonance spectroscopy (NMR) was used to analyze the anode exhausts of different alcohols on Pd/MWCNT (Multi Walled Carbon Nanotube) catalysts in galvanostatic experiments.[78] Ethanol was shown to be selectively oxidized to acetic acid, converted to acetate ion in the alkaline media of the reaction, while methanol yields carbonate, formate and CO_2. A much wider product distribution, including glycolate, glycerate, tartronate, oxalate, formate and carbonate, was obtained from the oxidation of glycerol. In another study, electrochemical Attenuated Total Reflection Surface Enhanced Infrared Reflection Absorption Spectroscopy (ATR-SEIRAS) and Infrared reflection–absorption spectroscopy (IRAS methods),[79] was used to explored the dissociation and electrooxidation of methanol on the surface of Pd electrodes. Methanol was shown to oxidize to either CO_2/bicarbonate or Co_{ad} and interfacial formate *via* a dual-pathway mechanism.

Hence, the reaction kinetics and mechanisms of all these oxidation reactions can be significantly affected by the structure, morphology and chemical composition of the materials. Therefore, it is an imperative to investigate the structure–property relationships of well characterized Pd based nanocomposites towards in order to gain a better understanding of their electrocatalytic properties and their oxidation mechanisms.

3.2 Palladium and its alloys for alcohol electrooxidation

In order to improve the activity and durability of the non-alloyed monometallic Pd electrocatalysts towards alcohols oxidation, it is important to have a good fundamental understanding of their electro-oxidation behavior and kinetics in alkaline media by understanding their structure–activity relationships. *Ab initio* Density Functional Theory (DFT) were used to calculate the electronic properties of different palladium surfaces and the interactions between an ethanol molecule and closed-packed Pd(111), the stepped Pd(110), and the open Pd(100) surfaces (Fig. 6A).[80] The reaction barriers of the first step dehydrogenation of an ethanol molecule on different palladium surfaces and complete electrooxidation pathways of ethanol on the most active palladium surface were calculated. It was estimated that the surface Pd–Pd bonds and electron energy density between atoms in surface, first and second layers – indicating that the atoms of Pd(100) allowed more interatomic interaction compared with those of Pd(110) and Pd(111) due to forceful electron coupling along the surface plane. Thus, the atoms in the surface layer of Pd(100) were deemed to be more sensitive in the electrocatalytic reaction of ethanol. Moreover, the binding energies calculated for the ethanol–Pd configurations showed that the interaction between the ethanol–Pd(100) was stronger than those of ethanol–Pd(110) and ethanol–Pd(111). However, there was only a small difference in binding energies, more specifically 0.03 eV, between ethanol–Pd(100) and ethanol–Pd(110). Their results also showed that the adsorbed ethanol species (CH_2CHOH)

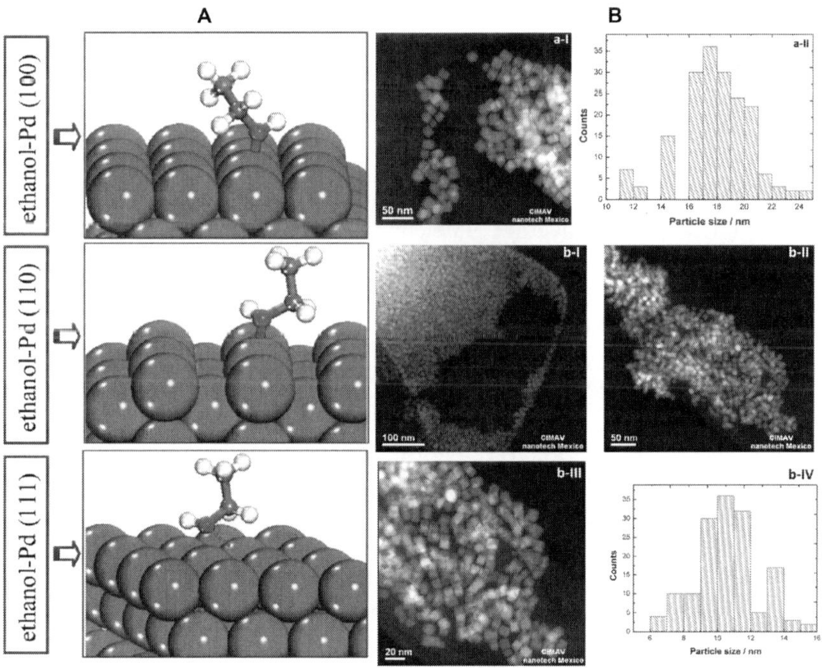

Fig. 6 (A) Optimized geometries of an ethanol molecule interacting with Pd(100), Pd(110), and Pd(111) surfaces. Reproduced with permission from E. D. Wang, J. B. Xu and T. S. Zhao, Density Functinoal Therory Studies of the Structure Sensitivity of Ethanol Oxidation on Palladium Surfaces, *J. Phys. Chem. C*, 2010, **114**, 10489–10497. Copyright 2010 American Chemical Society; (B) HR-TEM micrographs of Pd nanocubes synthesised with the support (a-I) and without the support (b-I to b-III), and their respective particle size distributions (a-II and b-IV, respectively). Reproduced from ref. 81 with permission from the Royal Society of Chemistry.

on the surface of Pd(100) were oxidized to acetic acid by the help of adsorbed OH^- species assisted in lowering the barrier energy.

Notable efforts have been carried out to design palladium-based catalysts and to investigate the electrocatalytic properties of ethanol oxidation by optimizing the structure of the nanoparticles. For instance, cubic shaped (200) Pd nanoparticles were obtained through a chemical reduction method in aqueous media by employing ascorbic acid, polyvinylpyrrolidone (PVP) and sodium bromide as the reducing agent, surfactant and additive, respectively (Fig. 6B).[81] Using cyclic voltammetry, the authors showed the current densities of the Pd nano-cubes towards ethanol and methanol electrooxidation exhibited a current density of 12 and 10 mA mg^{-1} cm^{-2}, respectively, which was about 4 folds higher than commercial Pd/Vulcan catalyst. However, it was noted that the surfactant PVP from was difficult to remove from the surface of the Pd nano-cubes, and an electrochemical method of repeated cycling in acidic and alkaline media was employed to remove the impurities. Unless eliminated completely, these polymeric surfactants can significantly affect the electrochemical accessible surface area of the nanoparticles, and in turn, their intrinsic activities.

One of the other methods implemented for enhancing the electrocatalytic activity of Pd nano-catalysts has been fabricating morphologically different structures by varying the shape or size of the Pd nanoparticles. For example, highly dispersed Pd nano-spheres were synthesized using polyoxypropylene–polyoxyethylene copolymer (P123) as a reducing agent and a stabilizer.[82] The size the Pd nano-spheres were controlled by varying the reaction temperature from 0–60 °C. Transmission electron micrographs showed that Pd nanoparticles synthesized at 0 °C (Pd0) were agglomerated to 200 nm size in diameter, whereas those synthesized at 30 and 60 °C had an average particle size of 7.5 nm (Pd30) and 15 nm (Pd60), respectively. Using cyclic voltammetry, it was shown that Pd nanospheres obtained at 30 °C (Pd30) had the highest catalytic performance, with a forward peak current density of 5.071 mA cm^{-2} for methanol oxidation in alkaline medium (1 M NaOH), compared to bulk Pd electrode, Pd0 and Pd60 nano-spheres. However, the surface area normalized activity of Pd60 nano-spheres (2.653 mA cm^{-2}) was lower than Pd0 (2.905 mA cm^{-2}), despite the presence of large agglomerates in the latter. The chronoamperometric studies showed that Pd30 nano-spheres were also the most stable. Moreover, EDS spectra of the Pd30 nano-spheres show a high carbon count, although no carbonaceous supports were used for the nano-spheres. This carbon count could be due to the P123 stabilizer still being present in the samples, which could affect nano-sphere stabilities and activities.

Formerly identified impediments accompanying subpar catalytic activity has usually originated from the existence of polymeric surfactants and their chemical bonding to the nanoparticle surface, which made them sensitive to poisoning, ultimately leading to deactivation. Recently, Pd nanoparticles were synthesized using a Soft Alcohol Reduction Method (SARM) where the Pd precursor $Pd(NO_3)_2{*}2H_2O$ was reducing using simple alcohols such as methanol (MeOH), ethanol (EtOH) and isopropyl alcohol (IPA) as reducing agents and deposited on 3D-Graphene nanosheets.[17] TEM particle size distribution analysis showed that when ethanol was used as a reducing agent (SARM-EtOH), the Pd nanoparticles had the smallest average particle size of 4.5 nm, which was on par with those reduced with IPA. However, those reduced using MeOH had Pd nanoparticles of ~30 nm in diameter (Fig. 7). The reduction in nanoparticle size was explained by the relative reduction power of the alcohols in the order of EtOH > IPA > MeOH. Moreover, the electroactivity of the Pd/3D-Graphene catalysts was studied by using cyclic voltammetry in 1 M KOH media. In both of the ethanol and methanol electrooxidation experiments, the most active material was derived from SARM-EtOH, which could be due to the by small size and the even dispersion of the Pd particles on the surface of the 3D-GNS. In a similar study, Pd nanoparticles were dispersed on 3D-Graphene nanosheets morphologically modified with different porosities using sacrificial silica templates for ethanol oxidation in alkaline media (Fig. 8).[83] It was shown that Pd nanoparticles deposited on thermally reduced 3D-Graphene nanosheets were the most stable, and had a peak current density of 1540 A/g(Pd), which was about 2.5 and 3 times greater than Pd nanoparticles supported on Vulcan (XC-72) and commercial Pt/C catalysts respectively. The results demonstrated that the nature of the supports – such as the templated pored of the 3D-Graphene matrix – can provide the underlying framework necessary for faster mass transport kinetics and hence, improved catalytic activity and durability.

Most of the studies have so far been focused on studying the effect of Pd morphology on electrochemical activity. However, the surface modification of Pd catalysts with organic ligands can also have an effect on electrochemical activity by modifying the nanoparticles core electronic structure and bonding properties. Recently, the surface of palladium nanoparticles were stabilized by the self-assembly of 1-octyne through metal–vinylidene interfacial bonding linkages (PdHC8).[84] The authors investigated the effects of surface ligands on electrochemical activity of ethylene glycol oxidation using TEM, Proton nuclear magnetic resonance (1H NMR) and FTIR. It was found that the average core size of PdHC8 was ~2.5 nm with average footprint of *ca.* 13 Å^2 per capping ligand on the nanoparticle surface. Using cyclic voltammetry in 0.1 M NaOH, it was shown that the PdHC8 nanoparticles displayed an enhancement of the electrocatalytic performance towards ethylene glycol oxidation, with an anodic peak *ca.* 70 mV more negative, and a mass specific peak current density of 1084.0 Ag^{-1}, which was twice as much in comparison with commercial Pt/C catalysts. The enhanced activity was ascribed to the electrochemically accessible surface area of 95.4 m^{2}g^{-1}, which was higher compared to Pd foams synthesized in another study.[85] This was indicative that that the organic ligand 1-octyne did not necessarily diminish the accessibility of the nanoparticle surface. On the other hand, the enhanced performance was also associated with the *partial removal* of the organic capping ligands, where it was speculated that the remaining molecules provided a hydrophobic environment for removal of water generated in the oxidative dehydration of adsorbed ethylene glycol.

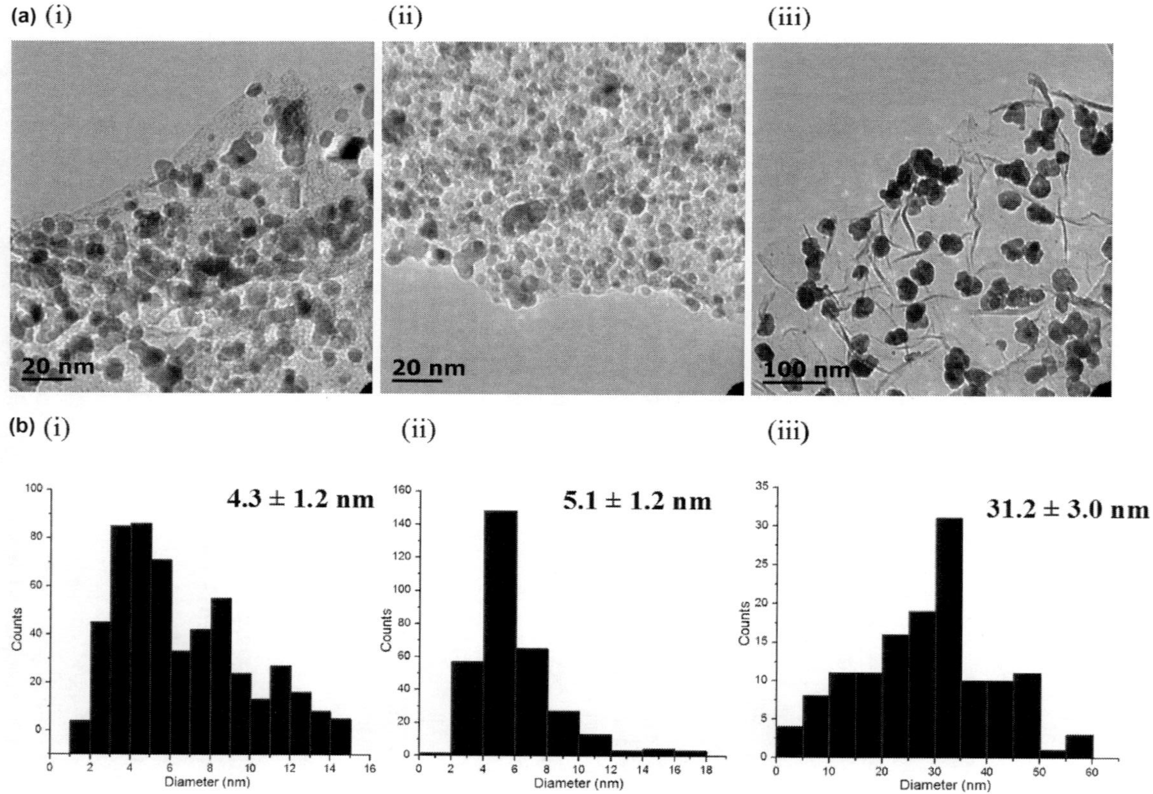

Fig. 7 (a) TEM micrographs and (b) particle size distribution of 30 wt% Pd/3D-GNS synthesized by Soft Alcohol reduction method with (i) Ethanol (ii) Isopropanol and (iii) Methanol as a solvent/reducing agents. Reproduced with permission from ref. A. Serov, N. I. Andersen, S. A. Kabir, A. Roy, T. Asset, M. Chatenet, F. Maillard and P. Atanassov, *J. Electrochem. Soc.*, 2015, **162**, F1305–F1309. Copyright the authors, 2015. Published under a Creative Commons Attribution 4.0 License (CC BY, http://creativecommons.org/licenses/by/4.0/).

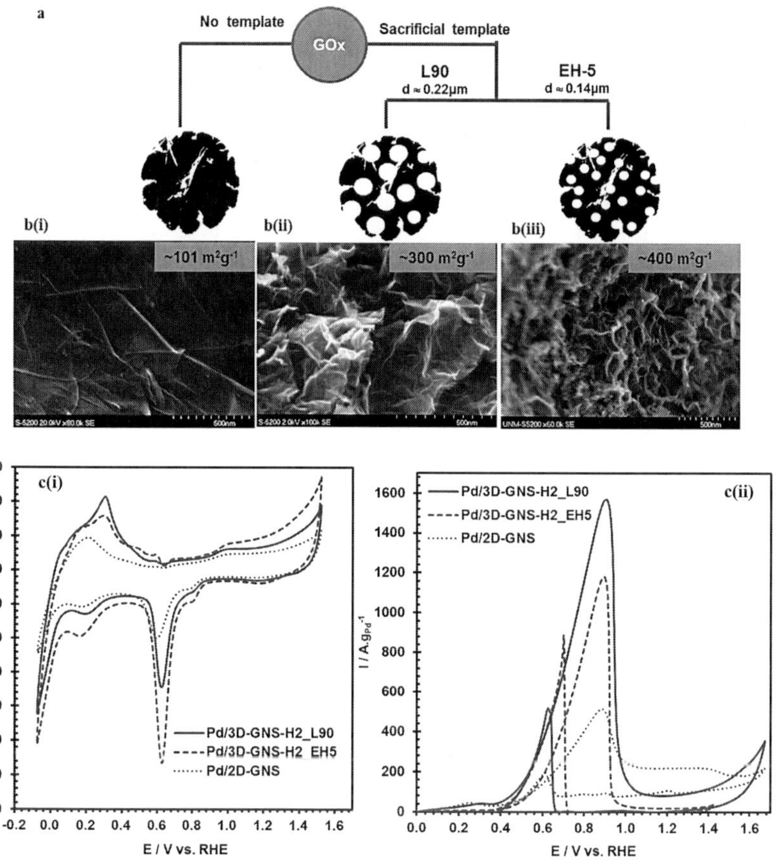

Fig. 8 (a) Schematic representation of how the morphology of graphene nanosheets (GNS) can be modified with silica templates; (b) SEM micrographs of thermally reduced GNS (i) unmodified (2D-GNS); and modified with (ii) L90 and (iii) EH5 sacrificial silica templates; (c) Cyclic voltammograms in N_2 purged (i) 1 M KOH and (ii) 1 M C_2H_5OH + 1 M KOH solution at a scan rate of = 20 mV s^{-1}. Reproduced from S. Kabir, A. Serov, K. Artyushkova and P. Atanassov, Design of Novel Graphene Materials as a Support for Palladium Nanoparticles: Highly Active Catalysts towards Ethanol Electro-oxidation, *Electrochim. Acta*, **203**, 144–153. Copyright 2016, with permission from Elsevier.

However, a comparative analysis of Pd nanoparticles of similar size synthesized without the 1-octyne capping agent was not provided to corroborate the effect of ligands towards modifying the EASA and activity of the nanoparticles.

In general, palladium nanoparticles are usually deposited on to carbonaceous supports to prevent them from sintering and to mitigate the dissolution of nanoparticles into the electrolyte. In particular, graphitized carbon nanomaterials such as multi walled carbon nanotubes, graphene oxide, nitrogen doped graphene *etc.* have been widely used as an excellent electrode support for fuel cells systems (micro and macro systems) due to their high conductivities, surface areas and stabilities in both alkaline and acid medium. Several studies have shown the enhanced performance of Pd catalysts toward alcohol oxidations by replacing the amorphous Vulcan support with graphitized carbons.

Pd nanocrystal-functionalized helical carbon nano fibers were (Pd-S-HCNF) were synthesized using a self-assembly method. The nanoparticles were found to 3–7 nm in diameter deposited in 60 nm HCNFs.[86] Using cyclic voltammetry in argon saturated 1 M KOH, it was shown that the Pd-S-HCNF catalysts had similar electrochemically active surface areas to commercial Pd/Vulcan catalysts. However, the peak ethanol and methanol oxidation currents for Pd-S-HCNF catalyst were found to be 1.5 and 2.3-fold higher than that of the Pd nanoparticles supported on commercial Vulcan XC-72. The enhanced activity of the catalysts could be attributed to the strong interaction of the catalysts with carbon support, although no experimental and spectroscopic evidence was provided to corroborate these metal-support interactions. Nethertheless, since both the Pd-H-CCNF and Pd/Vulcan catalysts had similar particle sizes and electrochemically active surface areas, the enhanced performance could be due to the uniform nature of helical carbon nanofiber support. A dispersity analysis of the Pd nanoparticles support on both Vulcan and HCNF supports could provide a better understanding of the supports properties towards enhancing electrocatalytic activities.

The fabrication of non-carbonaceous supports such as SiO_2 and TiO_2 can also be utilized to modify the structure and catalytic performance of Pd nanoparticles towards alcohol oxidation. Recently, Pd nanoparticles supported on 5–10 wt% TiO_2/C composites synthesized *via* an organic reductive colloid colloidal method.[87] XRD analysis of the synthesized composites showed no clear diffraction peaks corresponding to the presence of TiO_2 in the carbon powder. However, its presence was apparently verified using CV, where there was a decrease in cathodic and anodic currents as a function of TiO_2 concentration, possibly due to its semiconductor nature promoting less intense redox processes at relative higher overpotentials. Moreover, CO-stripping experiments of the Pd/C and Pd/TiO_2-C revealed that there was reduction in electrochemically accessible surface area (from 22–18 $m^2\,g^{-1}$) due to a possible formation of Pd–Ti alloy. And although the reaction order and reaction rate constants on both the Pd/C and Pd/TiO_2/C catalysts were of similar order, the electrocatalytic activity of the Pd/TiO_2/C catalyst-as a function of temperature and alcohol concentration – was higher towards methanol electrooxidation compared to Pd/C. Based on activation energy estimations, the increase in oxidation kinetics could be due to Pd-TiO_2/C's higher affinity to OH^-/OH_{ads} coupled with preferable adsorption at the TiO_2 sites.

The oxidation of primary and secondary alcohols in alkaline media involves many types of reaction intermediates, such as C1 species (most likely CO_{ads} and $CH_{x,ads}$), OH_{ads}, CH_3CO_{ads} and other forms of acetaldehyde in alkaline media. Therefore, a single type of metal catalysts may not be efficient enough to completely oxidize the intermediates further and provide the best performance for alcohol oxidation. Therefore, to further improve the activity and stability of the catalysts and to achieve the desired enhancement of their adsorption/dissociation properties, substantial efforts have been devoted towards modifying the Pd-based catalyst by alloying them with oxophilic ad-metals (M) through ligand effects manipulating the Pd-M bonding. The ability to fabricate and characterize homogeneous bimetallic alloys, where a less active adatom can enhance the activity of the more active metal, would be of broad interest in the field of electrocatalysis. Moreover, alloying Pd with a second or third ad-metals can also have an intrinsic effect by promoting the disassociation of water at lower potentials. The ad-metal can also promote a bifunctional effect, where the more oxophilic ad-metal forms oxygenated species from water, thereby reducing the operational potential of the anode. The second metal may diminish the electron density of the d-band of platinum/palladium and modify the electronic configuration, weakening the adsorption strength of poisonous species like CO and making them easier to oxidize and favoring the direct oxidation pathway. The presence of ad-atoms leads to significant changes on the amount and composition of strongly chemisorbed species, and on the electrooxidation mechanistic pathway. The ad-atom can also structurally alter the interatomic distances within the bimetallic alloy due to lattice strain, and expose more active sites. The electronic structure modification of the two interactive metallic components could be another important factor for improving catalytic activity.

For example, alloying palladium with silver (Ag) has been shown to increase the electrocatalytic performance of methanol and ethanol oxidation reactions due to reducing the poisoning effect by CO-like species and increasing adsorption properties of both alcohols and hydroxyls on the surface of electrocatalyst.[88,89] According density functional theory (DFT), there is an upward shift in d-band center toward the Fermi level *via* combination Ag overlayers or Ag impurities with Pd(111) surface compared to the parent metals.[90] This results in an increase of adsorbate binding energies which promotes the reactivity of Pd towards CO oxidation.

Some studies have also experimentally shown that the existence of ad-metals like silver (Ag) may alter the d-band center of the Pd metal.[91] For example, monodispersed bimetallic Pd–Ag nanoparticles with different compositions, *i.e.*, $Pd_{80}Ag_{20}$, $Pd_{65}Ag_{35}$ and $Pd_{46}Ag_{54}$ with an average size of 3.8 nm were synthesized *via* the thermal decomposition of palladium acetate and reduction of silver acetate in the emulsion-assisted EG ternary system with the presence of oleylamine (OAm) and oleic acid (OA) as the stabilizing agents.[92] A statistical analysis of TEM micrographs showed that there was a decrease in particle size and narrowing of the particle size distribution with increasing Ag content. XPS analysis of the PdAg alloyed materials showed an approximately 0.4–0.5 eV chemical shifts of the Pd 3d core level to higher binding energies, which was indicative of the strong interaction between Pd and Ag atoms. The nanoparticles were then loaded on to a carbon support (PdAg/C), and using cyclic voltammetry, the authors investigated and compared the catalytic performance of all PdAg/C for methanol oxidation reaction in alkaline media (1 M KOH). The results showed that the incorporation of silver in PdAg/C catalysts shifted the anodic onset potential by about 70 mV, suggesting that surface Ag significantly improved the kinetics of methanol oxidation. However, there was a decrease in anodic peak current densities with increasing Ag content. For example, the $Pd_{80}Ag_{20}/C$ catalyst had a peak current density of 691.6 mA mg_{Pd}^{-1}, which two times higher than the $Pd_{46}Ag_{54}/C$ (453.46 mA mg_{Pd}^{-1}). These results contradict the observation of decreasing particle size with increasing Ag content, where the $Pd_{80}Ag_{20}/C$ catalyst had a larger average particle size of 3.8 nm, compared to $Pd_{46}Ag_{54}$ with an average particle size of 3.3 nm. It therefore unclear as to why the alloys with (i) larger particles and (ii) lower Ag content shower better performance.

Hetero-structured Pd–Ag/C catalysts synthesized by polyol reduction method were shown to have a high activity of (0.956 A mg_{Pd}^{-1}) for the oxidation of ethylene glycol in alkaline media, was about 4.85 times higher than that of conventional Pd/C catalyst (0.197 A mg_{Pd}^{-1}).[93] In another study, Pd–Ag nanoparticles supported on reduced graphene oxide (Pd–Ag/RGO) nanocomposites were prepared by homogeneous deposition of Pd and Ag precursors on the surface of GO by hydrolysis of urea, followed by co-reduction of the precursors and GO with ethylene glycol.[94] TEM results showed that Pd–Ag nanoparticles deposited on MWCNT (Pd–Ag/MWCNT) were not highly uniformed and agglomerated, where as those on GO (Pd–Ag/GO) were more dispersed, indicating better interactions between the nanoparticle and support in the latter. The authors measured the electrochemical activities of the nanocomposites towards ethanol oxidation reaction in alkaline media using CV. Their results showed that Ag/RGO had no activity, whereas the activity of Pd_1Ag_1/RGO 0.090 A cm^{-2}, was about 4.1 times as high as that on Pd/RGO (0.022 A cm^{-2}) but significantly lower than the Pd_1Ag_2/RGO with higher Ag content. From the results obtained for Pd–Ag alloys, it seems to be that the addition of certain amount of Ag improved catalytic activity, but excessive Ag lead to lower activity. This could be due to the a competitive adsorption between hydroxyl and ethanol species on the surface of Pd–Ag alloys, and electrochemically accessible surface area analysis using either CO stripping and PdO reduction method can help explain this observation further. It is therefore imperative to optimize the Pd : Ag ratio in alloy to harness the best performance out of these nanocomposites.

Gold (Au) has also been shown to increase the catalytic activity of Pd–Au alloyed materials towards alcohol oxidation. A series of bimetallic PdAu/C catalysts were synthesized by the electroless deposition of Au on to Pd/C with increasing fractional coverage of Au on Pd.[95] XPS analysis of the bimetallic Pd–Au/C alloyed materials showed that there was an increasing shift to lower binding energies for the Au 4f5/2 and Au 4f7/2 peaks with decreased Au coverage, which was consistent with larger electronic effects on more dispersed Au atoms. Density functional theory calculation of the model systems suggested that the rate enhancement for ethylene glycol oxidation on the bimetallic surfaces compared to Pd was likely due to a decrease in coverage of strongly bound adsorbates, while the enhancement over Au was likely due to a decrease in the barrier for C–H scission. Hence by using a combination of DFT and XPS, a correlation between an increased electron density in Au core orbitals and more favorable thermodynamics for C–H scission as Au is added to Pd was established.

Other studies have also was shown that the addition of Au to Pd, for example 50–90 at.% of Au to Pd by electron beam irradiation – leads to more active and stable electrocatalysts for ethanol electro-oxidation.[96] *In situ* ATR-FTIR (Attenuated Total Reflectance Fourier Transform Infrared Spectroscopy) experiments identified acetate and acetaldehyde as the principal products of ethanol electro-oxidation PdAu/C electrocatalysts (Fig. 9), indicating the incomplete oxidation of ethanol.[96] It was also shown that the degree of alloying can – to some extent – have an effect on modifying the catalytic performance of Pd–Au alloys,[97] although a direct relationship between the two has not been established yet.

The promotional effects of lead (Pb) as an oxophilic metal is also being utilized towards ad-atom based Pd alloys to increase the rates of alcohol electrooxidation. In another study, by alloying Pb with Pd in a Pd_4Pb/C system and using various electrochemical techniques, it was demonstrated that the addition of lead facilitated the oxidative removal of adsorbed CO.[100] As discussed in previous sections, the promoting effect of lead was explained by the discussed bi-functional mechanism and d-band theory.[101,102] A recent study confirmed the observation that current density (2.02 mA cm^{-2}) and mass activity (0.51 A $mg_{Pd}{}^{-1}$) of "flower-like" Pd_3Pb intermetallic catalysts (prepared at low temperature by hydrothermal polyol solution based methods, Fig. 10a) were 1.5 times higher for ethanol oxidation than on Pd/C in alkaline media.[98] However,

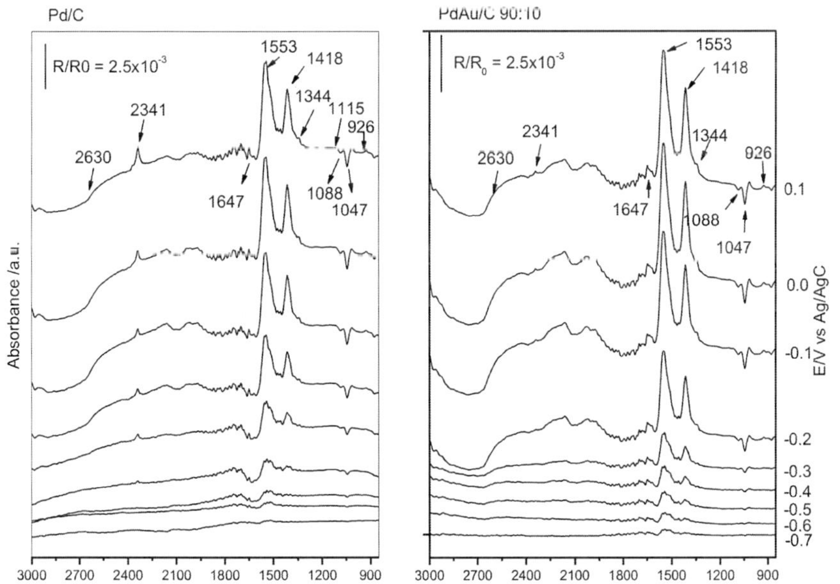

Fig. 9 *In situ* FTIR spectra of Pd/C and PdAu/C (90 : 10) electrocatalysts, taken in the potential range −0.70 to 0.20 V *vs.* Ag/AgCl, in 1.0 M ethanol in 1.0 M KOH. Reproduced from A. N. Geraldes, D. F. da Silva, E. S. Pino, J. C. M. da Silva, R. F. B. de Souza, P. Hammer, E. V. Spinacé, A. O. Neto, M. Linardi and M. C. dos Santos, Ethanol electro-oxidation in an alkaline medium using Pd/C, Au/C and PdAu/C electrocatalysts prepared by electron beam irradiation, *Electrochim. Acta*, **111**, 455–465. Copyright 2013 with permission from Elsevier.

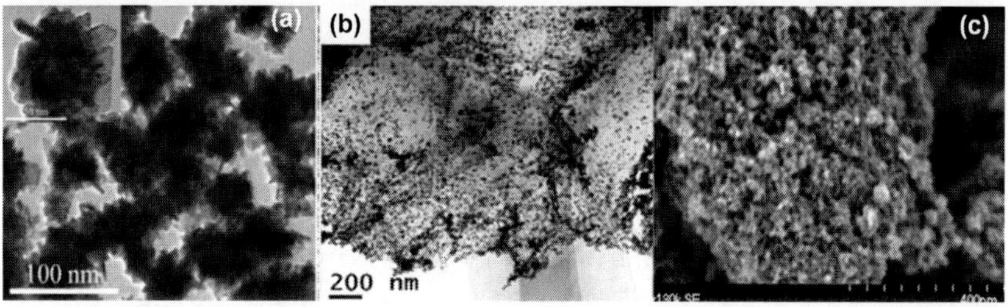

Fig. 10 (a) TEM image (inset shows magnified TEM image) of a single Pd_3Pb flower-like nanocrystals. Reproduced from R. Jana, U. Subbarao and S. C. Peter, Ultrafast synthesis of flower-like ordered Pd3Pb nanocrystals with superior electrocatalytic activities towards oxidation of formic acid and ethanol, *J. Power Sources*, **301**, 160–169. Copyright 2016 with permission from Elsevier. (b) TEM images of PdPb/NG(8 : 1.0). Reproduced from P. Wu, Y. Huang, L. Zhou, Y. Wang, Y. Bu and J. Yao, Nitrogen-doped graphene supported highly dispersed palladium-lead nanoparticles for synergetic enhancement of ethanol electrooxidation in alkaline medium, *Electrochim. Acta*, **152**, 68–74. Copyright 2014 with permission from Elsevier. (c) SEM micrograph of self-supported Pd_4Bi sample. Reprinted with permission from A. Zalineeva, A. Serov, M. Padilla, U. Martinez, K. Artyushkova, S. Baranton, C. Coutanceau and P. B. Atanassov, Self-Supported PdxBi Catalysts for the Electrooxidation of Glycerol in Alkaline Media, *J. Am. Chem. Soc.*, 2014, **136**, 3937–3945. Copyright 2014 American Chemical Society.

the electrochemically accessible surface area of the flower like Pd_3Pb (2.82 cm^2) catalyst was lower than Pd/C (2.85 cm^2), so it is unclear how the "follower like" morphology offered any additional morphological advantages, apart from its ordered structure.

In another study, PdPb nanoparticles (of different ratios) supported on N-doped graphene (PdPb/NG) synthesized by a one-step reduction method.[99] A statistical analysis of the TEM micrographs revealed that the PdPb nanoparticles were highly dispersed with on NG compared with an average diameter range of 3–15 nm, compared to Pd/G (2–100 nm) and Pd/NG (2–25 nm, Fig. 10b). However, using CO stripping, the electrochemically active surface areas for Pd/G, Pd/NG and PdPb/NG were calculated to be 35.1, 30.2 and 20.8 $m^2 g^{-1}$, suggesting that while the N doped graphene might help in narrowing down the particle size distribution of the nanoparticles, alloying Pd with Pb lowers results in lowered EASA, probably due to the amorphous nature of the Pb, as observed using XRD. Using CV, it was also demonstrated the enhanced electrocatalytic activity of the PdPb/NG catalysts towards ethanol oxidation. with the highest peak current density of 152.3 mA cm^{-2} for PdPb (8 : 1)/NG, which was 4.0 times higher than that of Pd/Graphene (38.0 mA cm^{-2}) and 2.9 times higher than that of Pd/C (51.9 mA cm^{-2}).[99] However, PdPb nanoparticles with a higher Pb ratio (*e.g.* PdPb/NG (8 : 3)) were shown to have a lower activity, suggesting that excessive Pb could lead to exceeded coverage of Pd active sites.

The activity and selectivity of Pd catalysts can also be tuned by modification with p-block atoms, such as bismuth.[16,103,104] Recently, the oxidation of glycerol on preferentially-shaped Pd(100) nanoparticles and Pd nano-spheres modified with bismuth (Bi) was investigated using cyclic voltammetry.[104] It was shown that bismuth modified palladium cubes with high surface structure [Pd(100)] had a higher activity compared Pd nano-spheres, which were found to have a low preferential domain orientation. However, experiments in 10^{-3} M glycerol showed that the increase of Bi surface coverage from 0 to 0.49 increased the oxidation peak currents. However further increase of Bi coverage decreased the intensity of the oxidation peaks, suggesting that the presence of Bi adatoms can lead to significant changes on the amount and composition of strongly chemisorbed species and the electrooxidation mechanism. In a similar study, self-supported, highly porous Pd_xBi catalysts (Fig. 10c) synthesized using the Sacrificial Support Method[16] also showed high activity toward glycerol oxidation compared to Pd/C catalysts. Addition of Bi to Pd shifted the onset oxidation potential by *ca.* 100 mV toward lower potentials and increased the oxidation current densities in the potential range from 0.5–1.2 V *vs.* RHE, according to the order in activity: $Pd_4Bi > Pd_6Bi > Pd_2Bi > Pd$. *In situ* Fourier transform infrared spectroscopy analysis of the materials showed that aldehyde and ketone were generated at low potentials, hydroxypyruvate at moderate potentials, and full oxidation with CO_2 production was achieved at high potentials. Similarly, these findings were corroborated in another study where carbon-supported Pd_xBi_{1-x}/C catalysts prepared by the water-in-oil microemulsion method. The addition of 10 at.% of bismuth to palladium led to the same activity as that obtained with a Pt/C catalyst. At low potentials, the oxidation of glycerol was shown to occur without activation of OH^- species which allowed the formation of aldehyde and ketone groups. However, further increase of Bi atomic ratio has no effect on the activity, which confirms that extra bismuth atoms are deposited on the carbon support and does not play any catalytic role.

Several studies have been focused on exploiting the effect of other p-group ad-atoms like tin (Sn) and towards enhancing the electrocatalytic oxidation of alcohols. For example, carbon supported Pd–Sn electrocatalysts prepared with a modified sequential polyol method showed high activities towards the electrooxidation of ethanol in alkaline media.[105] Using a combination of STEM-EELS, XRD, and XPS data, Pd–Sn/C catalysts were found to be homogeneous alloys, although Sn atoms on the surface were partially oxidized into SnO_2. Using CV, the activity and stability of the alloys followed the order: $Pd_{86}Sn_{14} > Pd_{75}Sn_{25} > Pd_{47}Sn_{53} >$ Pd/C (JM). In particular, $Pd_{86}Sn_{14}/C$ catalyst had the optimal electrocatalytic activity compared to the other alloys and Pd/C. However it also showed the least degree of alloy formation compared with the rest of two Pd–Sn catalysts. Therefore, an optimal Sn content must be alloyed with Pd to obtain best performance.

The electrocatalytic oxidation of ethylene glycol on Pd-based bimetallic alloys (PdSn and PdNi) synthesized using a microwave polyol process and supported on sulfonated multi-walled carbon nanotubes (SF-MWCNTs) have also been studied in alkaline media.[106] It was observed that the bimetallic nanocomposite mixtures (SF-MWCNT–PdSn/Ni) had higher electrochemically accessible surface areas and exhibited enhanced electrocatalytic oxidation of ethylene glycol compared to the Pd alone (*i.e.*, SF-MWCNT-Pd). Among the alloys, the SF-MWCNT–$PdSn_{mix}$ (51.9 mA cm^{-2}) electrode showed better electrocatalysis towards ethylene glycol oxidation (in terms of low onset and peak potentials, high current density, and lower CO poisoning) compared to the SF-MWCNT–$PdNi_{mix}$ (35.3 mA cm^{-2}) indicating that Sn exerts more influence on the EASA of the Pd, and hence the high electrocatalytic activity. Based on these results, Sn may be considered to be a more favored co-catalyst with Pd for, taking into account that nickel being 3d metal will have stronger electronic effect on the structured PdNi materials in contrast to p-metal tin.

The comparative analysis of several ad-atoms like antimony (Sb), tin (Sn) and lead (Pb), added to Pd nanoparticles was investigated for the electrochemical oxidation of alcohols like ethanol, ethylene glycol, *etc.* glycerol was done in alkaline media.[107] Among all the alloys, Pb seemed to have had the most pronounced effect in the electrooxidation activity compared to Sn and Sb. For example, CV results showed that the peak oxidation rate on Pd–Pb (5.00 mA cm^{-2}) was 2.9 times greater than on Pd (1.71 mA cm^{-2}). However, further comparisons were only made using chronoamperometry in order be relevant with steady state operations with fuel cells The authors speculated that enhanced performance of the PdPb alloy was due to lead's influence in inducing the least shift in Pd 3d5/2 binding energy (+0.16 eV) compared to Sb (+0.50 eV) and Sn (+0.40 eV), however a direct link between binding energy shifts of Pd and the rate at which key adsorbate intermediates interact with the palladium surface is still missing.

Platinum group metals like ruthenium (Ru) and rhodium (Rh) alloyed with Pd have also gained significant consideration for alcohol electro-oxidation due to their oxophilicity and intrinsic catalytic activity. Carbon supported Pd_xRu_y/C catalysts (with various Pd : Ru atomic ratios) synthesized by impregnation method were studied for electroreduction of ethanol.[108] XRD analysis showed that there was a decrease of the Pd (220) lattice constant with increasing Ru concentration. CV of the catalysts in blank electrolytes showed that the hydrogen adsorption/desorption region and the oxides formation/reduction of the Pd species overlapped with the redox of the RuO_xH_y species. This indicates some of Ru species in the Pd_3Ru/C catalyst existed in an amorphous form. Ru/C was also shown to be inactive for ethanol oxidation. However, the Pd_xRu_y/C catalyst showed a much lower onset potential (*e.g.* 0.2 V for Pd_3Ru/C) for EOR (Fig. 11). Additional CV experiments showed that all the Pd_xRu_y/C catalysts characterized by lower onset potential and peak potential for EOR than the Pd/C. However, using chronoamperometry, at 0.45 V, all of the Pd_xRu_y/C catalysts produced higher steady currents for EOR than the Pd/C catalyst with the order of performance following $Pd_3Ru/C > PdRu/C > Pd_2Ru/C > Pd_4Ru/C$. The steady current for EOR shows pseudo-volcano behavior as a function of the Ru atomic ratio *i.e.* activity of the Pd_xRu_y/C increased with increasing Ru content from 0 to 25 at.% (*i.e.* Pd : Ru atomic ratio of 3 : 1) but decreased further with the increasing of Ru content to 100% (*i.e.* Ru/C).

In another study, 20 wt% Pd_xRh/C ($x = 1/3$, 1, 3) electrocatalysts with a size of 3–4 nm were prepared by reduction of the precursor salts with $NaBH_4$ and studied for methanol oxidation reaction (MOR).[42] Cyclic voltammetry in blank electrolytes showed that the Rh oxide reduction

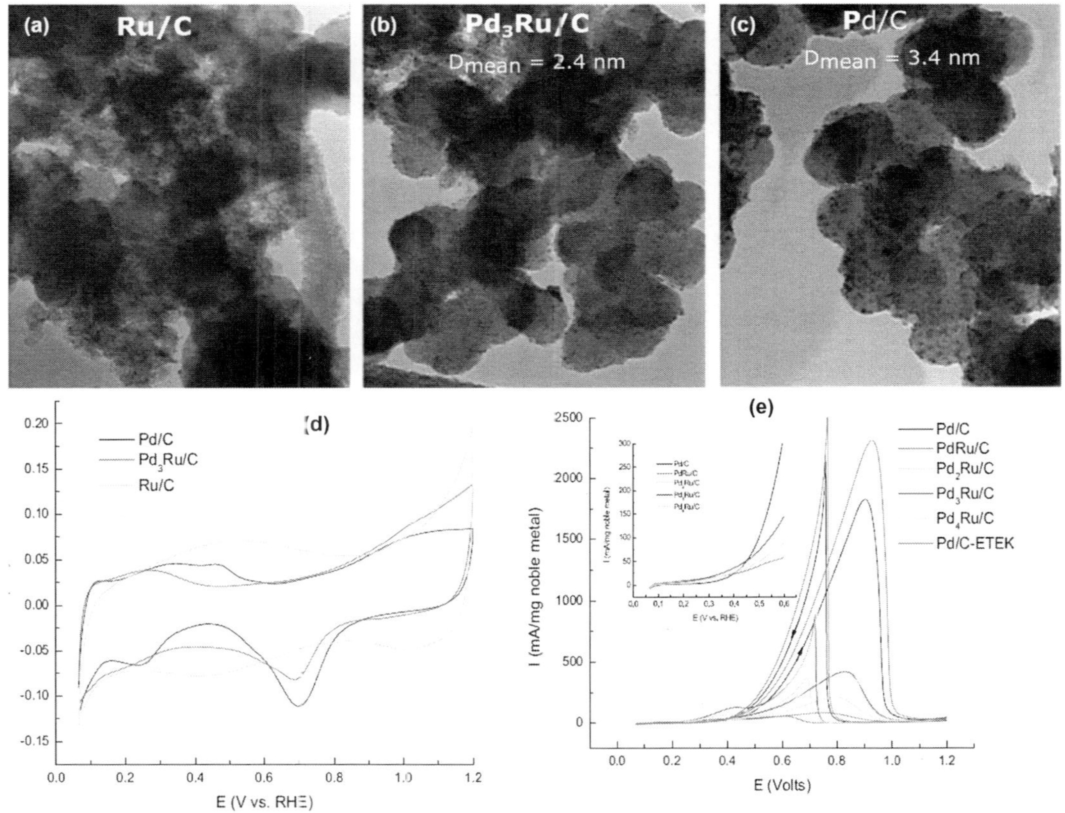

Fig. 11 TEM images of (a) Ru/C, (b) Pd_3Ru/C and (c) Pd/C; (d) CVs of Ru/C, Pd_3Ru/C and Pd/C electrocatalysts in an Ar-saturated 1 M NaOH solution; (e) CVs of Ru/C, PdRu/C, Pd/C and commercial Pd/C-ETEK electrocatalysts in an Ar-saturated 1 M NaOH + 1 M ethanol solution (insert: enlarged CVs). Reproduced from L. Ma, H. He, A. Hsu and R. Chen, PdRu/C catalysts for ethanol oxidation in anion-exchange membrane direct ethanol fuel cells, *J. Power Sources*, **241**, 696–702. Copyright 2013 with permission from Elsevier.

peak shifted negatively for Rh/C (0.337V *vs.* RHE) compared to Pt/C (0.697V *vs.* RHE), indicating that oxides are not only formed at lower potentials but also are more stable on the surface of Rh/C than on Pd/C or Pt/C (Fig. 12A). During the anodic scan metal oxide reduction peaks are shifting to lower potentials with rising Rh content in bimetallic catalysts. The oxide formation during the cathodic scan also occurs at lower potentials with rising Rh to Pd ratio, with alloys with Pd contents of 50 wt% and 75 wt% had the highest currents. The PdRh catalysts also had a 85–140 mV lower CH_3OH oxidation onset potential and higher mass current densities, with Pd_3Rh/C (955.7 mA mg_{Pd}^{-1}) having the highest catalysts compared with Pd/C (543.8 mA mg_{Pd}^{-1}) and Pt/C (669.9 mA mg_{Pd}^{-1}). A differential electrochemical mass spectrometry (DEMS) was used to investigate the product distribution during CH_3OH oxidation and to evaluate poisoning effects on Pd/C compared to PdRh/C and Rh/C. Analysis of the mass spectrometer (MS) signal of CO_2 ($m/z=44$) and methyl formate ($HCOOCH_3$, $m/z=60$) showed that the (i) Pt/C formed $HCOOCH_3$ at a higher extent than CO_2, and (ii) addition of Rh to Pd resulted in an higher conversion of methanol to CO_2 and enhanced CO_2 current efficiency compared to Pd/C or Pt/C.

Another class of add-metals that has gained significant consideration for alcohol electrooxidation are transition metals like Fe, Mn, Ni, Co, Cu *etc.*[109,110] For example, Cu_xPd_y/C catalysts with a 1 : 2 = Cu : Pd ratio were shown to be active, with anodic peak current densities for methanol and ethanol oxidation of 220 mA mg_{Pd}^{-1} and 520 mA mg_{Pd}^{-1} respectively.[110] HRTEM results showed the lattice of Cu_xPd_y nanoparticles were distorted and the d-spacing values got smaller with increasing Cu content. The compressive strain within the Pd layer, induced by the smaller lattice constant of Cu, can weaken the binding force of both reactants and intermediates Although the alloyed Cu_xPd_y/C catalysts showed higher forward current densities for EOR compared to MOR, the forward peak current density was lower than the reverse peak current density during EOR, whereas the opposite for true for MOR, indicating that the Cu_xPd_y surfaces were more likely poisoned with CO-like species generated during EOR.

A comparative analysis of PdCu and PdCo catalysts were made to demonstrate their activities towards the oxidation of ethanol in gas phase electrolyte.[70] The $Pd_{63}Co_{37}$ nanoparticles had an average size of (4.9 ± 0.7 nm), whereas the $Pd_{21}Cu_{79}$ nanoparticles were slightly bigger (~9.7 ± 0.6 nm), with some surface enrichment of Pd in a very thin layer of the alloyed PdCu surface. However, In the case of $Pd_{63}Co_{37}$, there was a clear surface enrichment of Co in a thick layer of the PdCo alloyed surface which did not seem to fully cover the surface of the nanoparticles. Moreover, the PdCu catalysts also displayed much higher mass activities for ethanol oxidation compared to PdCo catalysts due to the surface enrichment of Pd on the alloyed PdCu surface, in contrast to the surface enrichment of Co in the alloyed PdCo surface. Despite the fact that the catalytic conversion rate for CO_2 product was higher over Pd than PdCu, the catalytic activity over the nano-alloy PdCu catalyst was shown to suppress the formation of acetic acid, which is a significant portion of the

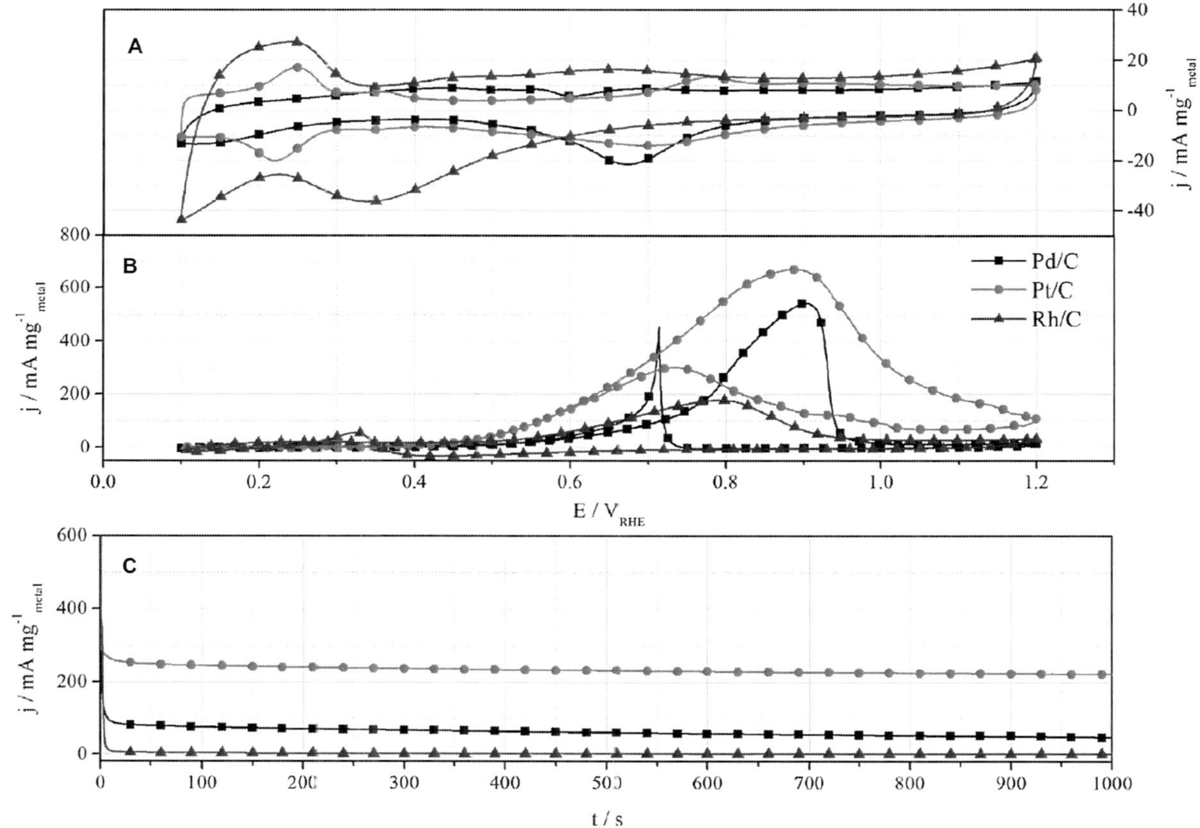

Fig. 12 Cyclic voltammetry curves for Pd_xRh/C electrocatalysts at a scan rate of 20 mV s^{-1} from 0.1 to 1.2 V RHE in (A) 0.5 M KOH and (B) 0.5 M CH_3OH + 0.5 M KOH solution and (C) chronoamperometry curves in 0.5 M CH_3OH + 0.5 M KOH solution at 0.7 V *vs.* RHE for 1000 s. Reproduced from T. Jurzinsky, R. Bär, C. Cremers, J. Tübke and P. Elsner, Highly active carbon supported palladium-rhodium PdXRh/C catalysts for methanol electrooxidation in alkaline media and their performance in anion exchange direct methanol fuel cells (AEM-DMFCs), *Electrochim. Acta*, **176**, 1191–1201. Copyright 2015, with permission from Elsevier.

product in the case of Pd catalyst. X-ray diffraction peaks for both the $Pd_{63}Co_{27}$ and PdCu alloys shifted to higher diffraction angles than those for Pd, indicating a reduction of lattice constant which were calculated to have similar magnitude (0.381 nm). The results in this study showed that for alloys with the same structural properties, the chemical composition of the alloys as well as degree of surface enrichment between the two alloy particles can play an important role in modifying electrochemical properties of the alloyed nanocomposites.

Alloying Pd with two or more oxophilic elements can also lower the electrocatalysts electronic binding energy by facilitating the adsorption of OH^- at lower potentials and promoting the oxidation of the organic species. A recent study reported the fabrication of ordered $Pd_{43}Ni_{10}Cu_{27}P_{20}$ bulk metallic glasses nanowires (Pd-BGM) using a CMOS (complementary metal oxide semiconductor) compatible method.[111] The nanowires used in this study had an average length and diameter of approximately 2–20 nm. The electrocatalytic activity of the Pd-BMG nanowires showed a decrease in onset potential for methanol and ethanol electro-oxidation in comparison to Pd flat electrodes, by 300 mV and 120 mV respectively, and a 300 mV lower onset potential for CO oxidation, indicating a decrease in the activation energy of the reaction. More specifically, the Pd-BMG nanowires activation energies were 17.11 $kJ\,mol^{-1}$ lower for methanol oxidation and 13.26 $kJ\,mol^{-1}$ lower for ethanol oxidation compared to Pd flat. This indicates a higher intrinsic activity for the Pd-BMG nanowires for methanol and ethanol oxidation. However, the synergistic effect of Ni and Cu present and P present in the structure of the Pd-BMG nanowires were not independently investigated. It would also be worthwhile to investigate the activity of the nanowires a function of length and/or diameter.

In another study, multi-walled carbon nanotubes supported PdCuSn(1 : 1 : 1) catalysts were synthesized by the borohydride reduction method (Fig. 13a–d).[112] It was shown that the incorporation of Sn to Pd with Cu decreased the average crystallite and particle size of the alloys, and activated the performance of the alloys towards ethanol oxidation. In particular, PdCuSn/CNTs showed highest mass activities of 395.94 and 872.70 $mA\,mg_{Pd}^{-1}$ toward methanol, and ethanol compared to it binary and non-alloyed counterparts. However, during the voltage-cycled activation, Cu was shown to be more prone to suffer leaching, while the dissolution of Sn occurred rarely. Moreover, for methanol oxidation, the promotion effect of Sn was comparable with that of Cu. For ethanol oxidation, however, the promotion effect of Sn was prominent, indicating that Sn could act as a more efficient promoter compared to Cu (Fig. 13e–f).

Partial substitution of Cu by Pd^{2+} ions and Bi^{3+} ion by self-adsorption processes were also recently used to fabricate core shell PdCuBi nanocatalyst. These tri-metallic catalysts showed catalytic activity of 171.1 $mA\,cm^{-2}$ towards ethylene glycol oxidation in alkaline medium in comparison to Pd/C.[113] However, bismuth can aggregate on the Pd particle surfaces *via* the spontaneous chemisorption process after its addition. The electrochemical surface area of Pd in the PdCuBi alloy was estimated to be 33.0 $m^2\,g^{-1}$, which was lower than Pd/Cu. In another paper,[114] hollow NiPdAu were prepared *via* a partly galvanic displacement

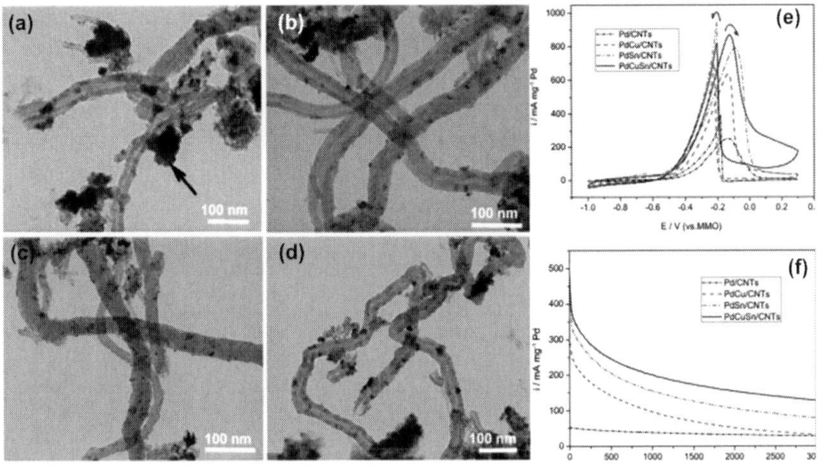

Fig. 13 TEM images of the (a) Pd/CNTs, (b) PdCu/CNTs, (c) PdSn/CNTs and (d) PdCuSn/CNTs catalysts. (e) CV curves of the activated Pd/CNTs, PdCu/CNTs, PdSn/CNTs and PdCuSn/CNTs catalysts in the N_2 saturated solution of 1.0 M KOH 0.5 M ethanol. Scan rate: 50 mV s^{-1}. (f) CA curves of the catalysts in solution of 1.0 M KOH + 0.5 M ethanol at a fixed potential of 200 mV (*vs.* MMO). Reproduced from F. Zhu, G. Ma, Z. Bai, R. Hang, B. Tang, Z. Zhang and X. Wang, High activity of carbon nanotubes supported binary and ternary Pd-based catalysts for methanol, ethanol and formic acid electro-oxidation, *J. Power Sources*, **242**, 610–620. Copyright 2013 with permission from Elsevier.

reaction between suitable metal precursors and Ni nanoparticles as a sacrificial template. The as-synthesized hollow NiPdAu nanoparticles of ~30.8 nm dispersed on a carbon substrate had a forward mass peak current of ($4.31\ \text{A}\,\text{mg}_{Pd}^{-1}$) which was 7.70 and 3.10 times that of commercial Pd/C ($0.56\ \text{A}\,\text{mg}_{Pd}^{-1}$) and Pt/C ($1.39\ \text{A}\,\text{mg}_{Pd}^{-1}$) and also higher than $Pd_{30}Au_{70}/C$ ($950.6\ \text{A}\ \text{mg}_{Pd}^{-1}$) catalyst. Additionally, a comparison between regular and hollow NiPdAu nanoparticles could offer additional insights into the morphological advantages offered by the latter. Moreover, analysis of diffraction peaks of the NiPdAu alloys at higher angles (from 80° to 145°) for estimating of lattice spacing could provide a better understanding of the alloyed structure in the nanoparticles.

4 Conclusions

The various methods implemented for enhancing the electro-catalytic activity of platinum and palladium based catalysts for the electro-oxidation of alcohols such as ethanol, methanol, ethylene glycol and glycerol were reviewed in this chapter. The electrooxidation kinetic of these alcohols were observed to be faster in alkaline media, and the removal of adsorbed poisoning species such as CO was much faster in alkaline media due to the availability of oxygen-containing species.[28] Additionally, the extent of Pd deactivation was smaller than that of Pt one.[65]

Numerous computational and fabrication methodologies have been implemented to investigate the effect of particle size, shape and morphology towards the electrooxidation of alcohols. Density functional theory studies demonstrated the activity and selectivity of ethanol oxidation on palladium are highly structure-sensitive, and that the atoms in the surface layer of Pd(100) were deemed to be more sensitive in the electrocatalytic reaction of ethanol.[80] Several studies also focused on optimizing the structure of the nanoparticles by varying their shape and morphology, for example, cubes,[81] and nanospheres,[82] using surfactants and reducing agents. However, polymeric surfactants have a tendency to attach themselves to the nanoparticles and unless eliminated completely, these can significantly affect the electrochemically accessible surface area of the nanoparticles, and in turn, their activities. Taking into account that physicochemical characteristics of these catalysts depend on the synthesis method, the method of preparation becomes a key factor regarding enhancement of electrochemical activity, and only simple reducing agents should be utilized during preparation of these nanoparticles.[17] Various graphitized carbon[17,99,109] and metal-oxide supports[87] can also be utilized to enhance the dispersion and stability of nanoparticles.

However, spectroscopic studies have shown that ethanol is usually selectively oxidized to acetic acid, or converted to acetate ion in the alkaline media, whereas methanol is oxidize to either CO_2/bicarbonate or Co_{ad} and interfacial formate *via* a dual-pathway mechanism.[79] A much wider product distribution, including glycolate, glycerate, tartronate, oxalate, formate and carbonate, was also obtained from the oxidation of glycerol.[78] So although the reaction kinetics and mechanisms of all these oxidation reactions can be significantly affected by the structure, morphology and chemical composition of the nanoparticles, a single type of metal catalysts may not be efficient enough to completely oxidize the intermediates further and giving the best performance for alcohol oxidation.

Several studies have shown improved electrocatalytic activities for alcohol oxidation through alloying Pd with a second or third ad-metals metals like Ag,[92] Pb[99,100] Bi,[104] Au,[95] Sn,[105] Ni,[106] Cu, Ru,[41] Sn[73] *etc.* These binary and ternary Pd-based catalysts have demonstrated current densities that were significantly higher than bare Pd/C. For alloys with same structural properties, the chemical composition and degree of surface enrichment between the two alloy particles plays an important role in modifying electrochemical properties. For example, the existence of admetals like silver may alter the d-band center of the Pd metal.[91] However, although the addition of certain amount of Ag improved catalytic activity, an excessive amount of Ag can lead to lower activity due to the competitive adsorption between hydroxyl and ethanol species on the surface of Pd–Ag alloys.[93] Other studies that used a combination of DFT and XPS showed a correlation between increased electron density in Au ad-atom core orbitals and more favourable thermodynamics for C–H scission towards glycerol electrooxidation.[95] Hence, the degree of alloyed Au can – to some extent – have an effect on modifying the catalytic performance of Pd–Au alloys,[97] although a direct relationship between the two has not been established yet. By alloying Pb with Pd in a Pd_4Pb/C system and using various electrochemical techniques, it was demonstrated that the addition of lead facilitated the oxidative removal of adsorbed CO.[100] However, PdPb nanoparticles with a higher Pb ratio were shown to have a lower activity, suggesting that excessive Pb could lead to exceeded coverage of Pd active sites.[99] The activity and selectivity of Pd catalysts can also be tuned by modification with p-group atoms, such as bismuth (Bi) and Sn,[16,103,104] although increase of Bi coverage can decrease the intensity of the oxidation peaks, and an optimal Sn content must be alloying with Pd to obtain best performance.[105] Adatoms like Pb, Sn and Sb can also enhanced performance alloy by inducing shifts in binding energies,[107] however a direct link between binding energy shifts of Pd and the rate at which key adsorbate intermediates interact with the palladium surface is still missing.

Platinum group metals like ruthenium (Ru) and rhodium (Rh) alloyed with Pd have also gained significant consideration for alcohol electro-oxidation due to their oxophilicity and intrinsic catalytic activity.[108] The addition of Rh to Pd can induce a higher conversion of methanol to CO_2 and enhanced CO_2 current efficiency compared to Pd/C or Pt/C.[42] However activity of the Pd_xRu_y/C increases with increasing Ru content up to a certain point (25%), and decreases with increasing Ru content.[108] Alloying Pd with transition metals like Co, or Cu can also introduce a compressive strain within the Pd layer, induced by the smaller lattice constant of Cu, can weaken the binding force of both reactants and intermediates. Surface enhancement of Pd nanoparticles with the Cu can also enhance electrochemical performances.[70] However, Cu can be more prone to suffer leaching in comparison to Sn.[112]

It is clear that alloying Pd with adatoms can improve electrocatalytic performances by (i) altering the interatomic distances within the bimetallic alloy due to lattice strain and exposing active sites; (ii) promoting the disassociation of water by formatted oxygenated species at lower potentials; (iii) diminishing the electron density of the d-band of platinum/palladium; (iv) modifying the electronic configuration and binding energy shifts; (v) weakening the adsorption strength of poisonous species like CO; (vi) changing on the amount and composition of strongly chemisorbed species, and on the electrooxidation mechanistic pathway.

But despite the increasing performance of these alloyed Pd-based catalysts for alcohol electrooxidation, the understanding of the catalyst shape, size, chemical composition, morphology, degree of alloying and electronic configurations structures for achieving C–C bond excision remains elusive. The optimum composition of the adatoms in the alloys are not well determined, and hence no direct correlations between degree of alloying and electrochemical performances are available. Bearing in mind that the properties of these catalysts depend method of fabrication – just as much as external factors such as temperature, concentration, and pH of the electrolyte – it is imperative to continue the investigation of structure–property relationships of Pd-based for the development of highly active catalysts for the electrooxidation of alcohols in fuel cells.

References

1 *Fuel Cells Bull.*, 2010, **2010**, 9–10.
2 *Fuel Cells Bull.*, 2014, **2014**, 5.

3 M. Warshay and P. R. Prokopius, *J. Power Sources*, 1990, **29**, 193–200.
4 J. R. Varcoe, P. Atanassov, D. R. Dekel, A. M. Herring, M. A. Hickner, P. A. Kohl, A. R. Kucernak, W. E. Mustain, K. Nijmeijer, K. Scott, T. Xu and L. Zhuang, *Energy Environ. Sci.*, 14, **7**, 3135–3191.
5 J. R. Varcoe, M. Beillard, D. M. Halepoto, J. P. Kizewski, S. Poynton and R. C. T. Slade, *ECS Trans.*, 2008, **16**, 1819–1834.
6 J. R. Varcoe and R. C. T. Slade, *Fuel Cells*, 2005, **5**, 187–200.
7 M. Wang, W. Zhang, J. Wang, D. Wexler, S. D. Poynton, R. C. T. Slade, H. Liu, B. Winther-Jensen, R. Kerr, D. Shi and J. Chen, *ACS Appl. Mater. Interfaces*, 13, **5**, 12708–12715.
8 Y. Wang, L. Li, L. Hu, L. Zhuang, J. Lu and B. Xu, *Electrochem. Commun.*, 2003, **5**, 662–666.
9 M. Alesker, M. Page, M. Shviro, Y. Paska, G. Gershinsky, D. R. Dekel and D. Zitoun, *J. Power Sources*, 2016, **304**, 332–339.
10 K. N. Grew and W. K. S. Chiu, *J. Electrochem. Soc.*, 2010, **157**, B327–B337.
11 J. R. Varcoe, *Phys. Chem. Chem. Phys.*, 2007, **9**, 1479–1486.
12 C. G. Arges, V. Ramani and P. N. Pintauro, *Electrochem. Soc. Interface*, 2010, 31.
13 U. Martinez, K. Asazawa, B. Halevi, A. Falase, B. Kiefer, A. Serov, M. Padilla, T. Olson, A. Datye, H. Tanaka and P. Atanassov, *Phys. Chem. Chem. Phys.*, 2012, **14**, 5512–5517.
14 E. A. Baranova, M. A. Padilla, B. Halevi, T. Amir, K. Artyushkova and P. Atanassov, *Electrochim. Acta*, 2012, **80**, 377–382.
15 A. Serov, K. Artyushkova, P. Atanassov, A. Aziznia, P. H. Benhangi and E. Gyenge, *J. Mater. Chem. A*, 2013, **1**, 14384–14391.
16 A. Zalineeva, A. Serov, M. Padilla, U. Martinez, K. Artyushkova, S. Baranton, C. Coutanceau and P. B. Atanassov, *J. Am. Chem. Soc.*, 2014, **136**, 3937–3945.
17 A. Serov, N. I. Andersen, S. A. Kabir, A. Roy, T. Asset, M. Chatenet, F. Maillard and P. Atanassov, *J. Electrochem. Soc.*, 2015, **162**, F1305–F1309.
18 G. J. K. Acres, *J. Power Sources*, 2001, **100**, 60–66.
19 R. Parsons and T. VanderNoot, *J. Electroanal. Chem. Interfacial Electrochem.*, 1988, **257**, 9–45.
20 K. Matsuoka, Y. Iriyama, T. Abe, M. Matsuoka and Z. Ogumi, *J. Power Sources*, 2005, **150**, 27–31.
21 P. J. Kulesza, I. S. Pieta, I. A. Rutkowska, A. Wadas, D. Marks, K. Klak, L. Stobinski and J. A. Cox, *Electrochim. Acta*, 2013, **110**, 474–483.
22 L. Jiang, A. Hsu, D. Chu and R. Chen, *Int. J. Hydrogen Energy*, 2010, **35**, 365–372.
23 Y. S. Li, T. S. Zhao and Z. X. Liang, *J. Power Sources*, 2009, **187**, 387–392.
24 S. C. S. Lai and M. T. M. Koper, *Phys. Chem. Chem. Phys.*, 2009, **11**, 10446–10456.
25 J. S. Spendelow and A. Wieckowski, *Phys. Chem. Chem. Phys.*, 2007, **9**, 2654–2675.
26 E. Antolini and E. R. Gonzalez, *J. Power Sources*, 2010, **195**, 3431–3450.
27 E. H. Yu, X. Wang, U. Krewer, L. Li and K. Scott, *Energy Environ. Sci.*, 12, **5**, 5668–5680.
28 M. Jing, L. Jiang, B. Yi and G. Sun, *J. Electroanal. Chem.*, 2013, **688**, 172–179.
29 S. P. S. Badwal, S. Giddey, A. Kulkarni, J. Goel and S. Basu, *Appl. Energy*, 2015, **145**, 80–103.
30 Z. Wang, C. Wang, S. Chen and Y. Liu, *Int. J. Hydrogen Energy*, 2014, **39**, 5644–5652.
31 Z. Zhang, L. Xin and W. Li, *Appl. Catal., B*, 2012, **119–120**, 40–48.
32 N. M. Aslam, M. S. Masdar, S. K. Kamarudin and W. R. W. Daud, *APCBEE Proc.*, 2012, **3**, 33–39.

33 X. Yu and P. G. Pickup, *J. Power Sources*, 2008, **182**, 124–132.
34 C. Rice, S. Ha, R. I. Masel and A. Wieckowski, *J. Power Sources*, 2003, **115**, 229–235.
35 E. Antolini, *J. Power Sources*, 2007, **170**, 1–12.
36 E. Antolini, *Energy Environ. Sci.*, 09, **2**, 915–931.
37 C. Bianchini and P. K. Shen, *Chem. Rev.*, 2009, **109**, 4183–4206.
38 Z.-Y. Zhou, Q. Wang, J.-L. Lin, N. Tian and S.-G. Sun, *Electrochim. Acta*, 2010, **55**, 7995–7999.
39 G. García, N. Tsiouvaras, E. Pastor, M. A. Peña, J. L. G. Fierro and M. V. Martínez-Huerta, *Int. J. Hydrogen Energy*, 2012, **37**, 7131–7140.
40 J. L. Cohen, D. J. Volpe and H. D. Abruna, *Phys. Chem. Chem. Phys.*, 2007, **9**, 49–77.
41 T. Jurzinsky, P. Kammerer, C. Cremers, K. Pinkwart and J. Tübke, *J. Power Sources*, 2016, **303**, 182–193.
42 T. Jurzinsky, R. Bär, C. Cremers, J. Tübke and P. Elsner, *Electrochim. Acta*, 2015, **176**, 1191–1201.
43 A. Santasalo-Aarnio, Y. Kwon, E. Ahlberg, K. Kontturi, T. Kallio and M. T. M. Koper, *Electrochem. Commun.*, 2011, **13**, 466–469.
44 A. V. Tripković, K. D. Popović, J. D. Momčilović and D. M. Draić, *J. Electroanal. Chem.*, 1996, **418**, 9–20.
45 A. V. Tripković, K. Đ. Popović, J. D. Momčilović and D. M. Dražić, *Electrochim. Acta*, 1998, **44**, 1135–1145.
46 S. Y. Shen, T. S. Zhao and Q. X. Wu, *Int. J. Hydrogen Energy*, 2012, **37**, 575–582.
47 J. Friedl and U. Stimming, *Electrochim. Acta*, 2013, **101**, 41–58.
48 J. Liu, J. Ye, C. Xu, S. P. Jiang and Y. Tong, *Electrochem. Commun.*, 2007, **9**, 2334–2339.
49 Z. X. Liang, T. S. Zhao, J. B. Xu and L. D. Zhu, *Electrochim. Acta*, 2009, **54**, 2203–2208.
50 L. Wang, H. Meng, P. K. Shen, C. Bianchini, F. Vizza and Z. Wei, *Phys. Chem. Chem. Phys.*, 2011, **13**, 2667–2673.
51 C. Lamy and J.-M. Leger, *J. Phys. IV*, 1994, **04**, C1-253–C1-281.
52 E. Santos and M. C. Giordano, *Electrochim. Acta*, 1985, **30**, 871–878.
53 C. B. Gorman, S. H. Bergens and G. M. Whitesides, *J. Catal.*, 1996, **158**, 92–96.
54 K. Miyazaki, T. Matsumiya, T. Abe, H. Kurata, T. Fukutsuka, K. Kojima and Z. Ogumi, *Electrochim. Acta*, 2011, **56**, 7610–7614.
55 L. Xin, Z. Zhang, J. Qi, D. Chadderdon and W. Li, *Appl. Catal., B*, 2012, **125**, 85–94.
56 M. Simoes, S. Baranton and C. Coutanceau, *ChemSusChem*, 2012, **5**, 2106–2124.
57 A. Villa, N. Dimitratos, C. E. Chan-Thaw, C. Hammond, L. Prati and G. J. Hutchings, *Acc. Chem. Res.*, 2015, **48**, 1403–1412.
58 A. Behr, J. Eilting, K. Irawadi, J. Leschinski and F. Lindner, *Green Chem.*, 2008, **10**, 13–30.
59 A. Serov, U. Martinez and P. Atanassov, *Electrochem. Commun.*, 2013, **34**, 185–188.
60 A. Zadick, L. Dubau, N. Sergent, G. Berthomé and M. Chatenet, *ACS Catal.*, 2015, **5**, 4819–4824.
61 F. E. López-Suárez, C. T. Carvalho-Filho, A. Bueno-López, J. Arboleda, A. Echavarría, K. I. B. Eguiluz and G. R. Salazar-Banda, *Int. J. Hydrogen Energy*, 2015, **40**, 12674–12686.
62 N. Erini, P. Krause, M. Gliech, R. Yang, Y. Huang and P. Strasser, *J. Power Sources*, 2015, **294**, 299–304.

63 X. Wang, C. Hu, Y. Xiong, H. Liu, G. Du and X. He, *J. Power Sources*, 2011, **196**, 1904–1908.
64 Z. Zhang, L. Xin, J. Qi, D. J. Chadderdon and W. Li, *Appl. Catal., B*, 2013, **136–137**, 29–39.
65 J. Wang, N. Cheng, M. N. Banis, B. Xiao, A. Riese and X. Sun, *Electrochim. Acta*, 2015, **185**, 267–275.
66 H. A. Gasteiger, S. S. Kocha, B. Sompalli and F. T. Wagner, *Appl. Catal., B*, 2005, **56**, 9–35.
67 K. Jiang, H. X. Zhang, S. Z. Zou and W. B. Cai, *Phys. Chem. Chem. Phys.*, 2014, **16**, 20360–20376.
68 L. Ma, D. Chu and R. Chen, *Int. J. Hydrogen Energy*, 2012, **37**, 11185–11194.
69 S. Kabir, A. Serov, A. Zadick, K. Artyushkova and P. Atanassov, *ChemElectroChem*, 2016, n/a–n/a.
70 J. Yin, S. Y. Shan, M. S. Ng, L. F. Yang, D. Mott, W. Q. Fang, N. Kang, J. Luo and C. J. Zhong, *Langmuir*, 2013, **29**, 9249–9258.
71 A. S. Aricò, S. Srinivasan and V. Antonucci, *Fuel Cells*, 2001, **1**, 133–161.
72 A. O. Neto, S. G. da Silva, G. S. Buzzo, R. F. B. de Souza, M. H. M. T. Assumpção, E. V. Spinacé and J. C. M. Silva, *Ionics*, 2015, **21**, 487–495.
73 A. N. Geraldes, D. Furtunato, da Silva, J. C. Martins, da Silva, O. Antonio, de Sá, E. V. Spinacé, A. O. Neto and M. Coelho dos Santos, *J. Power Sources*, 2015, **275**, 189–199.
74 M. Liu, R. Zhang and W. Chen, *Chem. Rev.*, 2014, **114**, 5117–5160.
75 M. M. Dimos and G. J. Blanchard, *J. Phys. Chem. C*, 2010, **114**, 6019–6026.
76 R. S. Ferreira Jr, M. Janete Giz and G. A. Camara, *J. Electroanal. Chem.*, 2013, **697**, 15–20.
77 A. Zalineeva, A. Serov, M. Padilla, U. Martinez, K. Artyushkova, S. Baranton, C. Coutanceau and P. B. Atanassov, *Appl. Catal., B*, 2015, **176–177**, 429–435.
78 V. Bambagioni, C. Bianchini, A. Marchionni, J. Filippi, F. Vizza, J. Teddy, P. Serp and M. Zhiani, *J. Power Sources*, 2009, **190**, 241–251.
79 Y.-Y. Yang, J. Ren, H.-X. Zhang, Z.-Y. Zhou, S.-G. Sun and W.-B. Cai, *Langmuir*, 2013, **29**, 1709–1716.
80 E. D. Wang, J. B. Xu and T. S. Zhao, *J. Phys. Chem. C*, 2010, **114**, 10489–10497.
81 N. Arjona, M. Guerra-Balcazar, L. Ortiz-Frade, G. Osorio-Monreal, L. Alvarez-Contreras, J. Ledesma-Garcia and L. G. Arriaga, *J. Mater. Chem. A*, 2013, **1**, 15524–15529.
82 X. W. Zhou, Y. L. Gan, Z. X. Dai and R. H. Zhang, *J. Electroanal. Chem.*, 2012, **685**, 97–102.
83 S. Kabir, A. Serov, K. Artyushkova and P. Atanassov, *Electrochim. Acta*, 2016, **203**, 144–153.
84 G. He, Y. Song, X. Kang and S. Chen, *Electrochim. Acta*, 2013, **94**, 98–103.
85 S. Cherevko, N. Kulyk and C.-H. Chung, *Nanoscale*, 2012, **4**, 103–105.
86 G. Hu, F. Nitze, H. R. Barzegar, T. Sharifi, A. Mikołajczuk, C.-W. Tai, A. Borodzinski and T. Wågberg, *J. Power Sources*, 2012, **209**, 236–242.
87 L. A. Estudillo-Wong, A. M. Vargas-Gómez, E. M. Arce-Estrada and A. Manzo-Robledo, *Electrochim. Acta*, 2013, **112**, 164–170.
88 Y. Wang, Z. M. Sheng, H. Yang, S. P. Jiang and C. M. Li, *Int. J. Hydrogen Energy*, 2010, **35**, 10087–10093.
89 K. Kakaei and M. Dorraji, *Electrochim. Acta*, 2014, **143**, 207–215.
90 A. Ruban, B. Hammer, P. Stoltze, H. L. Skriver and J. K. Nørskov, *J. Mol. Catal. A: Chem.*, 1997, **115**, 421–429.
91 D. A. Slanac, W. G. Hardin, K. P. Johnston and K. J. Stevenson, *J. Am. Chem. Soc.*, 2012, **134**, 9812–9819.

92 Z. Yin, Y. Zhang, K. Chen, J. Li, W. Li, P. Tang, H. Zhao, Q. Zhu, X. Bao and D. Ma, *Sci. Rep.*, 2014, **4**, 4288.

93 Y. Yang, W. Wang, Y. Liu, F. Wang, Z. Zhang and Z. Lei, *Int. J. Hydrogen Energy*, 2015, **40**, 2225–2230.

94 J. Liu, H. Zhou, Q. Wang, F. Zeng and Y. Kuang, *J. Mater. Sci.*, 2011, **47**, 2188–2194.

95 M. B. Griffin, A. A. Rodriguez, M. M. Montemore, J. R. Monnier, C. T. Williams and J. W. Medlin, *J. Catal.*, 2013, **307**, 111–120.

96 A. N. Geraldes, D. F. da Silva, E. S. Pino, J. C. M. da Silva, R. F. B. de Souza, P. Hammer, E. V. Spinacé, A. O. Neto, M. Linardi and M. C. dos Santos, *Electrochim. Acta*, 2013, **111**, 455–465.

97 Y.-H. Qin, Y. Li, R.-L. Lv, T.-L. Wang, W.-G. Wang and C.-W. Wang, *Electrochim. Acta*, 2014, **144**, 50–55.

98 R. Jana, U. Subbarao and S. C. Peter, *J. Power Sources*, 2016, **301**, 160–169.

99 P. Wu, Y. Huang, L. Zhou, Y. Wang, Y. Bu and J. Yao, *Electrochim. Acta*, 2015, **152**, 68–74.

100 Y. Wang, T. S. Nguyen, X. Liu and X. Wang, *J. Power Sources*, 2010, **195**, 2619–2622.

101 U. B. Demirci, *J. Power Sources*, 2007, **173**, 11–18.

102 J. Greeley and J. K. Nørskov, *Surf. Sci.*, 2005, **592**, 104–111.

103 C. Coutanceau, A. Zalineeva, S. Baranton and M. Simoes, *Int. J. Hydrogen Energy*, 2014, **39**, 15877–15886.

104 A. Zalineeva, S. Baranton and C. Coutanceau, *Electrochem. Commun.*, 2013, **34**, 335–338.

105 W. Du, K. E. Mackenzie, D. F. Milano, N. A. Deskins, D. Su and X. Teng, *ACS Catal.*, 2012, **2**, 287–297.

106 T. Ramulifho, K. I. Ozoemena, R. M. Modibedi, C. J. Jafta and M. K. Mathe, *J. Electroanal. Chem.*, 2013, **692**, 26–30.

107 A. Sadiki, P. Vo, S. Hu, T. S. Copenhaver, L. Scudiero, S. Ha and J. L. Haan, *Electrochim. Acta*, 2014, **139**, 302–307.

108 L. Ma, H. He, A. Hsu and R. Chen, *J. Power Sources*, 2013, **241**, 696–702.

109 L. M. Palma, T. S. Almeida, V. L. Oliveira, G. Tremiliosi-Filho, E. R. Gonzalez, A. R. de Andrade, K. Servat, C. Morais, T. W. Napporn and K. B. Kokoh, *RSC Adv.*, 2014, **4**, 64476–64483.

110 H. Mao, T. Huang and A. Yu, *Electrochim. Acta*, 2015, **174**, 1–7.

111 R. C. Sekol, M. Carmo, G. Kumar, F. Gittleson, G. Doubek, K. Sun, J. Schroers and A. D. Taylor, *Int. J. Hydrogen Energy*, 2013, **38**, 11248–11255.

112 F. Zhu, G. Ma, Z. Bai, R. Hang, B. Tang, Z. Zhang and X. Wang, *J. Power Sources*, 2013, **242**, 610–620.

113 Y. Huang, Y. Guo and Y. Wang, *J. Power Sources*, 2014, **249**, 9–12.

114 C. Shang, W. Hong, J. Wang and E. Wang, *J. Power Sources*, 2015, **285**, 12–15.

Newer polymer electrolytes and electrocatalysts for direct alcohol fuel cells

P. Sridhar,*[a] S. D. Bhat[b] and A. K. Sahu[b]
DOI: 10.1039/9781782622727-00102

1 Introduction

Fuel Cells are known for their higher energy conversion efficiency. Polymer Electrolyte Fuel Cells (PEFCs) are preferred in comparison with the other types of fuel cells due to their lower operating temperature, ease of electrolyte management *etc.* Hydrogen as a fuel in PEFCs has its own limitation in terms of generation and storage. Among several types of fuel cells, direct alcohol fuel cells are gaining tremendous interest as energy source for portable application, since liquid fuel without the external reforming system could stave off storage and safety issues.[1,2] Certain hydrogen-carrying organic fuels such as methanol, ethanol, propanol, ethylene glycol and diethyl ether are also considered for fuelling PEFCs directly. The use of methanol/ethanol as fuel has several advantages in comparison to hydrogen: it is a cheap liquid fuel, easily handled, transported, and stored, and with a high theoretical energy density.[3]

The specific energy of hydrogen compressed at 700 bar is 142 $MJ\,kg^{-1}$ while the specific energies for methanol and ethanol are 22.7 and 29.7 $MJ\,kg^{-1}$, respectively, while gasoline contains about 45 $MJ\,kg^{-1}$.[4] This sets the basis for sourcing methanol and ethanol as fuels in PEFCs. However, there are certain technical issues that are being addressed to overcome the obstacles for their commercialisation. The critical ones are (i) the polymer electrolyte and (ii) the catalysts used for the oxidation of alcohols at the anode and that for the reduction of oxygen at the cathode. Perfluorosulphonic acid membrane commercially available as Nafion manufactured by M/s DuPont Inc., USA is presently being widely used as the polymer electrolyte for Direct Methanol and Direct Ethanol fuel cells. Though thinner versions like Nafion 211, 212 are available, the preferred one being Nafion 117 on account of less methanol/ethanol permeability through the latter in comparison with the former. The ionic conductivity of these membranes are in the order of 10^{-2} $S\,cm^{-1}$ that makes them attractive inspite of higher alcohol permeability through these membranes. Higher permeability of the alcohol through the membranes results not only in loss of fuel available for oxidation, but interferes in the reduction of oxygen at the cathode resulting in the reduced conversion

[a]*M/s Mesha Energy Solutions Pvt. Ltd., 93A, Industrial Suburb, 2nd stage, Yeshwanthpur, Bengaluru 560 022, India. E-mail: psridhar@meshatech.com*
[b]*Central Electrochemical Research Institute (CECRI)-Madras Unit, CSIR Madras Complex, Chennai 600 113, India*

efficiency. This has made the researchers to look for alternative electrolytes. The polymer membrane electrolytes investigated for their application in Direct Alcohol Fuel Cells have been exhaustively reviewed by Horacio R. Corti.[5]

The real challenge is in designing a catalyst for methanol or ethanol oxidation, which involves transfer of six or twelve electrons for the oxidation to go to completion. Similar is the case in designing a catalyst for oxygen reduction that is insensitive for oxidation of alcohol. Designing such a catalyst will be helpful as permeability of alcohol through the polymer electrolyte may not be completely eliminated but could only be mitigated by proper design of the membrane electrolyte.

Alkaline Direct alcohol fuel cells (ADAFCs) have attracted increasing interest over the past decade because of their favourable reaction kinetics in alkaline media, higher energy densities achievable and the easy handling of the liquid fuels. Unlike in acidic media, the kinetics of both methanol oxidation reaction (MOR) and oxygen reduction reaction (ORR) in alkaline media is much faster allowing the use of non-precious metal catalysts to reduce the cost of the fuel cell.

Excellent review articles are available in the literature covering PEMs and electrocatalysts for Direct Ethanol Fuel Cell.[6–11]

The present chapter is organized in two major sections, each section dealing with the development of polymer membrane electrolyte and, anode and cathode catalysts for Direct Alcohol Fuel Cells. The section on Polymer Membrane Electrolyte focuses on recent advances that have been made under Nafion and sulfonated polyether ether ketone (sPEEK) composite membranes containing inorganic additives viz. Zeolites, Mesoporous materials, Clays and carbon nanostructures in the light of the potential that is seen in these composites for their ability to reduce the alcohol cross over in addition to improved ionic conductivity and durability and also on development of Anion Exchange Membrane (AEMs) for Alkaline Direct Methanol Fuel Cells (ADMFCs).

In the section under Electrocatalysts for DAFCs, the recent advances that have been made under methanol tolerant ORR catalysts for DMFCs and ADMFCs, methaol oxidation catalysts for DMFCs, ethanol tolerant ORR catalysts and ethanol oxidation catalysts for DEFCs are presented.

2 Proton conducting polymer electrolytes for DAFCs

In following sections, different types of proton exchange membranes (PEMs) based on Nafion and sPEEK blend and composites incorporated with inorganic additives are discussed for application in DAFCs.

2.1 Nafion based polymer electrolytes for DMFCs

2.1.1 Organic–inorganic composite polymer-electrolyte membranes for DMFCs. Organic–inorganic composite membranes comprising Nafion with inorganic materials such as silica, mesoporous zirconium phosphate (MZP) and mesoporous titanium phosphate (MTP) are fabricated and evaluated as PEM for DMFCs. For Nafion–silica composite membrane, silica is impregnated into Nafion matrix as a sol by a novel water hydrolysis process precluding the external use of an acid.[12] Instead, the acidic nature of Nafion facilitates *in situ* polymerization reaction with Nafion leading to a uniform composite membrane. The rapid hydrolysis and polymerization reaction while preparing zirconia and titania sols leads to uncontrolled thickness and volume reduction in the composite membranes, and hence is not conducive for casting membranes. Nafion–MZP and Nafion–MTP composite membranes are prepared by mixing pre-formed porous MZP and MTP with Nafion matrix. MZP and MTP are synthesised by co-assembly of a tri-block co-polymer, namely pluronic-F127, as a structure-directing agent, and a mixture of zirconium butoxide/titanium isopropoxide and phosphorous trichloride as inorganic precursors. The use of tri-block co-polymers as structure-directing agents enables the synthesis of mesoscopically ordered materials exhibiting a narrow pore-size distribution after surfactant removal. MZP and MTP reported here are surface-functionalized solid-superacid-proton conducting materials as well as inorganic fillers with high affinity to absorb water which helps fast proton-transport across the electrolyte membrane and makes it suitable for DMFCs. Specifically designed composite membranes increase water-uptake properties with reduced methanol permeability ameliorating the DMFC performance in comparison to a DMFC using commercially available Nafion-117 membrane while operating under identical conditions. Methanol release kinetics is also studied by volume-localized NMR spectroscopy (employing "point resolved spectroscopy", PRESS), the results clearly demonstrating that the incorporation of inorganic fillers in Nafion retards the methanol release kinetics under osmotic drag. Appreciable proton conductivity with reduced methanol permeability across the composite membranes leads to improved performance of DMFCs in relation to commercially available Nafion-117 membrane.

2.1.2 Diffusion of CH_3OH from membrane sac to 2 M CD_3OD in water. Permeation of methanol from membrane sacs – each saturated with 2 M CH_3OH in D_2O of Nafion, Nafion impregnated with 10 wt% SiO_2, Nafion with 5 wt% MTP and Nafion with 5 wt% MZP, to surrounding 2 M CD_3OD in water was recorded at different time intervals by PRESS spectroscopy.[13] It may be noted that 'shimming' the voxel of interest is facilitated by water in the surrounding medium, although this signal is finally suppressed by standard outer volume suppression and specific water signal suppression procedures. The medium surrounding the membrane sacs employed in these studies, namely 2 M CD_3OD in water, permits the accurate determination of small differences in the release kinetics pertaining to different membranes by slowing down the overall release kinetics in comparison to a system that has only water surrounding the membrane sac. The kinetics data are culled from the volume-localized spectral integral in the methyl region of the 1H spectrum; the spectra are presented as stack plots in Fig. 1(a)–(d), respectively.

The time at which each measurement commenced after dropping the sac in the solvent mixture is indicated to the right of each spectral trace.

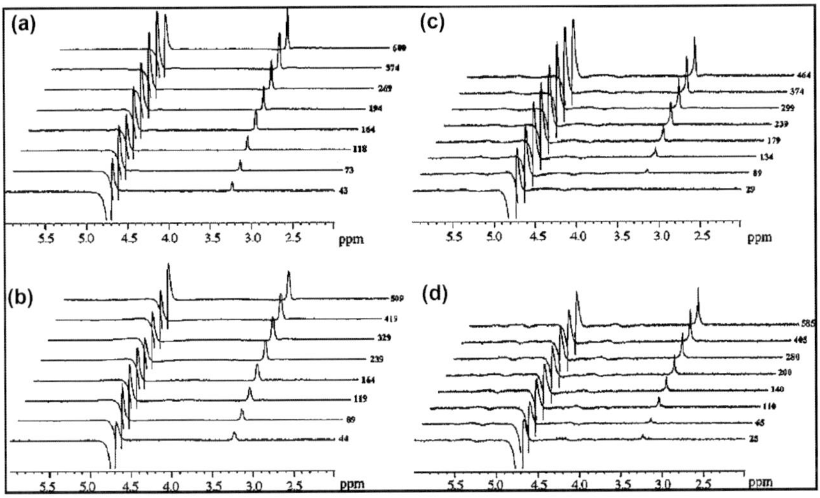

Fig. 1 Volume-localized spectra of methanol release from sacs of (a) Nafion, (b) Nafion + 10 wt% SiO_2, (c) Nafion + 5 wt% MTP and (d) Nafion + 5 wt% MZP, in each case after dropping the equilibrated sac in the solvent mixture. Reproduced with permission from A. K. Sahu, S. D. Bhat, S. Pitchumani, P. Sridhar, V. Vimalan, C. George, N. Chandrakumar and A. K. Shukla, Novel organic-inorganic composite polymer-electrolyte membranes for DMFCs, *J. Membr. Sci.*, **345**, 305–314. Copyright 2009 with permission from Elsevier B.V.

Eqn (1), corresponding to the Weibull function[14,15] has been used for fitting the curve.

$$I(t) = A - (A - B)\exp\left[-(kt)^d\right] \tag{1}$$

The integral I of the methyl signal in a voxel chosen to lie fully outside the sac has been plotted as a function of time in Fig. 2(a)–(d). It may be mentioned that the absolute values of the time-dependent methyl signal integral may not be compared between membranes, since they are not normalized to unit membrane mass, or to unit mass of absorbed methanol. Variations of the rate constant k are directly interpretable, while variations of the exponent d of (kt) lead to differences in short and long time behaviour. The methanol release kinetics measurement clearly demonstrates that Nafion with SiO_2 has the lowest methanol permeability, followed by Nafion with MZP. Under these conditions of permeation controlled by osmotic drag, Nafion and Nafion with MTP have similar methanol permeation characteristics (see Table 1). Differences in the exponent d between the various systems is also to be noted; further study is in progress to interpret this in terms of the pertinent release mechanism.[16]

The DMFC performance curves for the MEAs comprising Nafion–silica, Nafion–MZP and Nafion–MTP composite membranes are shown in Fig. 3(a–c), respectively. It is noteworthy that the DMFC performance increases with increase in filler content and reaches an intermediate value at the optimum performance. Excess filler in composite membrane hinders the proton conduction path which results in decreased DMFC performance. It is noteworthy that a peak power-density of 110 mW cm^{-2}

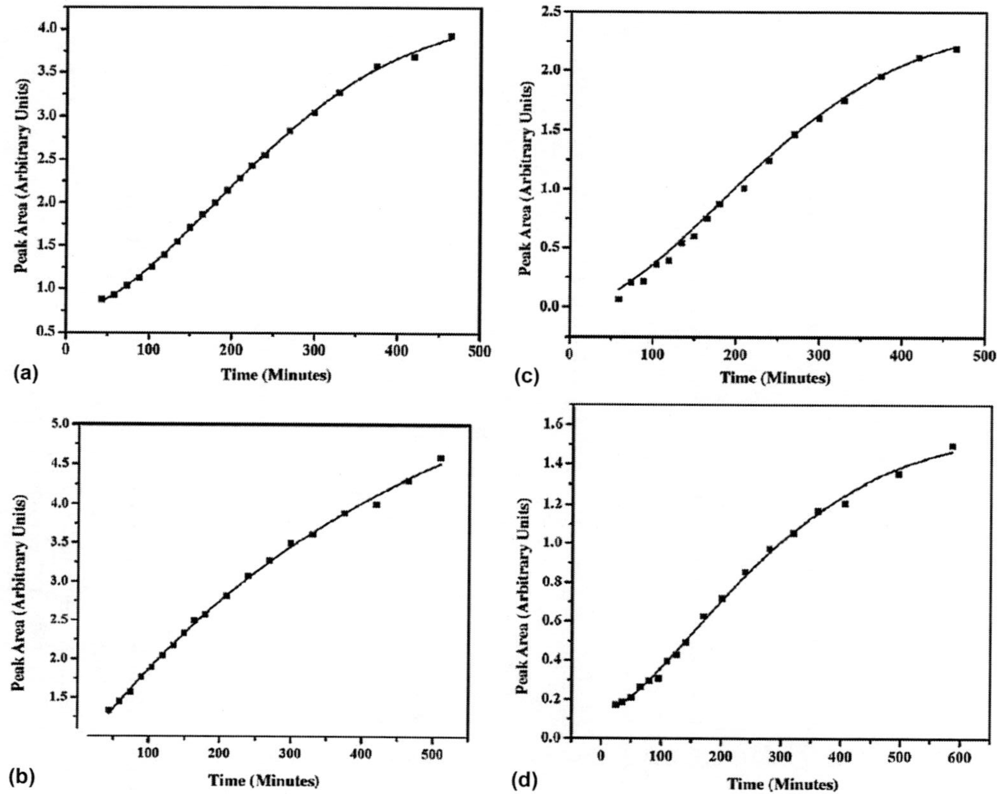

Fig. 2 Plots of the integral of the methyl signal of methanol *vs.* time for four different membrane sacs: (a) Nafion; (b) Nafion + 10 wt% SiO_2, (c) Nafion + 5 wt% MTP and (d) Nafion + 5 wt% MZP. All the plots were fitted with the Weibull equation: $I(t) = A - (A - B)\exp[-(kt)^d]$. Reproduced with permission from A. K. Sahu, S. D. Bhat, S. Pitchumani, P. Sridhar, V. Vimalan, C. George, N. Chandrakumar and A. K. Shukla, Novel organic-inorganic composite polymer-electrolyte membranes for DMFCs, *J. Membr. Sci.*, **345**, 305–314. Copyright 2009 with permission from Elsevier B.V.

Table 1 Methanol permeation in membrane sacs as measured by volume-localized NMR.

Fit equation $I(t) = A - (A - B)\exp[-(kt)^d]$					
Sample	A	B	d	k (min^{-1})	R^2
Nafion	4.22 ± 0.07	0.73 ± 0.03	1.76 ± 0.06	0.00355 ± 0.00007	0.99944
Nafion + SiO_2 (10 wt%)	6.26 ± 0.84	0.79 ± 0.15	1.01 ± 0.12	0.00223 ± 0.00053	0.99843
Nafion + MTP (5 wt%)	2.41 ± 0.15	0 ± 0.08	1.78 ± 0.21	0.00358 ± 0.00022	0.99541
Nafion + MZP (5 wt%)	1.57 ± 0.05	0.12 ± 0.02	1.51 ± 0.89	0.00321 ± 0.00015	0.99786

at a load current-density of 450 mA cm^{-2} is observed for the DMFC comprising Nafion-117 membrane. By contrast, peak power-densities of 140 mW cm^{-2}, 125 mW cm^{-2} and 140 mW cm^{-2} are achieved with Nafion–silica (10 wt%), Nafion–MZP (5 wt%) and Nafion–MTP (5 wt%) composite membranes, respectively, under identical operating conditions. It is obvious that the existence of inorganic fillers reduces the methanol crossover owing to the tortuous path around the filler material. It is noteworthy that the performance of DMFC is maximum for Nafion–silica (10 wt%) composite membrane, while for Nafion–MZP and Nafion–MTP composite membranes the optimized performance is obtained with 5 wt% filler. This may be due to impregnation of silica into Nafion matrix in the form of a sol by a novel water hydrolysis sol–gel process. The acidic character of Nafion facilitates *in situ* polymerization reaction in Nafion matrix leading to a uniform composite membrane without any phase separation. By contrast, the rapid hydrolysis and polymerization reaction while preparing titania and zirconia sol delivers uncontrolled thickness and volume reduction in the composite membranes and hence is inadaptable. As pre-formed MZP and MTP powders are adopted to mix with Nafion ionomer, phase separation is usually observed in the finished Nafion-MZP and Nafion-MTP composite membranes. Hence, beyond 5 wt% in the filler content the proton migration is hindered by the filler although this helps reducing methanol crossover. In the light of the foregoing, a compromise needs to be made between proton conductivity and methanol crossover in any composite membrane. In the present study, it is optimized at 10 wt% for silica, 5 wt% for MZP and 5 wt% for MTP content with respect to the Nafion matrix in realizing suitable composite membranes for DMFCs. Figure 3(d) shows the equivalent current for crossed-over methanol across Nafion-117, Nafion–silica (10 wt%), Nafion–MZP (5 wt%) and Nafion–MTP (5 wt%) composite membranes at varying load current-densities. It is seen that the methanol crossover decreases significantly for the composite membranes compared to pristine Nafion-117 membrane. Nafion–MTP composite membrane exhibits a lower methanol crossover in relation to other types of composite membranes, which could be due to the presence of relatively bigger particles and lower surface area of the filler.[17] The Nafion–MZP

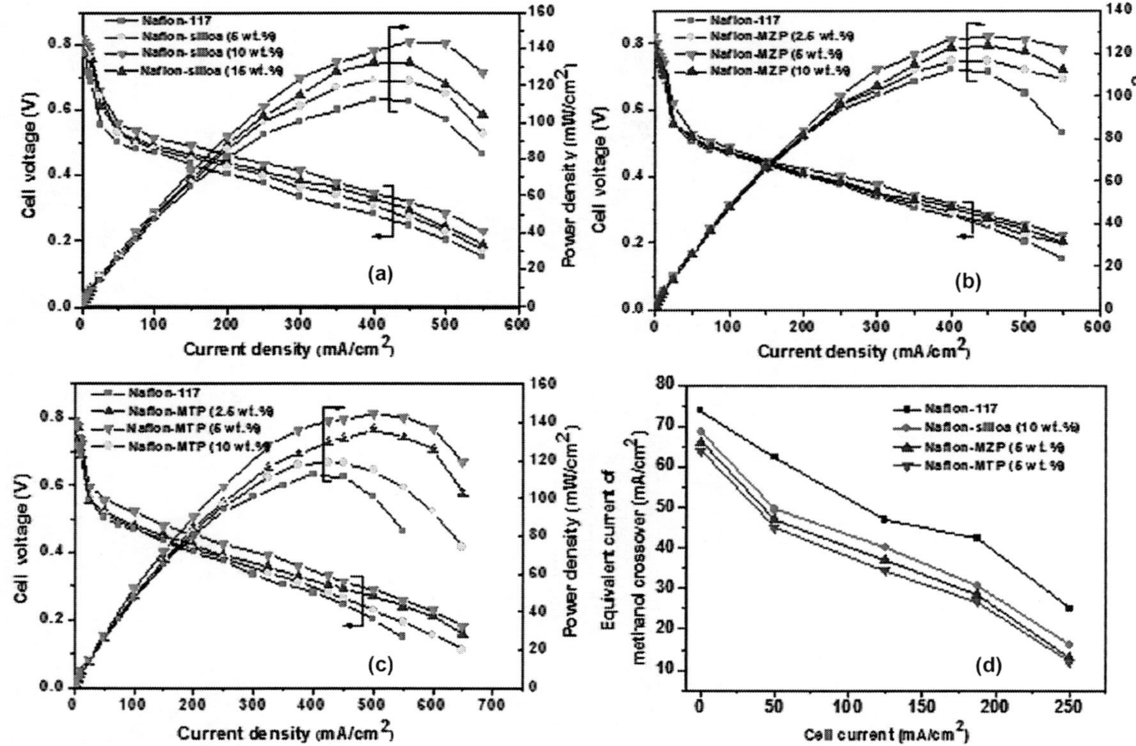

Fig. 3 Performance curves for DMFCs (a) Nafion–silica (b) Nafion–MZP and (c) Nafion–MTP composite membranes at 70 °C using methanol and oxygen reactants at atmospheric pressure. Anode: 2 M aqueous methanol with a flow rate of 2 sccm; cathode: oxygen with a flow rate of 200 sccm; (d) Equivalent currents for methanol crossover through Nafion-117, Nafion–silica, Nafion–MZP and Nafion–MTP composite membranes from anode to cathode in a $MeOH/O_2$ single cell at 70 °C. Reproduced with permission from A. K. Sahu, S. D. Bhat, S. Pitchumani, P. Sridhar, V. Vimalan, C. George, N. Chandrakumar and A. K. Shukla, Novel organic-inorganic composite polymer-electrolyte membranes for DMFCs, *J. Membr. Sci.*, **345**, 305–314. Copyright 2009 with permission from Elsevier B.V.

composite membrane shows an intermediate value compared to Nafion–silica and Nafion–MTP composite membranes. In general, the hydrophilic nature of silica, MZP and MTP particles in the composite membranes prevents the formation of non-selective voids for methanol transport without affecting the migration of protons as evidenced from the proton conductivity data.

2.1.3 Meso-structured silica and alumino-silicate-Nafion hybrid membranes for direct methanol fuel cells. Characteristics of Nafion-based hybrid membranes significantly differ with the nature of inorganic filler materials, namely their structural properties, composition and the method of impregnation to the host matrix. Porous materials are most effective due to their attractive textural and surface features, such as highly ordered-structures, high surface area, wide pore-range and pore-size distribution. It is noteworthy that the porous nature of the filler materials facilitates specific ionic-interactions when in contact with the host polymer matrix enhancing water-uptake and other related properties of the hybrid membranes.[18] In the light of the foregoing, meso-porous silica and aluminosilicate with varying structure and surface area is adopted for making hybrid with Nafion matrix and their possible exploration in the methanol fuel cell.[19,20] The fillers are acid functionalized to provide proton-conducting pathways between filler and matrix and obviously block the methanol pathway across the hybrid membrane. Silica nanopowder of surface area 60 $m^2 g^{-1}$, silica meso-structured cellular foam (MSU-F) of surface area 470 $m^2 g^{-1}$ and silica meso-structured hexagonal frame network (MCM-41) of surface area 900 $m^2 g^{-1}$ have been employed as potential filler materials to form hybrid membranes with Nafion framework.[19] In case of alumino-silicate, acid-functionalized mesostructured aluminosilicate with cellular foam framework (Al-MSU-F type) of surface area 463 $m^2 g^{-1}$, aluminosilicate molecular sieves (Al-HMS type) of surface area 651 $m^2 g^{-1}$ and meso-structured aluminosilicate with hexagonal network (Al-MCM-41 type) of surface area 799 $m^2 g^{-1}$ have been employed as potential filler materials to form hybrid membranes with Nafion.[20] The structural behavior, water uptake, proton conductivity and methanol permeability of these hybrid membranes have been thoroughly studied.

The aforesaid characteristics of the hybrid membranes could be exclusively attributed to the presence of pendant sulfonic acid groups in the filler, which provide fairly continuous proton-conducting pathways between filler and matrix in the hybrid membranes facilitating proton transport without any trade-off between its proton conductivity and methanol crossover.

From the aforesaid, excellent DMFC performance of Nafion-aluminosilicate is attributed to the strong Brønsted acidity and quasi-solid electrolytes properties of the aluminosilicate in which protons move along the hydrogen bonds.[21] Accordingly, the diffusion rate for protons in aluminosilicates is expected to be similar to the aqueous solutions. Furthermore, porous aluminosilicate materials have been found to be attractive due to their textural and surface features, namely highly ordered-structures, high surface-area, and wide pore-range/pore-size distribution.[22] Porous morphology of aluminosilicates facilitates specific ionic-interactions with the host matrix, and enhances water-uptake and other related properties of the composite membranes.

BET surface area data and the corresponding average pore size analysis for various silica and aluminasilicate powders before and after functionalization are summarized in Tables 2 and 3. No major change is noticed in the hysteresis behaviour and BET surface area for silica nano powder and silica MSU-F before and after functionalization but for a little reduction in the surface area. However, MCM-41 silica undergoes substantial structural distortion after acid functionalization. Before functionalization, BET surface area for silica MCM-41 is 900 $m^2 g^{-1}$ and the sorption isotherm is of type IV with H4 type hysteresis loop. After functionalization, the surface area reduces to 164 $m^2 g^{-1}$, suggesting substantial change in the structural behavior of MCM-41 with increased pore size distribution between 1 and 4 nm.[19] Similarly, BET surface area values for Al-MSU-F before and after functionalization are 463 $m^2 g^{-1}$ and 429 $m^2 g^{-1}$, respectively. The hysteresis behavior remains almost unaffected after functionalization. However, Al-HMS and Al-MCM-41 undergo structural distortion after acid functionalization.[20] Before functionalization, the BET surface area values for Al-HMS and Al-MCM-41 are 651 $m^2 g^{-1}$ and 799 $m^2 g^{-1}$, respectively. After functionalization, the BET surface area for the Al-HMS and Al-MCM-41 aluminosilicats are 385 $m^2 g^{-1}$ and 430 $m^2 g^{-1}$, respectively. The significant reduction in the surface area is an indication of structural distortion in Al-HMS and Al-MCM-41 with marginal increase in their pore size values.

Table 2 BET surface area and corresponding pore size data for various silica powders before and after functionalization.

Sample	BET Surface area ($m^2 g^{-1}$) (before functionalization)	BET Surface area ($m^2 g^{-1}$) (after functionalization)	BJH Average pore size (nm) (before functionalization)	BJH Average pore size (nm) (after functionalization)
SiO_2 Nano powder	60 ± 0.66	58 ± 0.58	15 ± 0.022	20 ± 0.0030
SiO_2-MSU-F	470 ± 3.49	429 ± 3.15	5.3 ± 0.008	6.5 ± 0.010
SiO_2-MCM-41	900 ± 10.12	164 ± 1.34	1.4 ± 0.002	2.5 ± 0.004

Table 3 BET surface area and corresponding pore-size of various aluminosilicates before and after functionalization.

Sample	BET Surface area ($m^2 g^{-1}$) (before functionalization)	BET Surface area ($m^2 g^{-1}$) (after functionalization)	BJH Average pore size (nm) (before functionalization)	BJH Average pore size (nm) (after functionalization)
Al-MSU-F	463	429	7	8
Al-HMS	651	385	1.5	1.8
Al-MCM-41	799	430	1.4	2

For further investigation, TEM images are recorded to corroborate the physical properties observed from N_2-sorption isotherm and pore-size distribution analysis. Figure 4 shows the TEM images for silica nanopowder, MSU-F and MCM-41 silica powders before and after functionalization. Electron diffraction patterns showed in the inset to Fig. 4 confirm the amorphous nature of silica nanopowder, MSU-F and MCM-41 silica powder. The amorphous nature of silica nanopowder, MSU-F and MCM-41 helps formation of – SO_3H groups in the silica matrix during functionalization. The nano-sized silica particles and cellular worm-hole like structural features are observed for silica nanopowders and silica MSU-F mesophases, respectively and the structural features for both silica

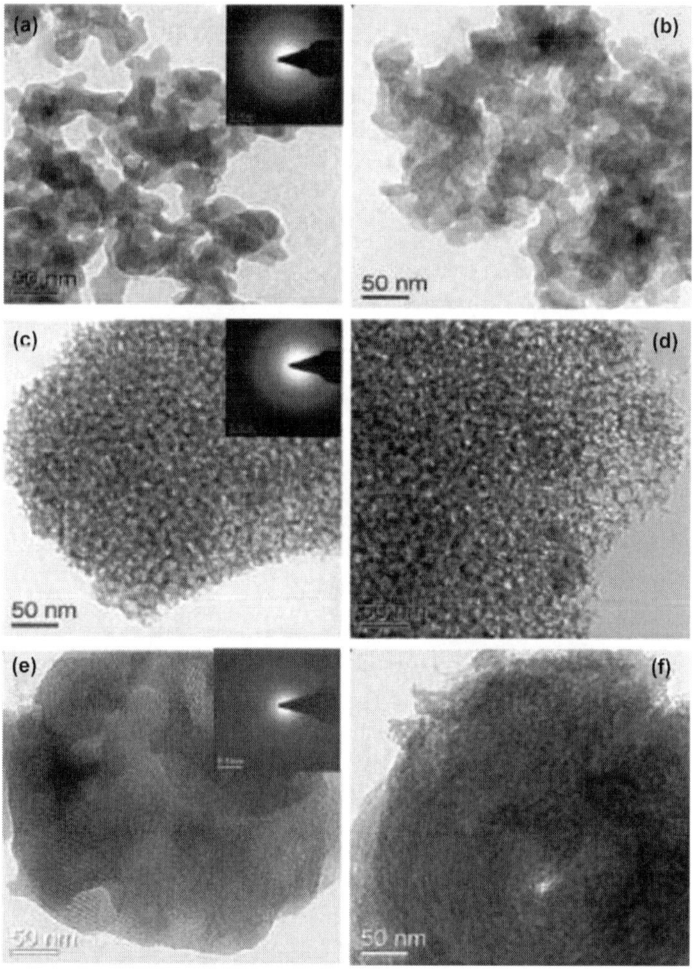

Fig. 4 TEM images for (a) Silica nanopowder, (c) MSU-F type silica powder and (e) MCM-41 type silica powders before functionalization, and (b) Silica nanopowder, (d) MSU-F type silica powder and (f) MCM-41 type silica powders after functionalization. The Scale bar in all the pictures is 50 nm. Reproduced with permission from A. K. Sahu, S. Meenakshi, S. D. Bhat, A. Shahid, P. Sridhar, S. Pitchumani and A. K. Shukla, Meso-Structured Silica-Nafion Hybrid Membranes for Direct Methanol Fuel Cells, *J. Electrochem. Soc.*, 2012, **159**(11), F702–F710. Copyright 2012, The Electrochemical Society.

nanopowder and MSU-F silica mesophase remain unchanged after acid functionalization which confirms their stability under acidic condition. By contrast, ordered hexagonal framework observed for MCM-41 undergoes severe structural collapse after acid functionalization. This further confirms that the hexagonal structure of MCM-41 is unstable under acidic condition during functionalization at 80 °C in agreement with the data observed from N_2 sorption isotherms.

TEM images are also recorded for all types of porous aluminosilicate materials shown in Fig. 5 before and after functionalization. The cellular

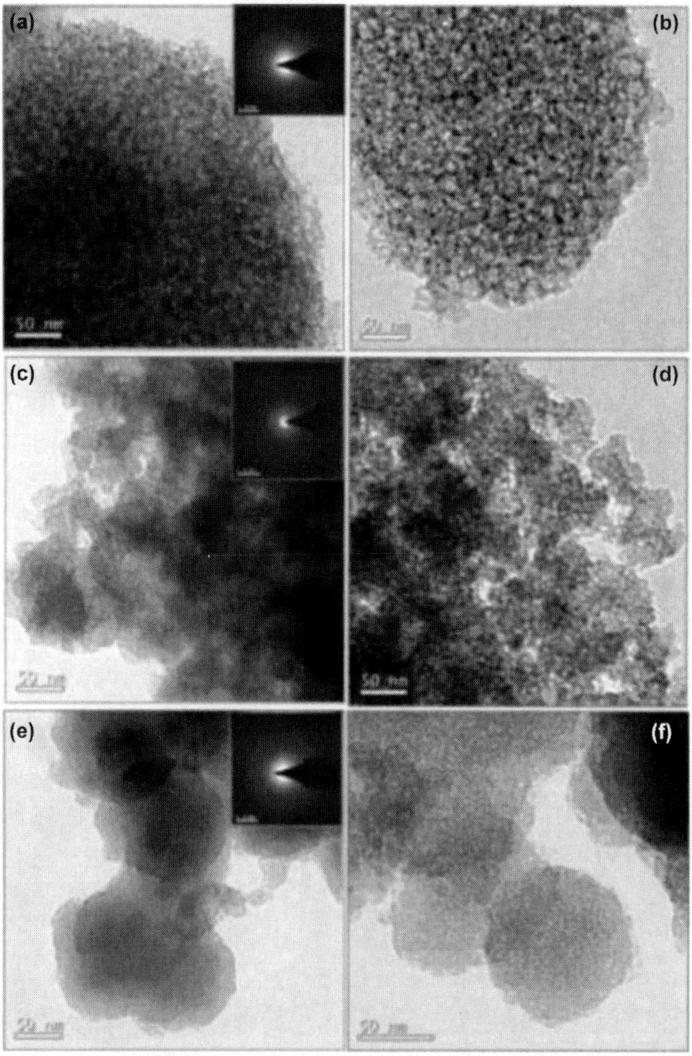

Fig. 5 TEM images for (a) Al-MSU-F, (c) Al-HMS and (e) Al-MCM-41 powders before functionalization and (b) Al-MSU-F, (d) Al-HMS and (f) Al-MCM-41 powders after functionalization. The scale bar in all the pictures is 50 nm. Reproduced with permission from S. Meenakshi, A. K. Sahu, S. B. D. Bhat, P. Sridhar, S. Pitchumani and A. K. Shukla, Mesostructured-aluminosilicate-Nafion hybrid membranes for direct methanol fuel cells, *Electrochim. Acta*, **89**, 35–44. Copyright 2012 with permission from Elsevier B.V.

foam framework like structural features are observed for Al-MSU-F type aluminosilicate meso-phase before and after functionalization suggesting that the structural features for Al-MSU-F type remain unchanged and stable under acidic condition. In the case of Al-HMS and Al-MCM-41, nano-sized worm-hole like structural features and ordered mesostructured hexagonal structure are observed respectively before functionalization. However, structure distortion is observed after acid functionalization as shown in Fig. 5, which is in agreement with the findings from N_2 sorption isotherm studies.

Performance curves for DMFCs comprising Nafion-silica hybrid membranes and Nafion-aluminosilicate hybrid membranes of optimized compositions are presented in Fig. 6a and b respectively. Presence of excess filler in the hybrid membrane hinders proton conduction path and reduce the proton conductivity and DMFC performance. Therefore, the filler contents in the hybrid membranes are optimized based on the proton conductivity data and are 5 wt%, 0.5 wt%, 0.25 wt% for silica nanopowder, silica MSU-F, silica MCM-41 additives in constituting Nafion-silica hybrid membranes. In case of Nafion-alumnosilicate hybrid membranes, the optimized wt% for Al-MSU-F, Al-HMS and Al-MCM-41 are 0.15 wt%, 0.5 wt% and 0.5 wt%, respectively. The peak power density of only 48 mW cm^{-2} was obtained for the DMFCs with pristine recast Nafion membrane. By contrast, peak power densities of 68, 127, 100, 156, 170 and 246 mW cm^{-2} are achieved for the DMFCs with Nafion-silica nanopowder (5 wt%), Nafion-silica MSU-F (0.5 wt%), Nafion-silica MCM-41 (0.25 wt%), Nafion-Al-MSU-F (0.15 wt%), Nafion-Al-HMS (0.5 wt%) and Nafion-Al-MCM-41 (0.5 wt%) hybrid membranes, respectively. Among silica additives, Nafion-silica MSU-F (0.5 wt%) hybrid membrane shows highest performance probably due to robust and worm-hole like mesostructures of silica MSU F type. The stable structure even after functionalization with sulfuric acid may as well contribute for dramatic reduction in methanol crossover and hence improved DMFC performance. Among aluminosilicate additves, Nafion-Al-MCM-41 hybrid membrane deliver peak-power densities of 246 mW cm^{-2} which is highest among all other hybrid membrane studied here.

2.1.4 Graphene based composite membranes.

Graphene oxide can be obtained by the exfoliation of graphite and subsequent oxidation. Graphene oxide contains carboxylic acid groups and the epoxy groups and is readily dispersible in many solvents due to the hydrophilicity imparted by the oxygen rich functional groups and attracted tremendous attention in many research fields due to its fascinating structural, physical and chemical properties.[23,24] Besides, ease of further functionalization with proton conducting groups like sulfonic acid which facilitate the proton transport has advantages towards its application in PEM fuel cells. In addition, significant improvement in mechanical properties can also be expected due to its high strength and when used as filler in PEM it reduce methanol crossover effectively.

DMFC research community has tried to take the advantage of GO and fabricated Nafion-GO/SGO composite membranes to solve the methanol

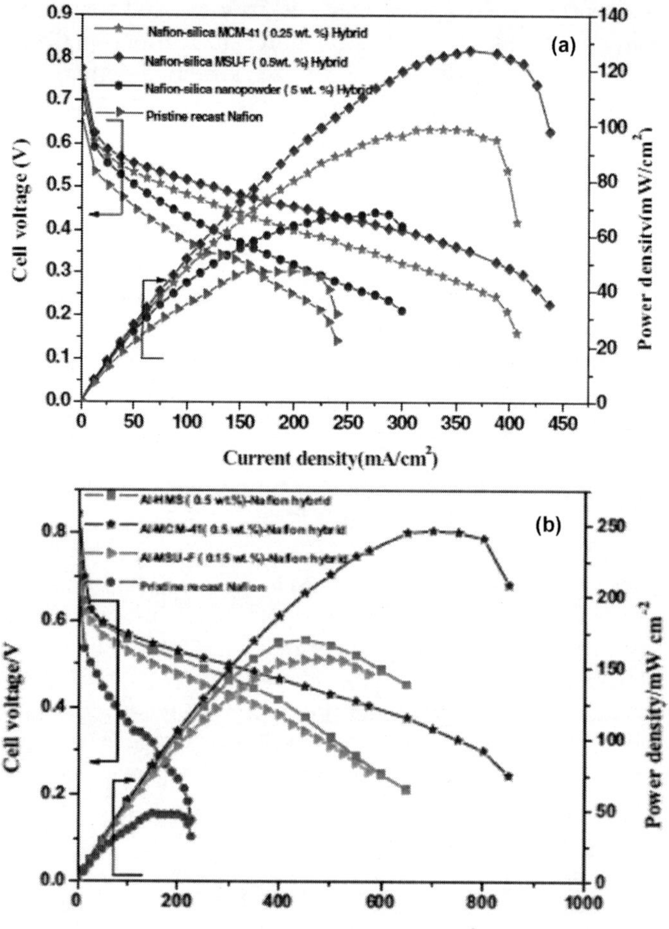

Fig. 6 (a) Performance curves for DMFCs with pristine recast Nafion, Nafion-silica nanopowder, Nafion-silica MSU-F and Nafion-silica MCM-41 hybrid membranes at 80 °C. Reproduced with permission from A. K. Sahu, S. Meenakshi, S. D. Bhat, A. Shahid, P. Sridhar, S. Pitchumani and A. K. Shukla, Meso-Structured Silica-Nafion Hybrid Membranes for Direct Methanol Fuel Cells, *J. Electrochem. Soc.*, 2012, **159**(11), F702–F710. Copyright 2012, The Electrochemical Society. (b) Performance data for DMFCs with pristine recast-Nafion, Al-MSU-F-Nafion, Al-HMS-Nafion and Al-MCM-41-Nafion hybrid membranes at 80 °C. Reproduced with permission from S. Meenakshi, A. K. Sahu, S. B. D. Bhat, P. Sridhar, S. Pitchumani and A. K. Shukla, Mesostructured-aluminosilicate-Nafion hybrid membranes for direct methanol fuel cells, *Electrochim. Acta*, **89**, 35–44. Copyright 2012 with permission from Elsevier B.V.

crossover issue. Bong Gill Choi *et al.*, reported the synthesis of Nafion-SGO and studied its transport properties such as proton conductivity and methanol permeability.[25] GO was functionalized with sulfonic acids by microwave treatment and incorporated in Nafion matrix to employ as PEM in DMFCs. Incorporation GO and SGO in Nafion decreases the average dimension of ionic cluster size due to shrinkage of ionic clusters through reorganization of Nafion matrix[26] as a result reduced methanol permeability was observed for Nafio-GO and Nafion-SGO composites in comparison to pristine Nafion.

C. W. Lin *et al.* fabricated GO laminated Nafion-115 and employed as PEM in DMFC, the authors specific claims are as follows: These membranes are effective in DMFCs operated with higher methanol concentration due to the effective restriction of methanol crossover by the laminated GO membranes compared to GO dispersed membranes. GO paper laminated on Nafion through transfer printing followed by hot press and the thickness of GO layer is about 1 micron. This membrane exhibited 70% depression in methanol permeability and shows 22% decrement in proton conductivity due to lack of conducting channels on GO. However this membrane shows higher selectivity compared to Nafion.[23]

Another research in this direction is by Kai Feng whose work includes the sulfonated GO-silica incorporated in Nafion.[24] It is noteworthy that at 0.5 wt% loading of SGO–SiO_2 in Nafion, the micro structure of Nafion is aligned in non-uniform fashion due to the interaction of sulfonic acid groups whereas at higher loading of SGO–SiO_2 (0.8 wt%), the compatibility effect of GO is dominant wherein aggregation of S–GO–SiO_2 takes place due to which concentration of sulfonic acid groups are lowered. In contrast to the reports by Bong Gill Choi *et al.* (Nafion–SGO), water uptake of the membrane greatly improved compared to pristine Nafion membrane due to the hygroscopic nature of SiO_2 and increased water content in the membrane promotes the protons resulting elevated proton conductivity for Nafion–SGO–SiO_2 composite membranes. It was reported that, at 0.7 wt% graphene loading, a PVA – graphene nano hybrid membrane exhibits a 76% increase in tensile strength and a 62% increase in Young's modulus; these results were attributed to effective load transfer to the graphene filler *via* interfacial hydrogen bonding.[27] Recently, Chien *et al.*[28] prepared various compositions of sulfonated graphene oxide (GO)–Nafion hybrid membranes for DMFC application and obtained improved proton conductivity with reduced methanol crossover. This was attributed to chemical interactions between the different kinds of oxygen functionalities of GO and Nafion.

Parthiban *et al.* explored sulfonic acid functionalized graphene (S-graphene) as a potential inorganic filler as well as a solid acid proton conducting medium to realize a hybrid membrane with Nafion for DMFC.[29] The simple, but effective, functionalization of graphene is performed by sulfonic acid containing aryl radicals to increase the number of sulfonate groups per unit volume of graphene domain (Scheme 1). Nafion-S-graphene hybrid membranes increase proton conductivity by 62% in comparison with recast Nafion. Further temperature dependence of proton conductivity is also evaluated. All the membrane samples exhibit an Arrhenius-type temperature dependence of proton conductivity, suggesting thermally activated proton conduction. The activation energy, which is the minimum energy required for proton transfer from one free site to another site. The high proton conductivity of the hybrid membrane is attributed to a Grötthus type mechanism, wherein reorganization of hydrogen bonds plays a vital role in hydrated graphene.[30]

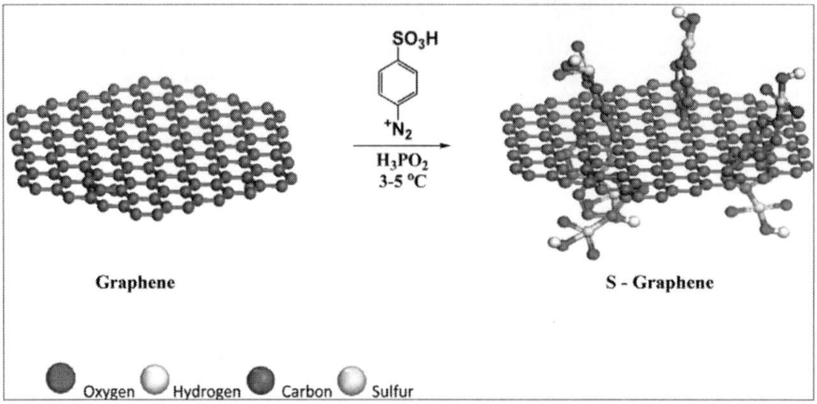

Scheme 1 Illustration for the preparation of sulfonated graphene. Reproduced with permission from V. Parthiban, S. Akula, S. Gouse Peera, N. Islam and A. K. Suha, Proton Conducting Nafion-Sulfonated Graphene Hybrid Membranes for Direct Methanol Fuel Cells with Reduced Methanol Crossover, *Energy Fuels*, 2016, **30**, 725–734. Copyright 2016 American Chemical Society.

The surface morphologies of recast Nafion and Nafion–S-graphene hybrid membranes are shown in Fig. 7(a,b) respectively. S-graphene is dispersed over the Nafion ionomer and tightly held in the polymer matrix due to strong interfacial interactions. Wrinkled features with agglomeration in some regions are also observed due to the presence of S-graphene in the hybrid matrix. Topography, phase image, and line profile of the pristine Nafion and Nafion–S-graphene hybrid membrane are studied by AFM, as shown in Fig. 7(c–h). The dark regions correspond to hydrophilic domains of membrane grafted with $-SO_3H$ groups, and the bright regions are accounted for by the polymeric chain. It is noteworthy that the surface of the pristine Nafion membrane is relatively smooth and the formation of bi-continuous ionic hydrophilic channels embedded in a hydrophobic matrix is seen. In the case of the Nafion–S-graphene hybrid membrane, the hydrophilic channels are widened randomly with the increase in the surface roughness. This appears to be advantageous toward more water uptake properties and high proton conductivity in the case of hybrid membranes as seen in the Table 4. Moreover, the increased surface roughness of the hybrid membrane is highly beneficial for more adhesion and compatibility with the electrodes during fabrication of membrane electrode assemblies.

The methanol permeability of the Nafion hybrid membrane with 1.5 wt% S-graphene is 2.88×10^{-7} mol s^{-1} cm^{-2}, which is 50% lower than that of the pristine recast Nafion membrane (5.53×10^{-7} mol s^{-1} cm^{-2}). The interaction of S-graphene in the polymer matrix increases the compactness of the membrane and blocks the pathway for methanol passage, which attributed to the lower methanol permeability. From the data (Table 4), it is observed that a compromise needs to be made between proton conductivity and methanol crossover in any hybrid membrane. In the present study, the S-graphene content is optimized to 1 wt% in the hybrid membrane to realize a good membrane for the DMFC applications. Figure 8(a) shows the polarization and performance curves for the

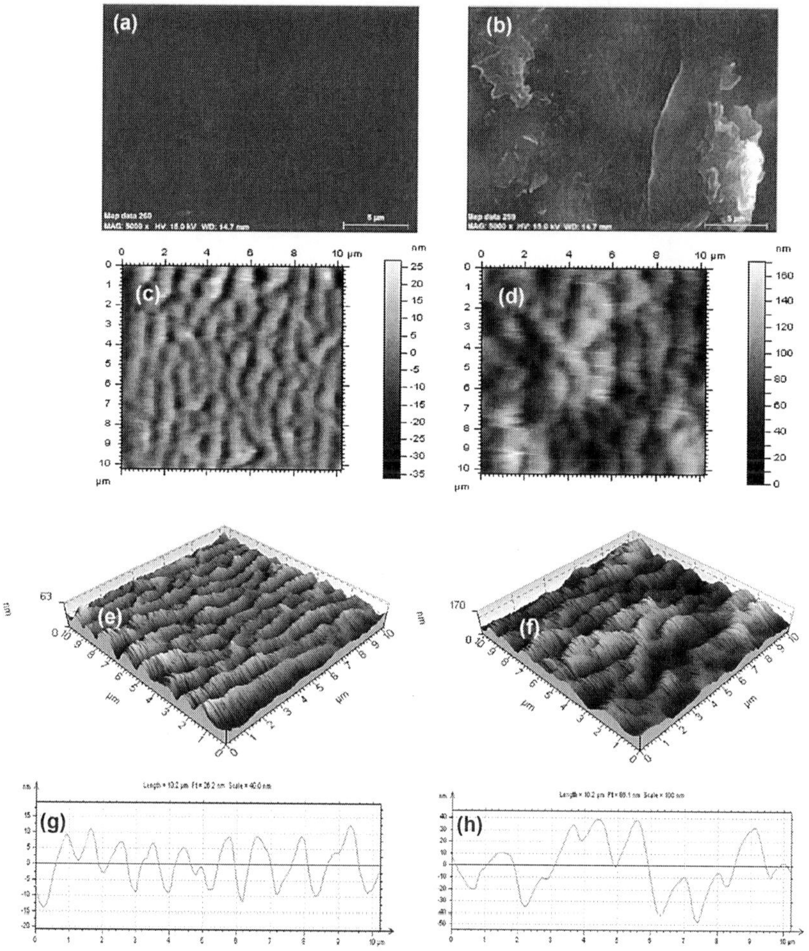

Fig. 7 SEM images for (a) pristine recast Nafion membrane and (b) Nafion–S-graphene hybrid membrane. (c) Topography of pristine recast Nafion membrane. (d) Topography of Nafion–S-graphene hybrid membrane. (e) Phase image of pristine Nafion membrane. (f) Phase image of Nafion–S-graphene hybrid membrane. (g) Line profile of pristine Nafion membrane. (h) Line profile of Nafion–S-graphene. Reproduced with permission from V. Parthiban, S. Akula, S. Gouse Peera, N. Islam and A. K. Suha, Proton Conducting Nafion-Sulfonated Graphene Hybrid Membranes for Direct Methanol Fuel Cells with Reduced Methanol Crossover, *Energy Fuels*, 2016, **30**, 725–734. Copyright 2016 American Chemical Society.

DMFCs comprising recast the Nafion membrane and Nafion–S-graphene hybrid membranes at 70 °C and ambient pressure. Hybrid membranes show higher open-circuit voltage (OCV) compared to recast pristine Nafion, indicating that the methanol crossover is less in the former (Fig. 8(b)). Higher OCV for the hybrid membranes also indicates no significant electronic conductivity effect on the presence of a small amount of graphene in the hybrid membrane, which would otherwise adversely affect the OCV. The hybrid membrane with optimized S-graphene content (1 wt%) delivers a peak power density of 118 mW cm^{-2}, which is much higher than that of the recast Nafion membrane, which delivers

Table 4 Water uptake, ion-exchange capacity, proton conductivity, methanol permeability, and peak power density values for different membranes.

Membrane type	Water uptake (%)	Ion-exchange capacity	Proton conductivity ($mS\,cm^{-1}$)	Methanol crossover ($\times 10^{-7}$ mol $s\,cm^{-2}$)	DMFC peak power density ($mW\,cm^{-2}$)
Recast Nafion	20.12 ± 0.5	0.88 ± 0.01	65.3 ± 0.5	5.53 ± 0.1	54
Nafion–S-graphene (0.5 wt.%)	24.56 ± 0.3	0.92 ± 0.01	82.0 ± 0.7	4.94 ± 0.2	80
Nafion–S-graphene (1 wt.%)	27.32 ± 0.7	0.96 ± 0.01	104 ± 1.0	3.72 ± 0.1	118
Nafion–S-graphene (1.5 wt.%)	29.17 ± 0.4	0.95 ± 0.01	94.0 ± 0.8	2.88 ± 0.3	95

only a peak power density of only 54 $mW\,cm^{-2}$. The high density of $-SO_3H$ groups on S-graphene acts as a solid acid proton conducting medium and helps the hybrid membrane to achieve higher proton conductivity. The presence of S-graphene in the hybrid membranes blocks the pathway for methanol passage and reduces methanol crossover.

2.2 A study of sPEEK based blend and composite polymer electrolyte membranes for direct methanol fuel cells (DMFCs)

2.2.1 Blend membranes fabricated from sulfonated poly(ether ether ketone) (sPEEK) and poly[bis(phenoxy)phosphazene] (POP) for DMFCs. Among the various sulfonated polymers studied in the literature, sPEEK has attracted considerable attention because of its good mechanical strength, high chemical stability, low cost and ease of preparation.[31] sPEEK is also resistant to methanol permeability, in comparison with Nafion in DMFCs, due to its narrow and more branched hydrophilic channels.[32] The proton conductivity, methanol permeability, water uptake and swelling ratio of sPEEK membranes are all dependent on the degree of sulfonation (DS). Although the sPEEK membrane with a low DS shows good performance in reducing permeability to methanol, there is an apparent reduction in ionic conductivity. It also becomes extremely difficult to process sPEEK with low DS due to its poor solubility in organic solvents. In contrast, a high DS results in excessive water uptake, increasing the swelling ratio and limiting the use of the sPEEK membrane.[33,34] The above factors may lead to a loss in the mechanical stability of the membrane making it unsuitable for long term operation in DMFCs. In order to address the above issues, different routes for modification of the sPEEK polymer have been explored. One of the most important routes is to fabricate a blend of sPEEK with different hydrophobic polymers to control water uptake and methanol permeability.[35–40] Polyphosphazenes are polymers which possess a backbone of alternating phosphorus and nitrogen atoms. The properties of polyphosphazenes are mainly controlled by the choice of side groups.[41] Polyphosphazenes with various

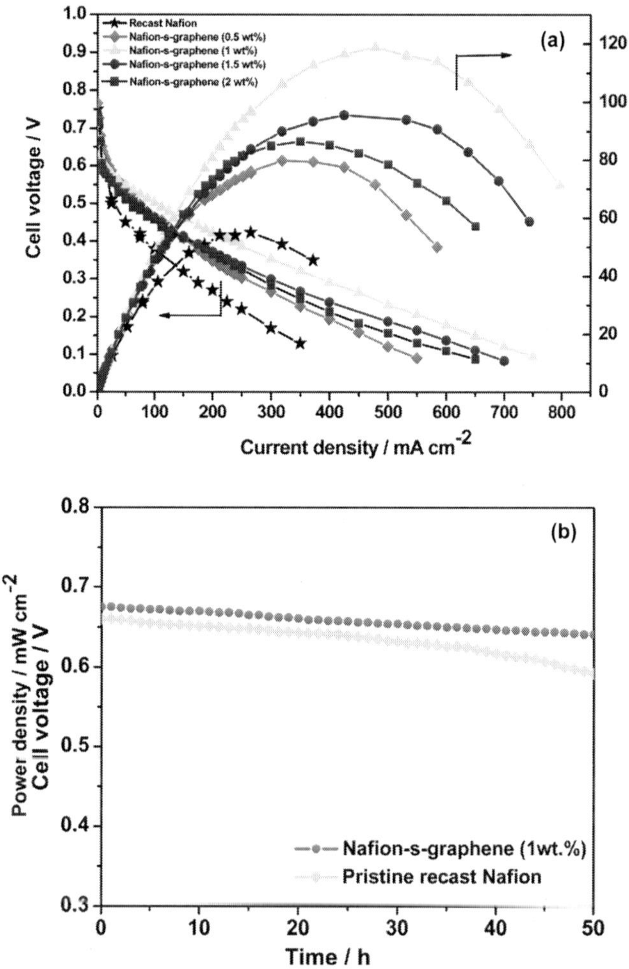

Fig. 8 (a) DMFC performance for recast Nafion and Nafion–S-Graphene composite membranes, (b) Open circuit voltage as function of time for the membranes. Reproduced with permission from V. Parthiban, S. Akula, S. Gouse Peera, N. Islam and A. K. Suha, Proton Conducting Nafion-Sulfonated Graphene Hybrid Membranes for Direct Methanol Fuel Cells with Reduced Methanol Crossover, *Energy Fuels*, 2016, **30**, 725–734. Copyright 2016 American Chemical Society.

properties can be synthesized by careful choice of side groups. Polyphosphazenes have several advantages over established hydrocarbon based polymers. One of their prime attributes is the thermal and chemical stability of the polymer backbone; both phosphorus and nitrogen are in their highest oxidation states resulting in a high degree of thermo-oxidative stability.[42] The polar nature of the bonding along the polymeric backbone inhibits chemical attack by free radicals.[43]

Gouse Peera *et al.* fabricated blend membranes from sulfonated poly(ether ether ketone) (sPEEK) and poly[bis(phenoxy)phosphazene] (POP).[44] The effect of POP content on the distribution of ionic channels is investigated by atomic force microscopy (AFM). The water uptake and methanol permeability for the blend membranes are also investigated.

The blend membranes are characterized in terms of their thermal and mechanical properties in conjunction with their ionic conductivity. The proton conductivity of the blend membranes slightly decreased with increasing POP content in comparison with the pristine sPEEK membrane. The hydrophobic nature of POP blocks the ionic channels in the sPEEK matrix, subsequently decreasing its water uptake and methanol permeability. The blend membranes showed higher power density compared to a pristine sPEEK membrane in direct methanol fuel cells (DMFCs).

Figure 9 show topological images of pristine sPEEK and sPEEK-POP blend membranes, wherein the light regions are assigned to the hydrophilic sulfonic acid groups and the dark regions are assigned to the hydrophobic polymer backbone.[45] All the images show even dispersions of POP, however, aggregations of the particles can still be observed. It is worth noting from the AFM that the connectivity and domain size of sPEEK varies depending on the content of POP, affecting its ionic channels. The hydrophilic and hydrophobic domains of the blend membranes decreased in size and were packed more densely. The microstructures of sPEEK vary significantly with the addition of POP, unlike the hydrophilic fillers, which may be due to the hydrophobic nature of POP. The decreasing hydrophilic domains may impact the change in water uptake, proton conductivity and methanol permeability properties of the membranes. Similar results were reported for polyaniline, poly(amide imide) and TiO_2 incorporated sPEEK matrices.[35,38,45,46]

Compared with pristine sPEEK, proton conductivities measured at 30 °C and 60 °C for blend membranes decreased with the increase in POP

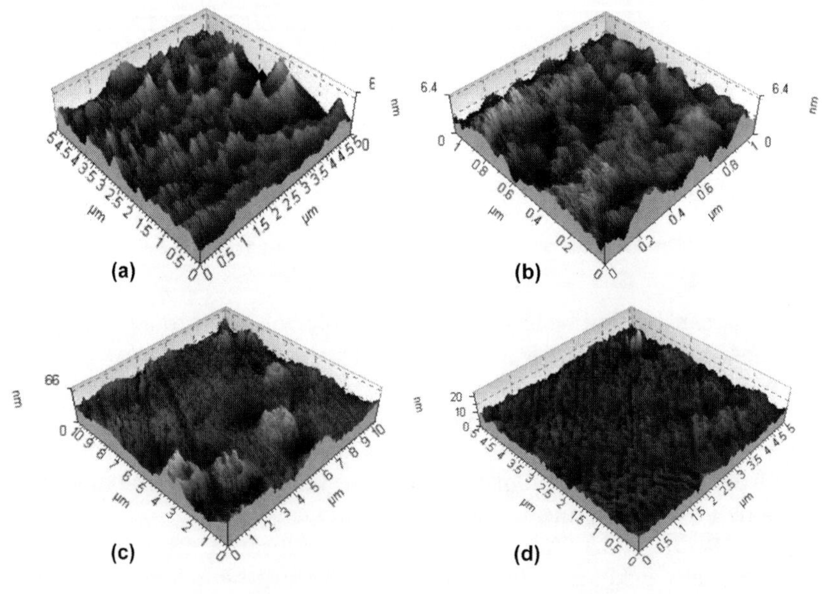

Fig. 9 AFM topological images for (a) pristine sPEEK, (b) sPEEK-POP (1 wt%), (c) sPEEK POP (2 wt%) and (d) sPEEK-POP (3 wt%) blend membranes. Reproduced from ref. 44 with permission from The Royal Society of Chemistry.

Table 5 Ionic Conductivity data for the membranes.

S. no	Membrane	Proton conductivity ($mS\,cm^{-1}$)	
		30 °C	60 °C
1	Pristine recast Nafion	51	86
2	Pristine sPEEK	38	50
3	sPEEK-POP (1 wt%)	35	44
4	sPEEK-POP (2 wt%)	32	41
5	sPEEK-POP (3 wt%)	26	30

content (Table 5). Proton transport in membranes requires well connected proton conducting channels formed from ionic clusters of hydrophilic sulfonated functional groups. The content and the diameter of the connected channels have a significant impact on the proton transport rate in membranes.[47,48] When the density of sulfonic groups is low, the hydrophilic sulfonic acid groups can form isolated ionic clusters in the continuous hydrophobic phase. Whereas, when the density of sulfonic acid groups increases, up to a certain limit, the isolated ionic clusters form crosslinked channels with good connectivity to allow facile transport of protons through these channels.[49,50] The introduction of hydrophobic polymer POP reduces the density of sulfonic acid groups of sPEEK in the membrane thereby slightly reducing the proton transport which is evident from the ionic conductivity data for the blend membranes as seen in Table 5.

The transport of methanol through the membrane also requires ionic channels with good connectivity. As discussed above, the content of well-connected channels in sPEEK decreased for blend membranes upon the introduction of POP. As shown in Fig. 10a, the methanol permeabilities of the sPEEK-POP blend membranes decreased with increasing POP content. Usually transport through the membrane is controlled by kinetic and equilibrium factors.[37] It is important to note that for porous membranes, methanol and/or water transport properties may be affected by the bulk water in pores, but sPEEK membranes, which are non-porous, have different transport properties due to the compact and smooth surface. In contrast, sPEEK-POP membranes have a dense entangled structure which may lead to low methanol permeability in comparison with the pristine sPEEK. Performance curves for DMFCs comprising pristine recast Nafion, pristine sPEEK and blend membranes of sPEEK-POP at 60 °C are presented in Fig. 10b. It is noteworthy that a peak power density of 47 $mW\,cm^{-2}$ is obtained for the DMFCs with the pristine recast Nafion membrane. Although the proton conductivity of the blend membranes is lower than that of pristine sPEEK, MEAs fabricated with blend membranes (1 wt% POP and 2 wt% POP) showed higher performance, which is attributed to the lower methanol cross-over and better electrochemical selectivity. Presence of excess POP (3 wt%) in the blend membrane hinders the proton conduction path as evidenced from the conductivity data, which results in decreased DMFC performance.

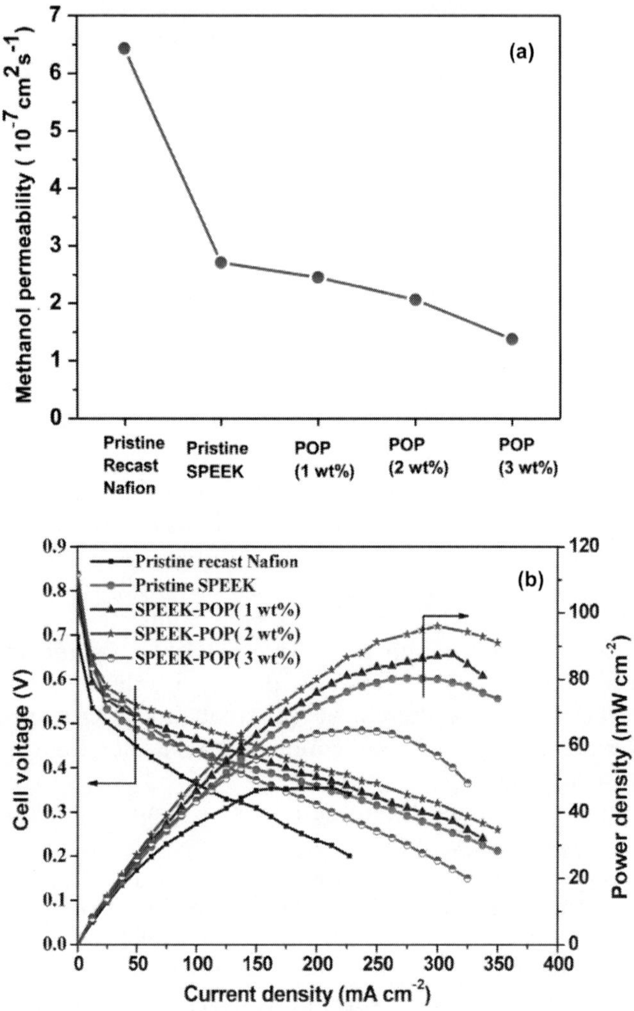

Fig. 10 (a) Methanol permeability of pristine recast Nafion, pristine sPEEK and blend membranes of sPEEK, (b) Performance curves for DMFCs with pristine recast Nafion, pristine sPEEK and blend membranes of sPEEK at 60 °C. Reproduced from ref. 44 with permission from The Royal Society of Chemistry.

2.2.2 Modified sulfonated poly(ether ether ketone) based mixed matrix membranes for direct methanol fuel cells. Several reports suggested fabrication of mixed matrix and composite membranes as electrolytes for DMFCs. These membranes also have restricted methanol permeability along with improved proton conductivity. In this regard, S. Meenakshi *et al.*[51] fabricated mixed matrix membrane from sPEEK blended with methane sulfonic acid (MSA) and incorporated with zeolite 4A for DMFCs. MSA is non-oxidizing and can act as a strong acid for surface modification reactions.[52] However, its use in fuel cell applications has not been explored so far in the literature. Hence in the present study, MSA is used to enhance the proton conductivity in sPEEK

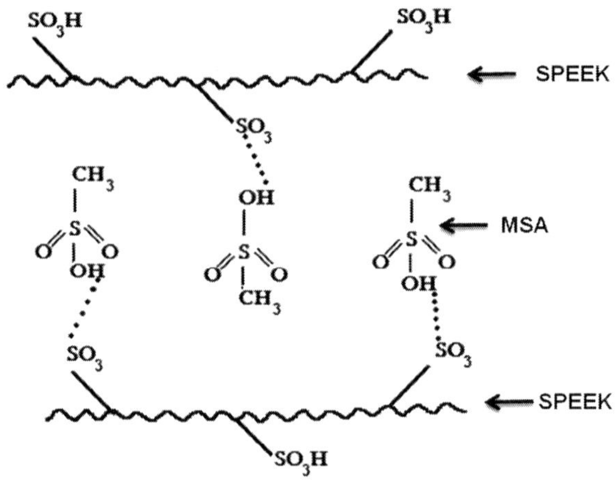

Scheme 2 Presumptive bridge interactions of MSA and sPEEK matrices. Reproduced with permission from S. Meenakshi, S. D. Bhat, A. K. Sahu, P. Sridhar and S. Pitchumani, Modified Sulfonated Poly(ether ether ketone) Based Mixed Matrix Membranes for Direct Methanol Fuel Cells, *Fuel Cells*, 2013, **13**(5), 851–861. Copyright 2013 Wiley-VCH Verlag GmbH & Co. KGaA, Weinheim.

for DMFCs and also to maintain the chemical stability. The presence of MSA in sPEEK enhances the proton conductivity due to the additional bridge provided by sulfonic acid groups present in MSA as presumed in Scheme 2.

In general, zeolites are alumino-silicates based on the infinitely extending three dimensional AlO_4 and SiO_4 tetrahedral framework structure, linked to each other by sharing all the oxygen.[53] 4A is one of the most hydrophilic zeolite and its crystal structure is definite with an effective pore dimension of 0.4 nm. Zeolite 4A has a distinct molecular sieving effect and selective adsorption capability along with its unique structural characteristics and hydrophilic nature.[54] Therefore, zeolite 4A can be extensively used in removal of volatile organic chemicals from air streams, separation of isomers, mixture of gases, shape-selective catalysis and ion exchange.[55,56] The aforesaid additive is incorporated to sPEEK-MSA blend to form a mixed matrix membrane in order to improve the overall stability of the membrane and also to reduce the methanol permeability in DMFCs.

Figure 11(a) showed the thermal stability for pristine sPEEK, sPEEK-MSA (20 wt%) blend and zeolite 4A (4 wt%)-sPEEK-MSA (20 wt%) mixed-matrix membranes studied by thermogravimetric analysis through thermal dehydration, decomposition, and degradation losses. Initial weight loss observed is due to the removal of absorbed water molecules from the membranes in the temperature range between 33 and 228 °C. First stage dehydration endotherm is observed at 85, 77, and 87 °C for pristine sPEEK, sPEEK-MSA (20 wt%) blend and zeolite 4A (4 wt%)-sPEEK-MSA (20 wt%) mixed matrix membranes, respectively as shown in DTG curves represented in inset of Fig. 11(a). The second weight loss between 228 and 423 °C is attributed to the loss of sulfonic acid groups of

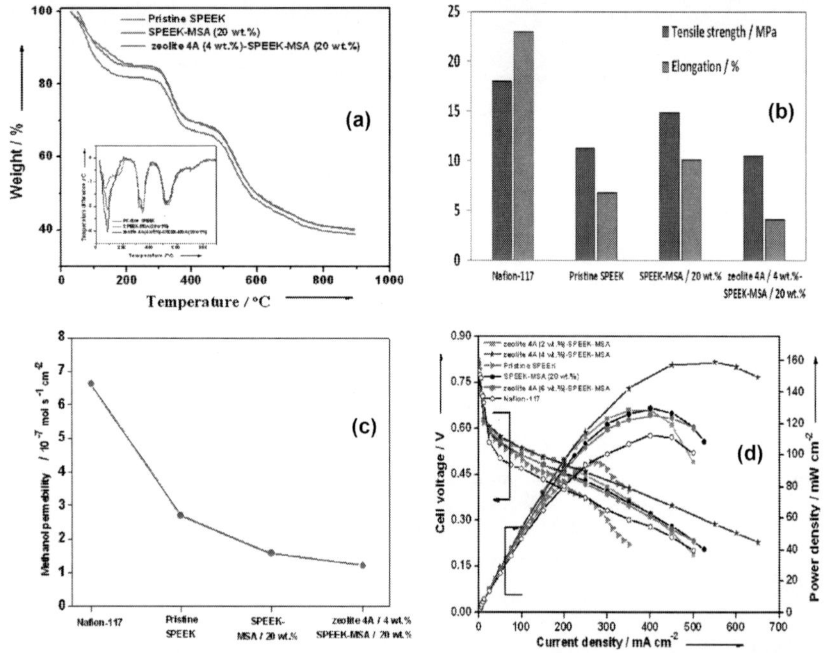

Fig. 11 (a) TGA analysis for pristine sPEEK, blend and its mixed matrix membranes (b) Tensile strength and Elongation for Nafion-117, pristine sPEEK, sPEEK-MSA (20 wt%) blend and zeolite 4A (4 wt%)-sPEEK-MSA (20 wt%) mixed matrix membranes, (c) Methanol permeability of Nafion-117, pristine sPEEK, sPEEK-MSA (20 wt%) blend and zeolite 4A (4 wt%)-sPEEK-MSA (20 wt%) mixed matrix membranes, (d) Performance curves for DMFCs with Nafion-117, pristine sPEEK and zeolite 4A (2, 4, and 6 wt%)-sPEEK-MSA (20 wt%) mixed matrix membranes at 70 °C. Reproduced with permission from S. Meenakshi, S. D. Bhat, A. K. Sahu, P. Sridhar and S. Pitchumani, Modified Sulfonated Poly(ether ether ketone) Based Mixed Matrix Membranes for Direct Methanol Fuel Cells, *Fuel Cells*, 2013, **13**(5), 851–861. Copyright 2013 WILEY-VCH Verlag GmbH & Co. KGaA, Weinheim.

sPEEK and MSA and decomposition of zeolite 4A chains. Second stage decomposition endotherm is observed at 351, 343, and 345 °C for the aforesaid membranes, respectively. The third weight loss between 423 and 675 °C corresponds to the degradation of the main polymer chain and structure collapse of zeolite 4A.[57,58] Third stage degradation endotherm is observed at 528, 521, and 536 °C for the aforesaid membranes, respectively. It is noteworthy that the mixed matrix membranes degrade at a relatively higher temperature in comparison to the blend as well as pristine membranes, suggesting the overall improved thermal stability of mixed matrices as a result of incorporation of MSA and zeolite 4A molecular sieves into the cluster of sPEEK chain.

The mechanical properties for Nafion-117, pristine sPEEK, sPEEK-MSA (20 wt%) blend and zeolite 4A (4 wt%)-sPEEK-MSA (20 wt%) mixed-matrix membranes in sorbed conditions are shown in Fig. 11(b). It is noteworthy that both tensile strength and percentage elongation are higher for Nafion-117 membrane because of its flexible chain mobility. However, tensile strength and elongation are lower for sPEEK membranes because of the restriction in chain mobility. In contrast, for sPEEK-MSA (20 wt%) blend membranes, tensile strength, and percentage elongation increased due to the flexible chain mobility induced by MSA. The membrane rigidity increased by addition of 4A molecular sieves to the sPEEK-MSA blend thereby slightly decreasing its tensile strength and percentage elongation.

Methanol permeabilities for the pristine sPEEK, sPEEK-MSA blend and zeolite 4A-sPEEK-MSA mixed matrix membranes measured at OCV conditions are shown in Fig. 11(c). The blend and the mixed matrix membranes possess lower methanol permeability in relation to the pristine sPEEK membrane. It is noteworthy, that zeolite 4A-sPEEK-MSA mixed matrix membrane exhibits lowest methanol permeability in comparison with blend, pristine sPEEK, and Nafion-117. This is due to the fact that zeolite 4A has pores with a dimension larger than water molecule but similar to aqueous methanol molecule resulting in a tortuous path for methanol transport are reported in our earlier studies.[17,59]

Cell polarization for DMFCs comprising MEAs with pristine sPEEK membrane and different weight percentage of sPEEK-MSA-zeolite 4A mixed matrix membranes are explored in Fig. 11(d). It is noteworthy that a peak power density of 159 mW cm^{-2} is observed for the mixed matrix membrane comprising zeolite 4A (4 wt%)-sPEEK-MSA (20 wt%) due to the active ionic sites in the zeolite 4A crystal lattice that could possibly form strong interactions with the water molecules, acting as sorption, and transport sites.[60] It is worthwhile, that the well-defined micropore structure of the zeolite 4A cage could help in a partial molecular sieving of the large solvent molecules like methanol thereby restricting the methanol permeability and enhancing the performance. The higher ionic conductivity and lower methanol permeability of mixed matrix membranes attributed to achieve high electrochemical selectivity in DMFCs.

There are also some important reports on clay composite membranes used for DMFC applications. Z. Gaowen *et al.*[61] used intercalated structures like organic-montmorillonite (OMMT) to modify sPEEK matrix suiting to DMFC applications and found that methanol cross-over was lower for these nanocomposite-sPEEK membrane electrolyte. Y. F. Lin *et al.*[62] explored the incorporation of sulfonated montmorillonite into Nafion® to form composite membranes for DMFCs which showed improved stability, reduced methanol cross-over and enhanced DMFC performance. Also D. J. Kim *et al.*[63] prepared sulfonated poly(arylene ether sulfone)/laponite-SO_3H composite membrane for DMFCs which exhibited good mechanical stability and excellent proton conductivity.

In continuation of the above studies, S. Sasikala *et al.*[64] explored to modify sPEEK matrix through functionalized Bentonite clay and studied its effect as membrane electrolyte for DMFCs. In general, Bentonite clay is an aluminium-phyllosilicate layered structure[65–67] which comprises rings of tetrahedron, linked by shared oxygen in a two dimensional plane forming a sheet-like layer connecting to each other. Bentonite clay has been widely studied as ion exchanger and/or adsorbent due to its unique framework and has large surface area.[68–70] S. Sasikala *et al.*[64] studied potassium form of Bentonite (K^+-Bentonite) functionalized by introducing organo sulfonic acid (HSO_3^-) groups through silane condensation to induce the acidity. The functionalization of bentonite by silane

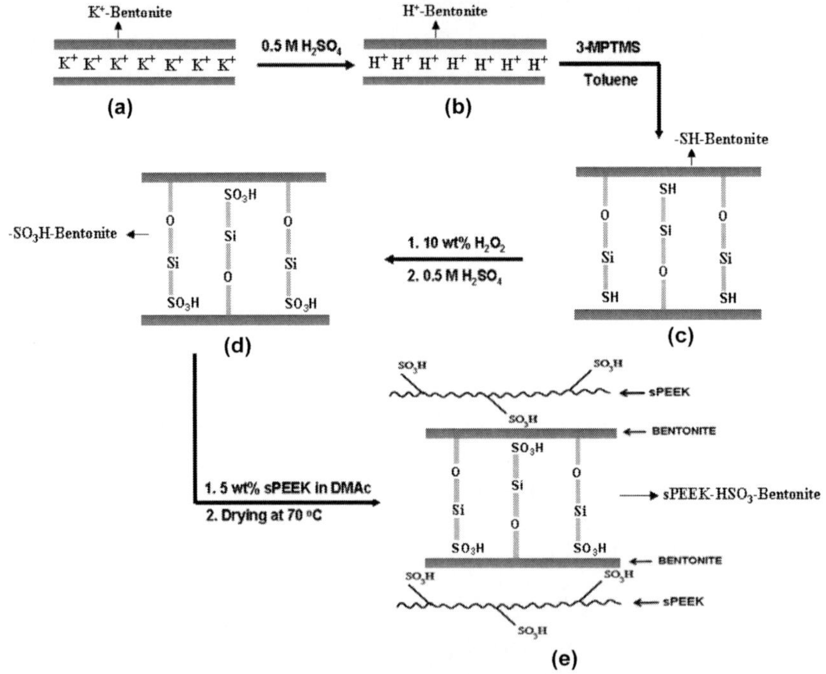

Scheme 3 Functionalization of bentonite through silane condensation (a–d); interaction of sPEEK-HSO_3-BEN composite (e). Reproduced with permission from S. Sasikala, S. Meenakshi, S. D. Bhat and A. K. Sahu, Functionalized Bentonite clay-sPEEK based composite membranes for direct methanol fuel cells, *Electrochim. Acta*, **135**, 232–241. Copyright 2014 with permission from Elsevier B.V.

condenstation was performed.[71] The silane condensation of the bentonite is represented as Scheme 3(a–d).

The resulting functionalized bentonite is characterized by solid state Magic Angle Spinning (MAS) NMR study. The ^{29}Si MAS and ^{27}Al MAS NMR spectra for K^+-BEN, H^+-BEN and HSO_3-BEN particles are represented in Fig. 12(a and b). The ^{29}Si resonance is found in the range from −90 to −160 ppm. The Q_3 peak at around −133 ppm corresponds to the tetrahedral layers of the Bentonite. The Q_3 peak is slightly shifted to −131 ppm and −127 ppm for H^+-BEN and HSO_3-BEN confirming the presence of functional groups on the surface of the BEN. The new peak observed at −90 ppm (T) for HSO_3-BEN confirms the silane condensation. A peak at −150 ppm (H^+-BEN) represents Si in Q_4 arrangements. However, the peak shifted to −155 ppm for HSO_3-BEN indicating the grafting of organo-sulfonic acid groups on the surface of BEN. The trend is similar to the studied literature elsewhere.[70] ^{27}Al MAS NMR as shown in Fig. 12(b) for BEN particles highlights the significant change in the peak position. Peak at Octahedral Al of 18 ppm (K^+-BEN) is shifted to 15 ppm (H^+-BEN) presumably suggesting the effect of ion exchange and 14 ppm (HSO_3-BEN) confirming the surface functionalization of BEN by grafting organo sulfonic acid groups. Interestingly, a broad peak at 35 ppm (K^+-BEN) corresponding to tetrahedral Al also shifted to 33 ppm (H^+-BEN) and 30 ppm (HSO_3-BEN) confirming the grafting on the surface of BEN.

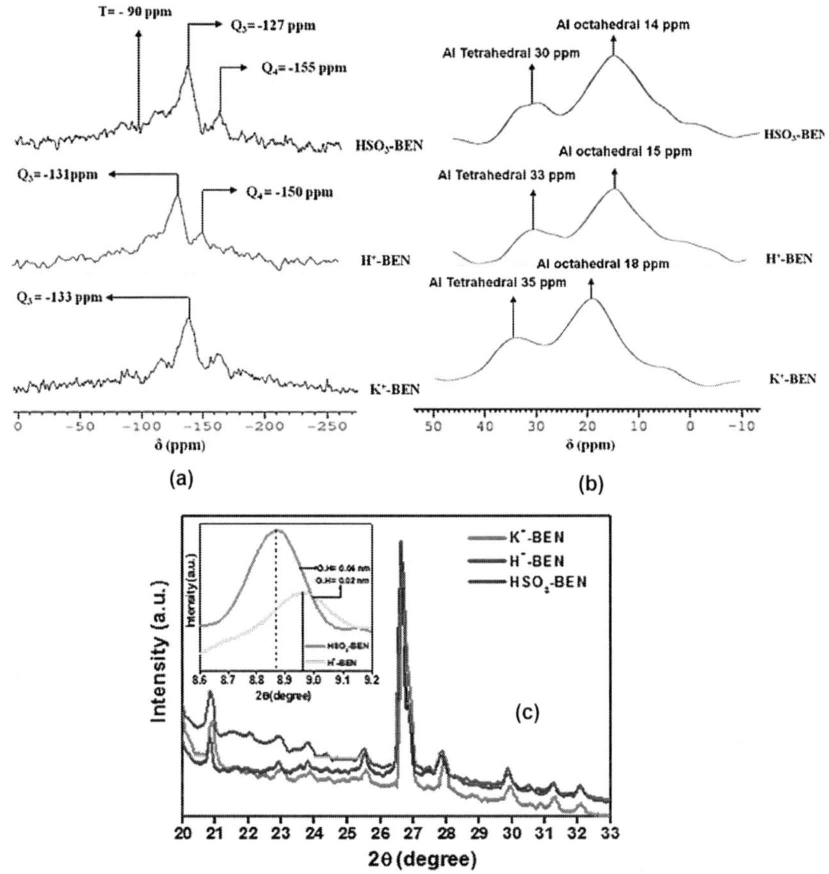

Fig. 12 (a) ^{29}Si and (b) ^{27}Al-MAS NMR spectra for BEN particles (c) X-ray diffraction pattern for BEN powder samples. Inset to Fig. 12(c) suggests the gallery height for H^+-BEN and HSO_3-BEN powders. Reproduced with permission from S. Sasikala, S. Meenakshi, S. D. Bhat and A. K. Sahu, Functionalized Bentonite clay-sPEEK based composite membranes for direct methanol fuel cells, *Electrochim. Acta*, **135**, 232–241. Copyright 2014 with permission from Elsevier B.V.

These bentonite powders further analysed by XRD analysis and the gallery height of bentonite is calculated from Bragg's equation. Figure 12(c) shows the XRD pattern for K^+-BEN, H^+-BEN and HSO_3-BEN particles. Inset to the Fig. 12(c) reveals the gallery height for the XRD pattern for the H^+-BEN and HSO_3-BEN particles. It is noteworthy, that the peak at $2\theta = 8.97^\circ$ observed for H^+-BEN is gradually shifted to 8.86° for HSO_3-BEN confirming the grafting of organo sulfonic acid groups.[72] The gallery height increased from 0.02 nm observed for H^+-BEN to 0.04 nm for HSO_3-BEN suggesting the grafting of organo-sulfonic acid groups in the internal surface of BEN. This may be due to the interaction of surface hydroxyl (–SiOH) groups of BEN and 3-MPTMS along with oxidation of –SH to $-SO_3H$ group.

These functionalized bentonite powder is incorporated with sPEEK matrix to form a composite membrane. These sPEEK based modified BEN composite showed high proton conducting as well as methanol

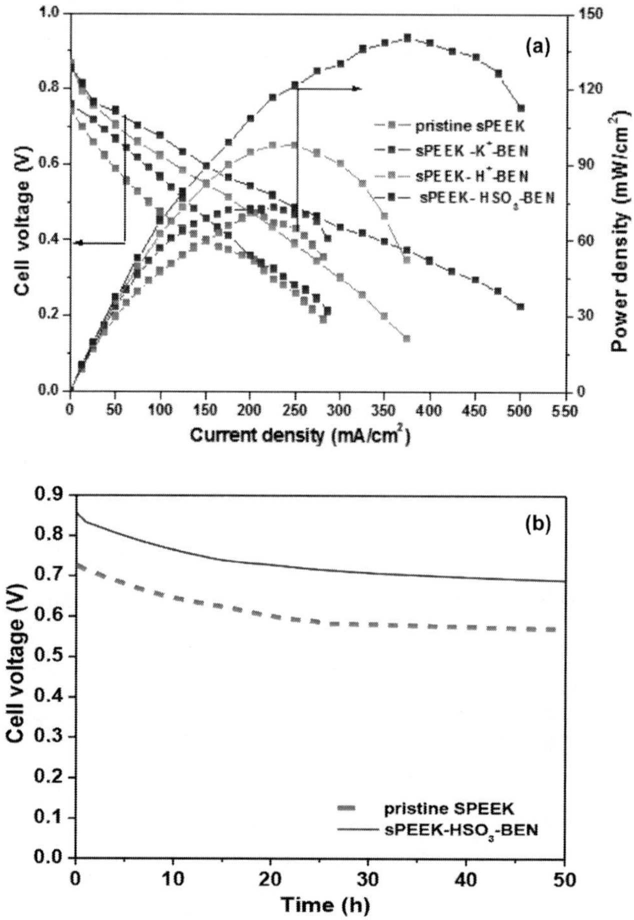

Fig. 13 (a) Cell polarization studies for DMFCs with pristine sPEEK and composite membranes at 70 °C, (b) Stability test for pristine sPEEK and sPEEK-HSO_3-BEN under OCV in relation to time. Reproduced with permission from S. Sasikala, S. Meenakshi, S. D. Bhat and A. K. Sahu, Functionalized Bentonite clay-sPEEK based composite membranes for direct methanol fuel cells, *Electrochim. Acta*, **135**, 232–241. Copyright 2014 with permission from Elsevier B.V.

restricting behaviour thereby resulting in better electrochemical selectivity. The hydrogen bonding interaction between the sPEEK and BEN composite leads to enhanced overall DMFC cell performance. The optimized sPEEK-HSO_3-bentonite clay (90 : 10) composite membrane shown higher peak power density of 140 mW Cm^{-2} in comparison with pristine sPEEK membrane of 73 mW Cm^{-2} at 70 °C as shown in Fig. 13a.

The stability test for the membranes shown in Fig. 13b under OCV condition in relation to time,[73] suggest that there is an initial slight drop in OCV for both the MEAs comprising pristine sPEEK and sPEEK-HSO_3-BEN and then after stabilization, the drop is negligible suggesting no degradation for 50 h. It is noteworthy that, sPEEK-HSO_3-BEN composite membrane has shown better stability in comparison with pristine sPEEK.

2.2.3 Carbon nanomaterials as additives in PEMs.

Carbon nanotubes (CNTs) are known for their unique structural and physical properties like high strength due to the strong sp^2 bond and high surface area and low density.[74,75] Hence CNTs were chosen as an additive material in polymer electrolyte membranes to overcome the methanol permeability and mechanical strength issues in DMFCs. However it is difficult to disperse pristine CNTs in a polymer matrix due to the Van der Waals interactions of the bare tubes. This situation can be overcome by modifying the surface of CNTs by functionalizing it before dispersing in polymer matrix.[76,77]

Sang Hoon Joo *et al.* prepared nanocomposite membranes using functionalized CNTs as additives in sulfonated poly(arylenesulfone) composite membranes. In this study, two types of functionalized CNTs are explored viz sulfonated CNTs and Pt–Ru supported CNTs. The functionalized CNTs homogeneously dispersed within a sPAS matrix, primarily *via* π–π interactions. Among these, Pt–Ru/CNT-sPAS exhibits a low Ohmic resistance and high open-circuit voltage compared with a sPAS membrane.[77] Weihua Zhou *et al.* reported composite membranes consisting of sulfonated carbon nanotubes (sCNTs) and sulfonated poly(ether sulfone ether ketone ketone). The composite membranes shoes improved mechanical strength and restricts methanol permeability acting as barrier for methanol.[78]

Considering the advantages and challenges of CNTs, a novel composite membrane for direct methanol fuel cells (DMFCs) was prepared by the incorporation of polystyrene sulfonic acid (PSSA) functionalized multi-walled carbon nanotubes (PSSA–MWCNTs) into sulfonated polyether ether ketone (sPEEK) matrix.[79] Functionalization of CNTs was performed using poly(styrene sulfonic acid) (Scheme 4), wherein Van der Waals

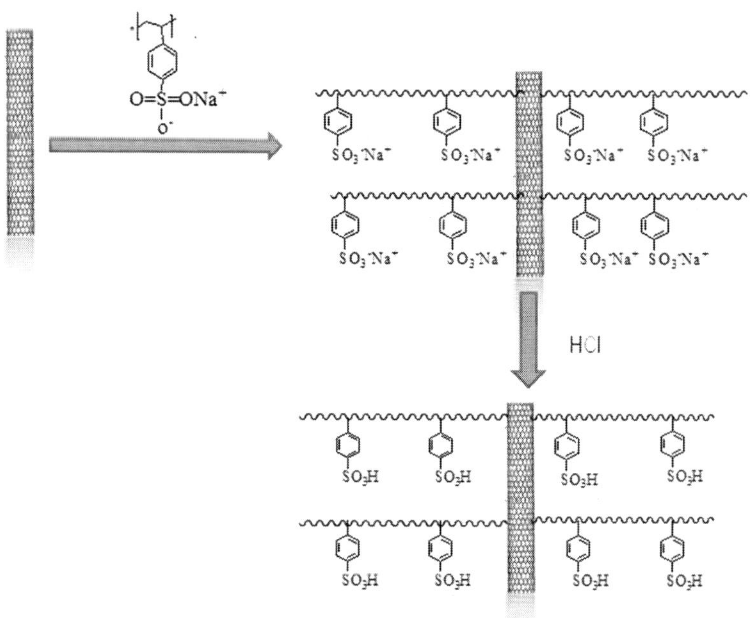

Scheme 4 Sulfonation of CNTs.

attractive forces are balanced by the electrostatic interactions inturn improving the dispersion of PSSA-CNTs in the polymer matrix.

Uniform dispersion of PSSA-CNTs in sPEEK improves the mechanical and thermal properties due to the interfacial interactions between PSSA-CNTs and sPEEK matrix.[80] Ionic conductivity of the composite membrane is enhanced due to restructuring of hydrogen bonding between PSSA and sPEEK which will facilitate the continuous pathways for proton conduction thereby enhancing the ionic conductivity for sPEEK-PSSA-CNTs composites in relation to pristine sPEEK membrane as shown in Table 6. Methanol crossover for recast Nafion, pristine sPEEK and composite membranes were carried out under OCV condition at 60 °C and shown in Table 6. It is noteworthy that methanol permeability of sPEEK-PSSA-CNTs composites decreased in comparison with pristine sPEEK membrane and recast Nafion leading to higher electrochemical selectivity as shown in Fig. 14. As the content of PSSA-CNTs is increased

Table 6 Proton conductivity and methanol permeability for the membranes.

Membrane type	Proton conductivity ($mS\ cm^{-1}$) at 30 °C	Proton conductivity ($mS\ cm^{-1}$) at 60 °C	Methanol permeability ($10^{-7}\ cm^2\ s^{-1}$) at 60 °C
Recast Nafion	53	82	6.43
Pristine sPEEK	44	62	2.71
sPEEK-PSSA-CNT (0.25 wt%)	62	89	2.52
sPEEK-PSSA-CNT (0.5 wt%)	68	101	2.17
sPEEK-PSSA-CNT (1 wt%)	65	95	1.61

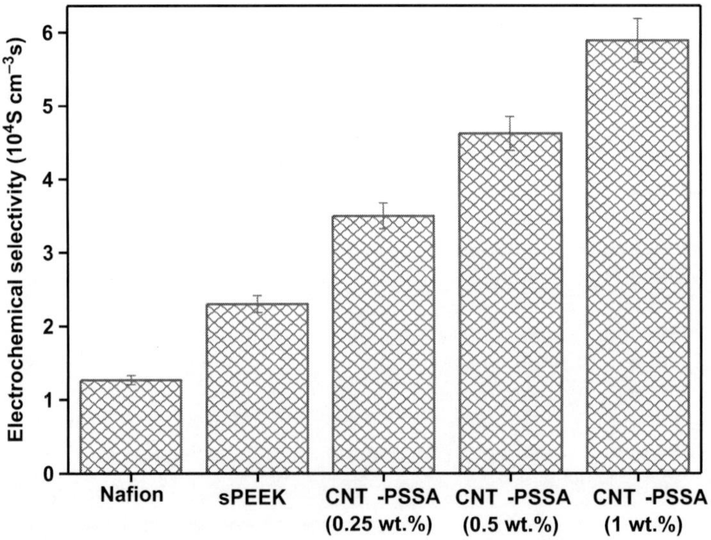

Fig. 14 Electrochemical selectivity for the membranes. Reproduced with permission from G. Rambabu and S. D. Bhat, Simultaneous tuning of methanol crossover and ionic conductivity of sPEEK membrane electrolyte by incorporation of PSSA functionalized MWCNTs: A comparative study in DMFCs, *Chem. Eng. J.*, **243**, 517–525. Copyright 2014 with permission from Elsevier B.V.

from 0.25 to 1 wt% the methanol permeability decreased due to the presence of PSSA-CNTs on the pores of sPEEK. In addition, better dispersion of CNTs in sPEEK due to the presence of PSSA provides the tortuous pathways for the methanol transport thereby restricting the methanol crossover.

The composite membranes were employed in DMFC where in the optimized composite membrane delivers peak power density of 93 mW cm^{-2} which is considerably higher than pristine sPEEK and on par with Nafion-117 membraneas represented in Fig. 15a. In addition, Fig. 15b shows stability for the membranes in OCV condition wherein the composite membranes shows higher OCV due to low methanol permeability and show on par stability with recast Nafion.

Another class of carbon nanomaterial Fullerenes are also identified as potential filler material in the membranes due to their unique

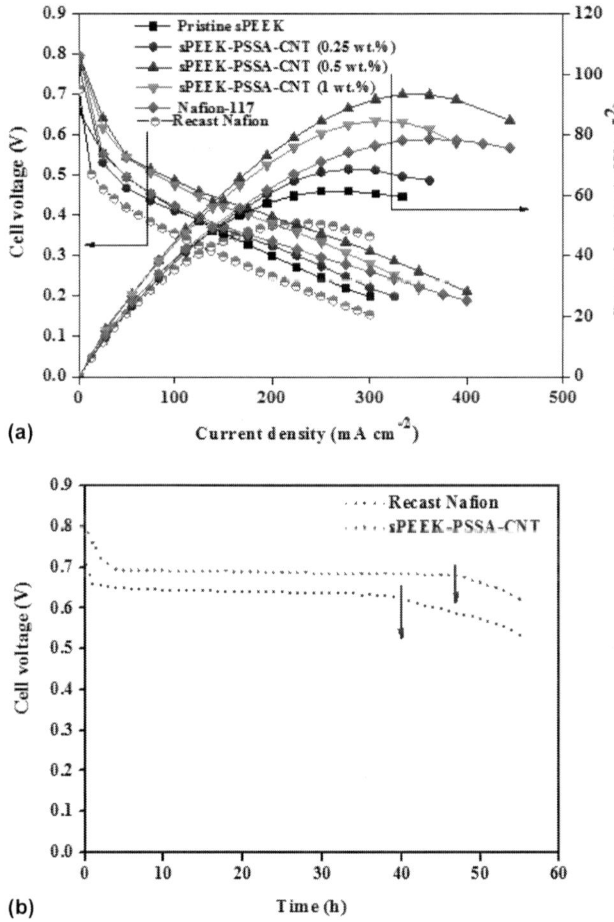

Fig. 15 (a) DMFC performance for sPEEK and sPEEK-PSSA-CNT composite membranes, (b) Open circuit voltage as function of time for the membranes. Reproduced with permission from G. Rambabu and S. D. Bhat, Simultaneous tuning of methanol crossover and ionic conductivity of sPEEK membrane electrolyte by incorporation of PSSA functionalized MWCNTs: A comparative study in DMFCs, *Chem. Eng. J.*, **243**, 517–525. Copyright 2014 with permission from Elsevier B.V.

characteristics such as the high electron affinity, the high volumetric density of the functional groups and its radical scavenging behavior.[81,82] Considering these advantages of fullerene, few research groups have explored fullerene based composite membrane electrolytes for DMFCs. Saga *et al.* prepared composite polyelectrolyte membranes from sulfonated polystyrene and fullerene.[82] The additive effect of the fullerene on the membrane properties, electric resistance, mechanical strength, oxidation resistance, and methanol permeability were measured. The addition of fullerene improved the oxidation resistance, and reduced the methanol crossover. The mechanical strength of the fullerene-composite membrane, on the other hand, was not improved. The direct methanol fuel cell (DMFC) based on a 1.4 wt% fullerene-composite membrane showed the highest power density of 47 mW cm^{-2} at the current density of 200 mW cm^{-2}.

Tasaki *et al.* reported the preparation of Fullerene–Nafion composite membranes. The fullerenes used for the composites included C_{60} and polyhydroxy fullerene (PHF) $C_{60}(OH)_n$,[81] The miscibility of the hydrophobic fullerene, C_{60}, in the Nafion matrix was further improved by a new fullerenedispersant, poly[tri(ethylene oxide)benzyl] fullerene. The solution-cast fullerene composites also demonstrated a significant improvement in the physical stability relative to the fullerene-doped Nafion composites through a better integration of the fullerene into the Nafion matrix. Furthermore, increased loadings of the fullerene in Nafion were made possible through the new solution-casting method, compared to doping method.

Motivated by the above studies, Rambabu *et al.* extended the beneficial characteristics of fullerene to prepare composite membrane of sPEEK. Different from the previous studies, fullerene was sulfonated in a simple method and incorporated in sPEEK matrix to improve its electrolyte properties in DMFCs.[83] Prior to sulfonation, fullerene was oxidised using 6 M H_2SO_4 and HNO_3 (1 : 1 ratio) at 130 °C for 8 h to activate the surface for further functionalization.[84] Oxidized fullerene was sulfonated using 4-benzene diazonium sulfonic acid precursor formed by diazotization reaction.[85] The Ball-stick model view of the sulfonation process is represented in Scheme 5.

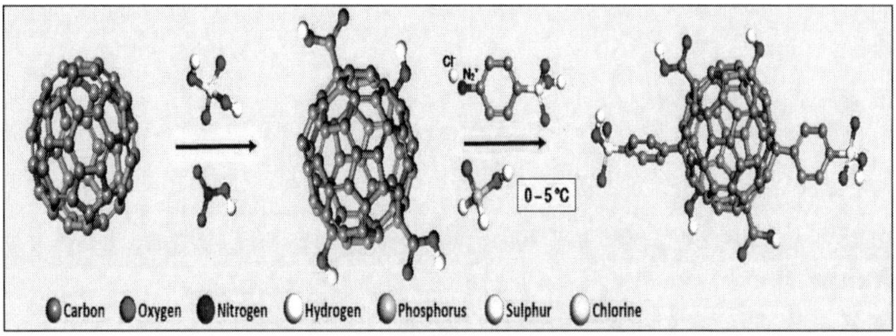

Scheme 5 Oxidation followed by sulfonation of fullerene. Reproduced with permission from G. Rambabu and S. D. Bhat, Sulfonated fullerene in sPEEK matrix and its impact on the membrane electrolyte properties in direct methanol fuel cells, *Electrochim. Acta*, **176**, 657–669. Copyright 2015 with permission from Elsevier B.V.

Composite membranes were prepared by varying the amount of Sfu in sPEEK matrix (0.25 to 1 wt%). These composite membranes show improved IEC and proton conductivity due to the additional sulfonic acid groups present on Sfu. The effect of methanol concentration on tensile strength and sorption was investigated as shown in Fig. 16a. It is noteworthy that the composite membranes are stable even at higher methanol concentration. On the other hand, compared to water sorption, water–methanol sorption is found to be higher due to the sorption of both water and methanol through the ionic clusters of the membranes. At higher methanol concentration, Nafion-117 and pristine sPEEK show higher sorption which may be the reason for depletion of tensile strength. In contrast, as Sfu amount is increased in sPEEK, composite membranes show less affinity towards methanol compared to water and the effect of methanol concentration is minimal suggesting the stability of the composite membranes in methanol environment. Figure 16b shows concentration profile which is used to determine the methanol permeability for the membranes wherein Nafion-117 shows higher methanol permeability due to its larger ionic clusters and sPEEK-Sfu (1 wt%) shows lowest methanol permeability due to the effective blockage of methanol by Sfu in the membrane.

In addition, oxidative stability for the membranes was carried out to evaluate the stability of the membranes in fuel cell environment and the date is presented in Fig 16c. It is noteworthy that sPEEK-Sfu composite membranes have better oxidation resistance than pristine sPEEK with the weight loss of only 7% for sPEEK-Sfu (1 wt%) while it is 30% for pristine sPEEK. Higher oxidative stability for composite membranes is due to the effective capturing of HO and HOO radicals formed by the Fenton's solution during experiments. DMFC comprising the optimized membrane shows peak power density of 105 mW cm^{-2} which is 57% higher than pristine sPEEK as shown in Fig. 16d. The improvement in peak power density is attributed to enhanced ionic conductivity, methanol barrier properties and its oxidative stability of the composite membranes.

2.3 A comparative study of Nafion and sPEEK polymers incorporated with meso and microporous additives for direct ethanol fuel cells (DEFCs)

At present methanol has been considered as the most promising fuel because of its efficient oxidation in comparison with other alcohols. However, environmental and safety issues are associated with methanol due to its higher toxicity.[86] In addition, methanol diffuses from the anode to cathode through the membrane, reduces the available sites for oxygen reduction and creates a mixed potential at cathode leading to lower fuel cell efficiency.[87] On the other hand, as far as other alternate liquid fuels are concerned, ethanol is the preferred one as it has higher energy density (8.2 kWh kg^{-1}) than methanol (6.1 kWh kg^{-1}). In addition, it can be produced in large quantities from biomass through fermentation process of renewable resources like sugarcane, wheat, corn or straw.[88] These features make ethanol more attractive fuel source than methanol

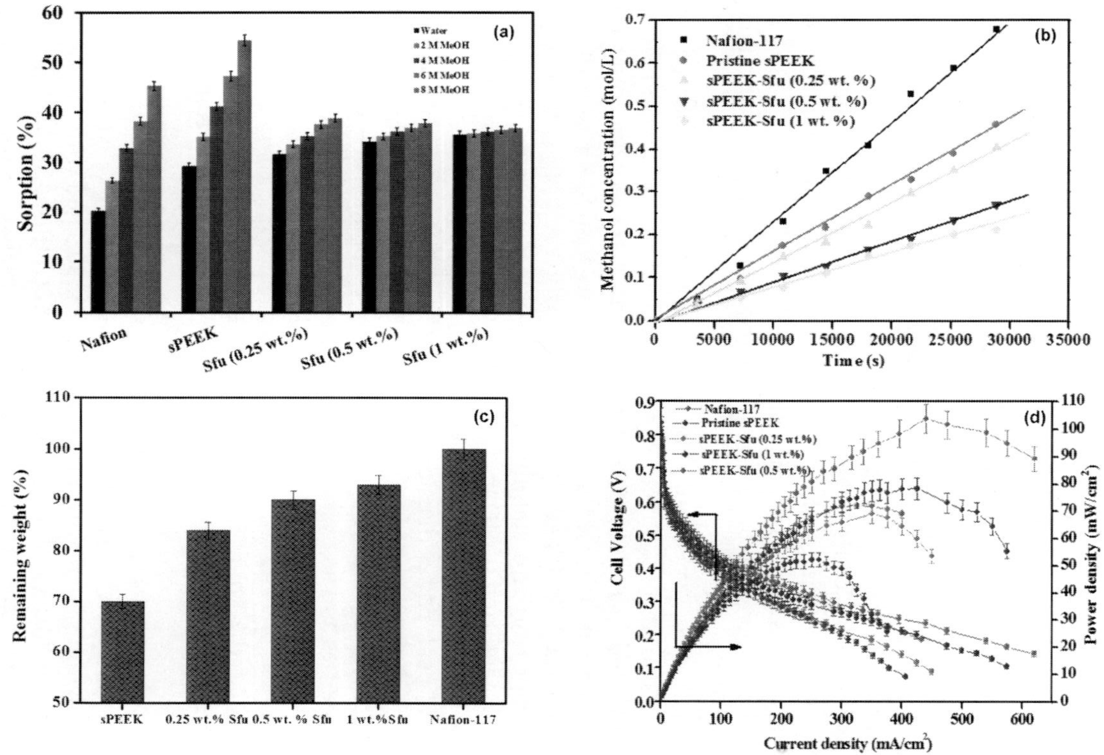

Fig. 16 (a) Water–methanol sorption for the membranes, (b) Methanol concentration profile pertaining to the methanol permeability for the membranes (c) Oxidative stability for the membranes, (d) DMFC performance for sPEEK and sPEEK-SFu composite membranes. Reproduced with permission from G. Rambabu and S. D. Bhat, Sulfonated fullerene in SPEEK matrix and its impact on the membrane electrolyte properties in direct methanol fuel cells, *Electrochim. Acta.*, **176**, 657–669. Copyright 2015 with permission from Elsevier B.V.

for direct alcohol fuel cells. To accomplish the desired performance of DEFCs, efforts are being pursued in developing catalyst with lower ethanol permeability.[8]

In this context, several researchers have investigated Nafion based composite membranes for DEFCs to reduce the ethanol permeability and to obtain higher performance. Blasi *et al.*[89] fabricated composite Nafion membrane with incorporation of silica and found increased water retention property with reduced ethanol permeability in turn affecting the fuel cell performance. Similarly, Barbora *et al.*[90] reported Nafion–TiO_2 composite membranes with reduced ethanol permeability. However, these composite membranes are not evaluated for fuel cell studies.

As far as non-fluorinated polymer electrolytes are concerned, sulfonated polyether ether ketone (sPEEK) based composite materials are widely studied in the literature.[91] K. S. Roelofs *et al.*[92,93] have studied sPEEK based composite membranes with incorporation of various silica materials and found lower ethanol permeability with good electrochemical selectivity in DEFCs. Also, A. R. Tan *et al.*[94] explored nanocomposite membranes based on heteropolyacid (HPA) sealed silica matrix with sPEEK/PES blend showing improved proton conductivity and lower ethanol sorption. The ethanol permeability and selectivity of sPEEK/PVDF/PWA composite membranes was also studied by S. Xue *et al.*[95]

In this regard, Meenakshi *et al.*[96] studied the ethanol permeability behaviour of two different membranes based on Nafion-mesostructured molecular sieves and zeolite-4A-sPEEK-MSA hybrid and evaluated their overall impact on the DEFC cell performance. In this study, aluminosilicate with mesostructured hexagonal network MCM-41, silica mesostructured cellular foam (MSU-F) and zeolite-4A additives were incorporated into Nafion and sPEEK matrix respectively, to form composite membranes which showed higher IEC and proton conductivity, lower ethanol permeability with better mechanical stability.

The proton conductivity for the membranes depends on the proton mobility through the ion-cluster pores and the ionic channels present in the polymer backbone.[97] The proton conductivity data for the pristine and hybrid membranes are presented in Table 7 at 30 °C and 70 °C. It is found that the proton conductivity for Nafion-Si-MSU-F and Nafion-Al-MCM-41 hybrid membranes is higher than that of the pristine recast

Table 7 Properties for the membranes.

Membrane type	Proton conductivity (mS cm^{-1}) at 30 °C	at 70 °C	Ethanol permeability (10^{-7} cm^2 s^{-1}) at 30 °C	Ion exchange capacity (meq g^{-1}) at 30 °C
Pristine recast Nafion	51	70	0.90	8.13
Nafion-Si-MSU-F hybrid	137	175	0.92	4.30
Nafion-Al-MCM-41 hybrid	185	236	0.96	0.60
Pristine sPEEK	36	49	1.68	7.41
Zeolite 4A-sPEEK-MSA hybrid	65	137	1.75	1.60

Nafion membrane. This corresponds may be due to the incorporation of $-SO_3H$ groups along the interior pore walls of mesoporous Si-MSU-F and Al-MCM-41 that increases the hydrophilicity of the membrane thereby increasing the proton-conduction pathways.[19,20] However, in case of zeolite 4A-sPEEK-MSA hybrid, higher proton conductivity may be due to the facile pathways provided by the micropores of 4A and ionic channels of sPEEK-MSA for proton transport in comparison to the torturous pathways of pristine sPEEK membrane.[52]

Nyquist plots are shown in Fig. 17(a–d). It can be clearly seen that the membrane resistance is lower for the hybrid membranes in comparison with pristine matrices. It is interesting to note that the resistance values obtained at 30 °C from Nyquist plots for sPEEK and Zeolite 4A-sPEEK-MSA hybrid membranes are 2.1 and 1.8 ohm, respectively closely following the trend observed in IEC values. Similarly at 70 °C, the resistance values obtained from Nyquist plots for the pristine sPEEK membrane and Zeolite 4A-sPEEK-MSA are 2.0 and 0.68 ohm, respectively.

Figure 18 shows the variation in ethanol concentration for different membranes as a function of time. The time interval is varied from 1 to 20 h and ethanol concentration were measured in both compartments. Concentration profiles from the compartment 2 in Fig. 18(a and b) for the aforesaid membranes is a clear indication that hybrid membranes show lower ethanol permeability than the pristine recast membranes.

These aforesaid properties lead to the enhanced DEFC cell performance for these membranes in comparison with recast Nafion. The DEFC performance curves for Nafion and sPEEK based hybrid membranes are represented in Fig. 18(c and d).

3 Anion exchange membranes for alkaline DMFC

There has been prodigious concern towards the development of anion exchange membranes (AEMs) for alkaline direct methanol fuel cells (ADMFCs) which replaces the proton conducting DMFC.[98] In alkaline DMFC, anodic methanol oxidation kinetics will be faster than acidic media, thus allowing cost effective non-noble metals like Ni, Fe, and Ag to be used as electro-catalysts.[99] Furthermore, the methanol crossover will be curtailed by the electro-osmotic movement of ions (OH^-) transported from cathode to anode, thus opposing the flow of methanol. Since the ion transport is from the cathode to anode, water is now produced at the anode and consumed at the cathode thereby making the water management potentially simplified. AEMs are the positively charged polyelectrolyte acting as the charge carrier (usually quaternary ammonium type) grafted on the polymer backbone with dissociated anions for ionic conductance.[100] Polymer electroneutrality is maintained by attaching a mobile counter-ion to each ionic functional group. The choice of AEM to be used as solid electrolyte demands high OH^- conductivity with excellent methanol tolerance and good chemical stability.

Poly (phenylene oxide) (PPO) is one of the widely used thermoplastic for the preparation of ion exchange membranes because of its easy accessibility, good solubility, high chemical and thermal stability.

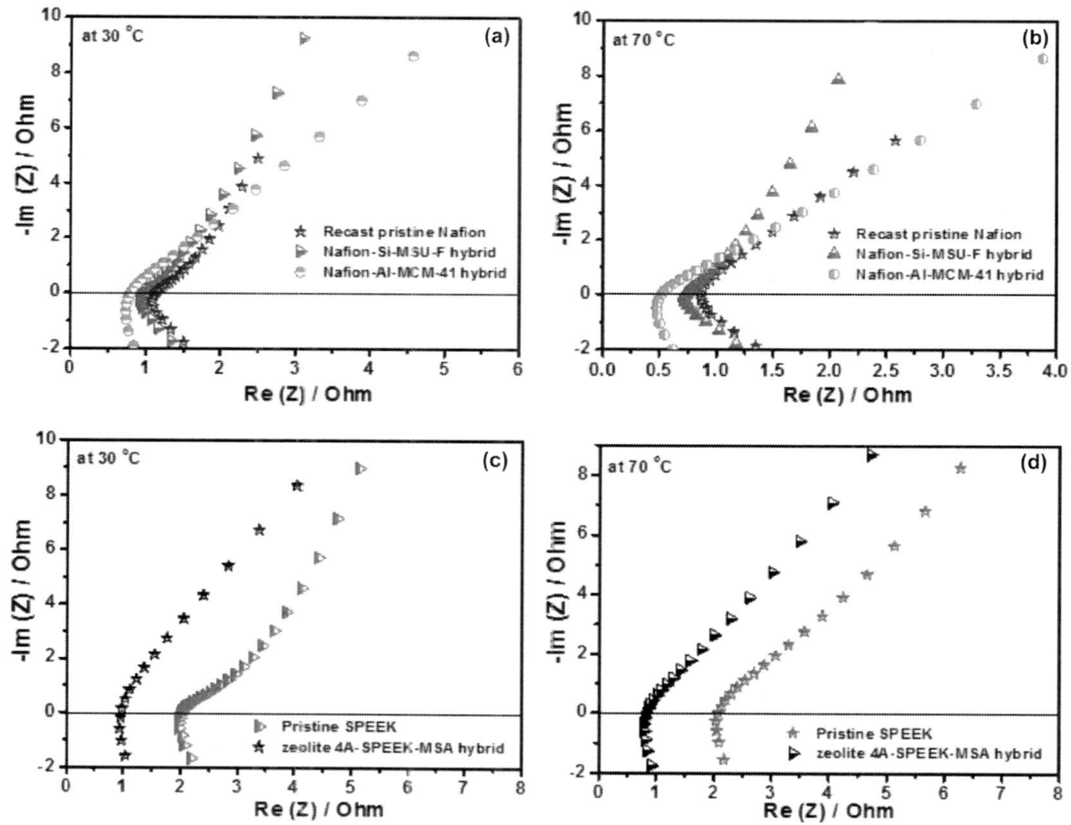

Fig. 17 Nyquist plots for the membranes at 30 °C and 70 °C. Reproduced with permission from S. Meenakshi, A. Manokaran, S. D. Bhat, A. K. Sahu, P. Sridhar and S. Pitchumani, Impact of Mesoporous and Microporous Materials on Performance of Nafion and sPEEK Polymer Electrolytes: A Comparative Study in DEFCs, *Fuel Cells*, 2014, **14**, 842–852. Copyright 2014 WILEY-VCH Verlag GmbH & Co. KGaA, Weinheim.

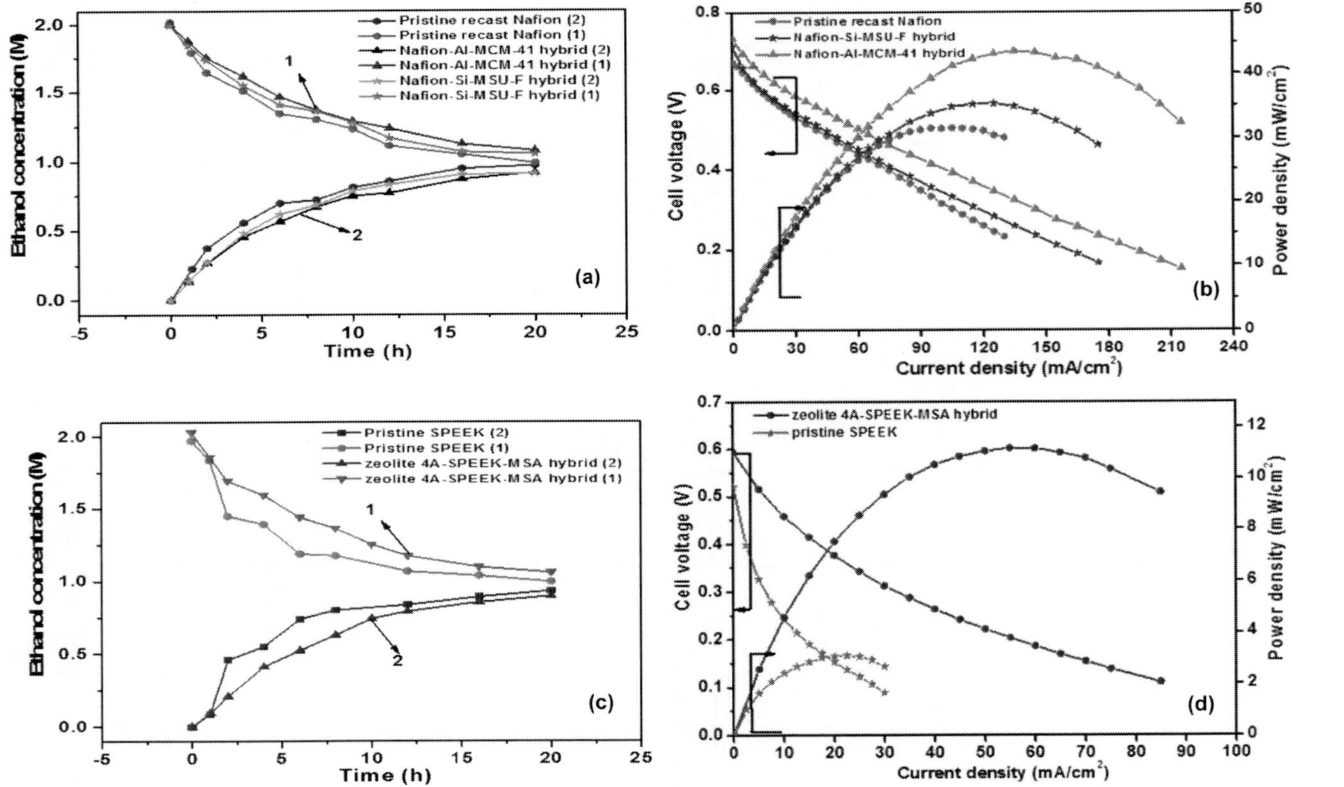

Fig. 18 Time *vs.* Ethanol concentration for (a) pristine recast Nafion, Nafion-Si-MSU-F hybrid and Nafion-Al-MCM-41 hybrid, (b) Pristine sPEEK and zeolite 4A-sPEEK-MSA hybrid membranes, (c) Cell polarization curves of DEFCs for pristine recast Nafion, Nafion-Si-MSU-F and Nafion-Al-MCM-41 hybrid membranes (d) pristine sPEEK and zeolite-4A-sPEEK-MSA hybrid membranes at 70 °C. Reproduced with permission from S. Meenakshi, A. Manokaran, S. D. Bhat, A. K. Sahu, P. Sridhar and S. Pitchumani, Impact of Mesoporous and Microporous Materials on Performance of Nafion and sPEEK Polymer Electrolytes: A Comparative Study in DEFCs, *Fuel Cells*, 2014, **14**, 842–852. Copyright 2014 WILEY-VCH Verlag GmbH & Co. KGaA, Weinheim.

Generally, PPO is used as a precursor for the preparation of various functional polymers by chemical modification either in their aryl or benzylic positions for its use as solid polymer electrolyte.[100,101]

Katzfuß *et al.*[102] developed an AEM from 1,5-dimethyl polyphenylen-oxide by using N-bromosuccinimide (NBS) as brominating reagent. The brominated PPO (BrPPO) was further cross-linked with DABCO to form a covalently cross-linked membrane which leads to its higher thermal and chemical stability. Properties like ion exchange capacity, ion conductivity, methanol uptake and single cell performance of the membranes were investigated. The alkaline stability was investigated at 90 °C in 1 M KOH solution, where the membranes showed excellent stabilities. The covalent cross-linking caused an additional steric hindrance for the attack of hydroxyl ions, and therefore this membrane seems to be reasonably stable.

Janarthanan *et al.*[101] investigated the performance of poly (phenylene) based anion exchange membrane in a direct methanol fuel cell with different anode gas diffusion layers. A poly(phenylene) AEM with cation consisting of six methylene spacers attached to a trimethyl ammonium group demonstrated good performance with a maximum current density of 11.8 $mA\,cm^{-2}\,mg$ in a KOH free fuel. They also established that a hydrophilic anode GDL and hydrophobic cathode GDL gives the best performance. MEA with a hydrophilic Zoltek gas diffusion layer on the anode showed the highest performance (226 $mA\,cm^{-2}$ and 53.8 $mW\,cm^{-2}$) in the presence of KOH solution, suggesting that hydrophilic gas diffusion layer plays a significant role in improving the fuel cell performance.

Hari Gopi *et al.*[103] fabricated a series of AEMs from poly(phenylene oxide) with quaternary ammonium moiety for application in ADMFCs. In the first step, poly(phenylene oxide) (PPO) was chloromethylated by substituting chloromethyl groups in the aryl position of polymer and the influence of reaction time on the chloromethylation reaction was investigated. It is evident that, the degree of substitution (DS) increases as the reaction time is increased. The degree of substitution for different time intervals was calculated from the ^1H NMR spectra as represented in Fig. 19a. In the second step, the chloromethylated PPO (CPPO) was homogeneously quaternized *via* Menshutkin reaction and finally ion-exchanged to form an anion exchange membrane (AEM) as given in Fig. 19b.

From the second step, series of AEMs are prepared (QPPO 1, QPPO 2, QPPO 3 & QPPO 4) by varying the mole ratio of amine with relation to CPPO. AEMs with different degree of quaternization are evaluated for its ionic conductivity, ion exchange capacity (IEC) methanol permeability and single-cell DMFC performance. During cell polarisation measurements, the variation of metal loading in the catalyst layer and membrane durability at OCV were also investigated.

It is well known that the conductivity of an AEM is determined by many factors such as the degree of chloromethylation, degree of quaternization, basicity of the tethered cationic group, and micro-morphology of membrane. Figure 20a and b represents the ionic conductivity of QPPO membranes prepared with different mole ratio of amine as a function of

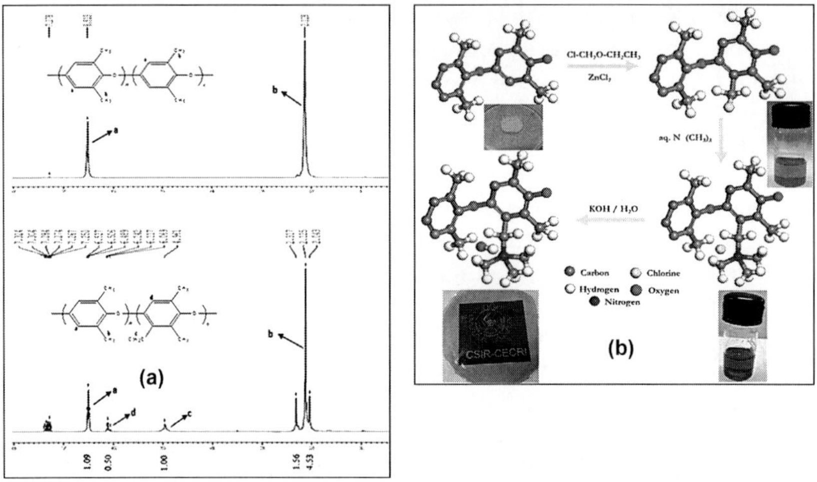

Fig. 19 (a) ^{1}H NMR spectrum of PPO and CPPO in $CDCl_3$, Reproduced with permission from K. Hari Gopi, S. Gouse Peera, S. D. Bhat, P. Sridhar and S. Pitchumani, Preparation and characterization of quaternary ammonium functionalized poly (2,6-dimethyl-1,4-phenylene oxide) as anion exchange membrane for alkaline polymer electrolyte fuel cells, *Int. J. Hydrogen Energy*, **39**, 2659–2668. Copyright 2014 with permission from Elsevier B.V. (b) Ball & Stick model scheme for the preparation of QPPO. Reproduced with permission from K. Hari Gopi, S. D. Bhat, A. K. Sahu and P. Sridhar, Quaternized poly(phenylene oxide) anion exchange membrane for alkaline direct methanol fuel cells in KOH-free media, *Appl. Polym. Sci.*, 2016, **133**(29), 43693. Copyright 2016 Wiley Periodicals, Inc.

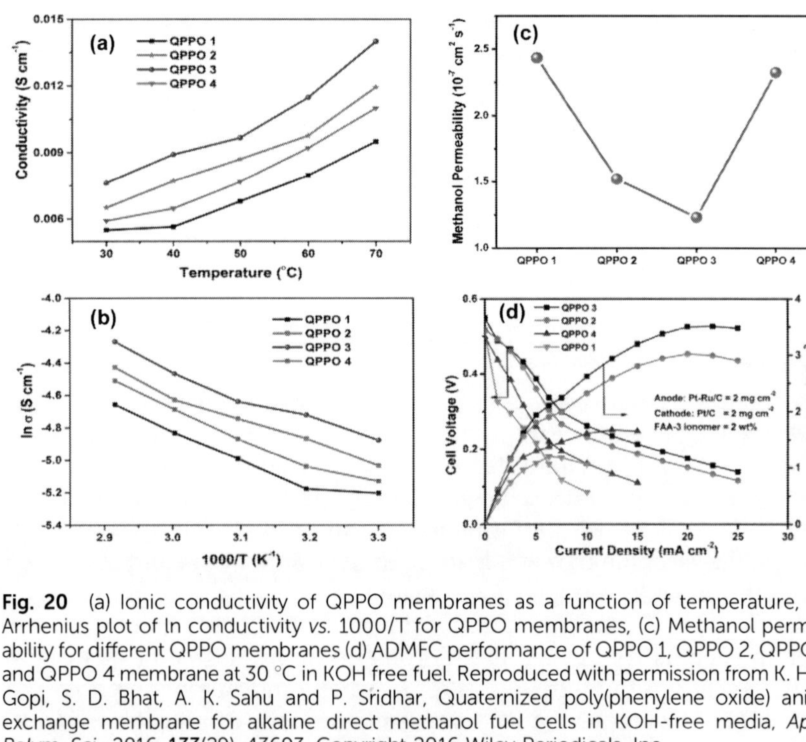

Fig. 20 (a) Ionic conductivity of QPPO membranes as a function of temperature, (b) Arrhenius plot of ln conductivity *vs.* 1000/T for QPPO membranes, (c) Methanol permeability for different QPPO membranes (d) ADMFC performance of QPPO 1, QPPO 2, QPPO 3 and QPPO 4 membrane at 30 °C in KOH free fuel. Reproduced with permission from K. Hari Gopi, S. D. Bhat, A. K. Sahu and P. Sridhar, Quaternized poly(phenylene oxide) anion exchange membrane for alkaline direct methanol fuel cells in KOH-free media, *Appl. Polym. Sci.*, 2016, **133**(29), 43693. Copyright 2016 Wiley Periodicals, Inc.

Table 8 Effect of amination on the properties of different QPPO membrane.

Membranes	CPPO : amine (mol : mol)	Degree of quaternization (%)	IEC (mmol g^{-1})	Ionic conductivity at 30 °C (mS cm^{-1})
QPPO 1	1 : 4	4.20	0.83	5.5
QPPO 2	1 : 6	4.25	1.13	6.5
QPPO 3	1 : 8	4.31	1.38	7.6
QPPO 4	1 : 10	4.22	0.95	5.9

temperature. From the figure it is seen that, as the quantity of added amine increases, there is a substantial increase in ionic conductivity. The ionic conductivity of membranes at 30 °C increased from 5.5 mS cm^{-1} (QPPO 1) to 7.6 mS cm^{-1} (QPPO 3) by varying the amount of amine as shown in Table 8.

The conductivity of QPPO membranes at a given temperature gradually increase in line with IEC, with the increase in the concentration of active sites for anion transport. Also, the methanol permeability for these membranes was carried out at OCV for 5 h under identical operating conditions. As seen from Fig. 20c, there is corresponding decrease in methanol permeability as the amount of added amine increases. Among them, the QPPO 3 membrane has the lowest methanol permeability of 1.23×10^{-7} cm^2 s^{-1}. But at higher and lower ratio of amine (QPPO 4 & QPPO 1) methanol permeability was increased probably due to their low ionic conductivity, as the total number of hydroxide ions opposing the methanol will be reduced.

The fabricated membranes are subjected to cell polarization studies in ADMFCs as shown in Fig. 20d, wherein QPPO 3 (CPPO : amine ratio of 1 : 8) membrane exhibited higher peak power density of 3.5 mW cm^{-2} when compared with the other ratios of CPPO : amine in the absence of KOH solution. As the quantity of the added amine increases, the quaternary ammonium groups grafted on to the polymer is increased due to the high degree of quaternization, which has a pronounced effect on cell polarisation. Also, the chemical stability at high pH (4 M KOH solution) and at 75 °C for the optimized membrane was studied for 300 h.

Electrocatalysts for DAFCs

As already mentioned in the Introduction, in this section mainly anode catalysts for oxidation of alcohols and alcohol-tolerant cathode catalysts for oxygen reduction with respect to methanol and ethanol will be discussed.

Methanol-tolerant ORR catalysts for DMFCs

Noble metal catalysts, such as platinum, which when dispersed in particulate form on carbon exhibit high activity for oxygen reduction, however, these catalysts show little methanol tolerance. The addition of a base metal, such as iron, cobalt, nickel, and chromium, to platinum improves its methanol tolerance toward oxygen reduction at the cathode. However, the alloy catalysts exhibit poor long-term stability due to the dissolution of the base metal. Recently, the introduction of a second platinum group metal (PGM), such as palladium and gold, to platinum has found considerable attention due to their high oxygen reduction activity, methanol tolerance, and better durability.[104–108] Among these, gold happens to be the most attractive catalyst because of its inertness in the bulk state and high catalytic activity at nanoscale;[109–113] gold when alloyed with platinum provides enhanced activity toward oxygen reduction reaction (ORR).[114–120] Besides, gold is known to have a stabilizing effect on Pt even under highly oxidizing conditions. In general, a bimetallic substance differs in properties from its single metal counterparts, as the presence of another metal atom in the parent lattice affects its composition, crystal structure, and electronic nature. These factors are known to influence the catalytic behavior of a bimetallic alloy and are primarily driven by the experimental conditions used for its preparation. DFT (density functional theory) calculations have been employed to analyze the electronic structure of Pt–Au alloys and their ameliorating catalytic activity as compared to the bare Pt nanocluster. Linear sweep voltammetry experiments suggest clear methanol-tolerant behavior for the Pt–Au catalyst toward ORR in relation to Pt, and the performance of DMFC employing a (2 : 1) Pt–Au/C cathode catalyst is superior to the DMFC employing a Pt cathode. DFT calculations on a model cluster also reflect electron transfer from Pt to Au within the Pt–Au alloy to be responsible for the synergistic promotion of the oxygen-reduction reaction on the Pt–Au electrode.[108] However, the use of Pt–PGM alloys is cost intensive.

It has been reported that Pt and Pd exhibit similar reactivity towards ORR, but different electrochemical reactivity towards methanol oxidation reaction (MOR) as Pd is completely inactive towards methanol electro-oxidation in an acidic medium.[121,122] The stability of the Pt–Pd catalyst is also improved due to the stability of Pd in the acidic environment. Another advantage of Pt–Pd catalyst is its high selectivity towards ORR in presence of methanol that makes it an attractive methanol-tolerant-cathode catalyst.[106,123,124] For methanol oxidation reaction (MOR) to proceed, three Pt sites are initially required for the dissociative chemisorptions of methanol. It appears that such a situation is hindered with Pt–Pd (2 : 1) catalyst.[125]

Recent studies show that it is necessary to improve the Pd core catalytic activity for which several approaches have been developed such as incorporation of a second element namely, Fe, Co *etc.*[126–129] However, incorporation of such elements into Pd may lead to dissolution within the Pd nanostructure under fuel cell operating conditions akin to that observed in the case of Pt-based alloy catalysts, such as Pt–Fe, Pt–Co, Pt–Ni *etc.* As a consequence, the dissolution of the non-noble metal components limits the life of DMFCs.[130] Instead of incorporating a second element, we have adopted a different synthetic method of encapsulating the Pd nanostructure on carbon with Pt. It is also reported that monolayer Pt deposited on Pd(111) could prolong ORR activity with minuscule Pt loading compared to Pt–M (M: base metals) catalysts. Development of Pt-based core–shell nanoparticles as electro-catalysts for polymer electrolyte membrane fuel cells has been an emerging research area as this class of material can offer great scope for cost reduction as well as electrochemical stability.[126–129] The core–shell construction can lead to the use of a low-cost metal core and a noble metal over layer to reduce the cost of the catalyst. It has been reported that the strain caused by the lattice mismatch between the surface and core components may be used to modify the electronic properties of the surface metal atoms resulting in an improvement in the catalytic activity toward ORR. The enhancement in the activity was accounted for electron transfer from Pd to the Pd-modified Pt surface and the contraction of the Pt lattice in a related computational study.[128,129] Pt was deposited on the surface of Pd/C nanoparticles by a spontaneous displacement reaction. Pt@Pd/C electro-catalyst was found to have enhanced ORR activity and excellent stability. Pt@Pd/C showed much higher methanol tolerance than Pt/C suggesting that it could be an alternative cathode electro-catalyst for DMFCs.[131]

Recently, stable metal oxides, such as Pt–MO_x M = Ru, W, or Ti, have also been investigated for ORR and have been found to exhibit higher activity than Pt.[121,132–143] Among these Pt–TiO_2 catalysts, being cost effective and acid stable are particularly attractive. Brewer and Wengert[144] reported that the hypo d-electron character of titanium oxide facilitates its interaction with noble metals, like Pt, changing the catalytic activity of the noble metal. The change in the catalytic activity is explained using Hammer and Norskov's concept,[145] according to which the reactivity alters through the changes in the adsorbate interaction energy due to the shift in the local d-band position relative to the Fermi level. In the literature,[135–143] the suitability of TiO_2 as a catalyst support material has been studied extensively. Although TiO_2 shows higher durability in relation to conventional carbon supports, its electronic conductivity is relatively lower, which results in increased ohmic resistance for the cell.[135–137] Incorporation of TiO_2 in carbon-supported Pt ameliorates the electrocatalytic activity of Pt toward oxygen reduction with better methanol tolerance. The DMFC employing a Pt–TiO_2/C heat treated at 750 °C with a Pt to Ti atomic ratio of 2 : 1 [750 Pt–TiO_2/C (2 : 1)] exhibits better performance in relation to Pt/C.[146]

The altervalent features of transition metal have been utilized to improve the catalytic activity of electrocatalysts through its dynamic spillover effect.[147–150] In particular transition metal oxides like TiO_2 and WO_3 have been used. The working principle of such metal oxide systems is based on the hydrophilic behavior due to water molecules trapped inside the oxide network turning to hydrous network, which substantially behaves as a continuous undisturbed reversible membrane mechanism of the altervalent changes resulting in OH^- transfer within the system with consequent spillover of primary oxide over metallic catalyst particle. Based on this lead, Platinum–Yttrium hydroxide based catalyst supported on carbon is developed and evaluated for ORR activity in presence and absence of methanol. Desired atomic ratio of Pt to Y for ORR in presence and absence of methanol is also optimized through half-cell and cell polarization studies. Among the various Pt–$Y(OH)_3$/C catalysts, the one with Pt to Y in 3 : 1 atomic ratio shows the highest activity for ORR in aqueous $HClO_4$ solution without methanol while the one with Pt to Y in 2 : 1 atomic ratio shows the maximum activity for ORR in presence of methanol.[151]

Pt–Au supported on carbon as a methanol tolerant cathode catalyst for DMFCs

A Pt–Au alloy catalyst of varying compositions was prepared by codeposition of Pt and Au nanoparticles onto a carbon-support to evaluate its electrocatalytic activity towards oxygen reduction reaction (ORR) with methanol-tolerance in Direct Methanol Fuel cells (DMFCs). The optimum atomic weight ratio of Pt to Au in the carbon-supported Pt–Au alloy (Pt–Au/C) as established by linear-sweep voltammetry (LSV) and cyclic voltammetry (CV) and cell polarization studies, and illustrated in Fig. 21(a–c) is determined to be 2 : 1. A DMFC comprising

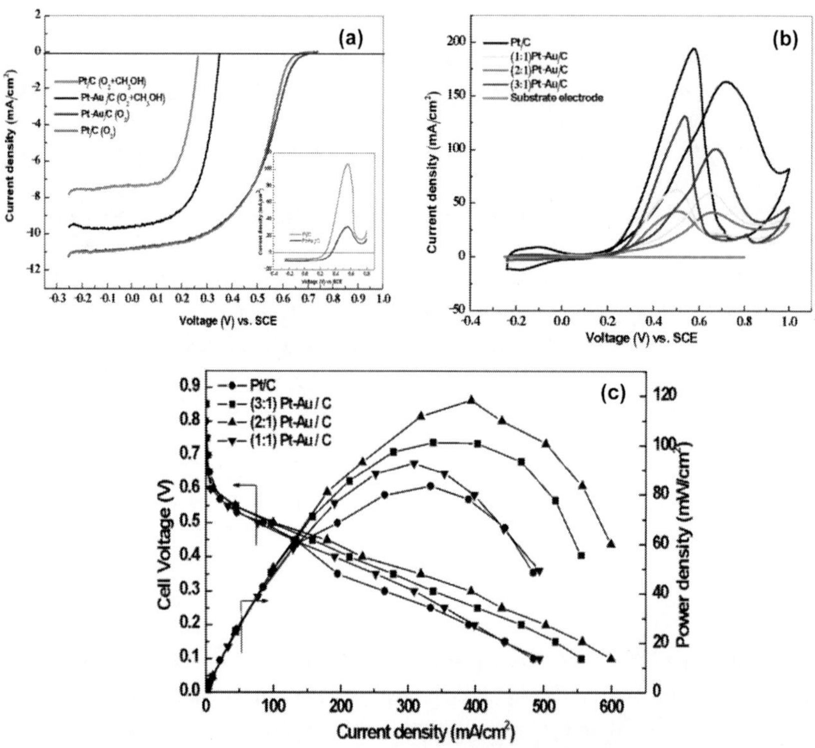

Fig. 21 (a) Linear sweep voltammetry data for ORR on Pt/C and (2 : 1) Pt–Au/C catalysts in O_2-saturated aq 0.5 M H_2SO_4 in the presence and absence of methanol at 1 mV s^{-1} scan rate (electrode rotation rate: 1500 rpm). The inset shows the LSV data for ORR on Pt/C and (2 : 1) Pt–Au/C catalyst in O_2-saturated aq 0.5 M H_2SO_4 + 0.5 M CH_3OH. (b) Cyclic voltammograms for Pt/C and bimetallic Pt–Au/C catalyst with various atomic ratios in N_2-saturated aq 0.5 M H_2SO_4 + aq 0.5 M CH_3OH solution at a scan rate of 50 mV s^{-1}. (c) Steady-state performances of DMFCs (CH_3OH & O_2) for Pt/C and Pt–Au/C with varying atomic ratios. Reproduced with permission from G. Selvarani, S. V. Selvaganesh, S. Krishnamurthy, G. V. M. Kiruthika, P. Sridhar, S. Pitchumani and A. K. Shukla., A Methanol-Tolerant Carbon-Supported Pt–Au Alloy Cathode Catalyst for Direct Methanol Fuel Cells and Its Evaluation by DFT, *J. Phys. Chem. C*, 2009, **113**, 7461–7468. Copyright 2009 American Chemical Society.

carbon-supported Pt–Au (2 : 1) alloy as the cathode catalyst delivers a peak power density of 120 mW cm^{-2} at 70 °C in contrast to the peak power density value of 80 mW cm^{-2} delivered by the DMFC with carbon-supported Pt catalyst operating under identical conditions.

Pt–Au alloy shows high ORR with better methanol tolerance in relation to bare Pt catalyst. Generally, the rate determining step for ORR is breaking of O–O bond to form water. The kinetics of the reaction depends on the degree of interaction of oxygen with catalyst adsorption sites. The enhanced electrocatalytic activity and methanol tolerance for Pt–Au/C alloy can be explained by the electronic factor, namely, the change of the d-band vacancy in Pt upon alloying and/or by a geometric effect. Both the effects may enhance the reaction rate for oxygen adsorption and cleavage of O–O bond during the reduction reaction. It is known that the presence of Au in Pt affects charge transfer from Pt to Au because of the higher electronegativity of Au (2.54) in relation to Pt (2.2).[152–154] The Au $4f_{7/2}$ binding energy obtained from the XPS (83.88 eV for Au foil and 83.50 eV for Pt–Au catalyst) supports this argument. The reduction in binding energy and Full Width at Half Maximum (FWHM) values for Au in Pt–Au alloy which is attributed to the higher electronegativity of gold in relation to Pt results in the increase of d-orbital vacancies in Pt metal. Density Functional Theory (DFT) calculations on a small model cluster reflect electron transfer from Pt to Au within the alloy to be responsible for the synergistic promotion of the oxygen-reduction reaction on Pt–Au electrode. Au_{20} cluster is tetrahedral with its faces resembling the fcc (111) structure on nanoscale.[155] All the Au atoms in Au_{20} cluster were replaced by the Pt atoms to obtain pure Pt_{20} cluster. $Pt_{10}Au_{10}$ (1 : 1) and $Pt_{15}Au_5$ (3 : 1) alloy clusters were obtained by substitution of Au into Pt_{20} cluster. For the alloy composition, Pt atoms in Pt_{20} were replaced by Au atoms to realize $Pt_{10}Au_{10}$ (1 : 1) and $Pt_{15}Au_5$ (3 : 1) clusters with Au atoms at varying atomic positions leading to about 20 starting geometries for each composition. These clusters were then optimized to obtain the most stable Au–Pt alloy geometry for each composition. The electronic properties, such as energy gap between the Highest Occupied Molecular Orbital (HOMO) and Lowest Unoccupied Molecular Orbital (LUMO), atomic charges and charge density, were analyzed for resultant geometry. Subsequently, the results were compared with homogeneous Pt and Au clusters.

Methanol-oxidation activity for Pt/C catalyst is higher than all Pt–Au/C catalysts. Among the Pt–Au/C catalysts, (2 : 1) Pt–Au/C and (1 : 1) Pt–Au/C exhibit lower methanol-oxidation current. It is known that the methanol adsorption-dehydrogenation process requires at least three neighboring Pt atoms with appropriate crystallographic arrangement.[118] In the Pt–Au/C samples, the probability of finding three-neighboring Pt atoms on the surface decreases with increasing Au content in the alloy. Accordingly, the methanol oxidation current is smaller for (2 : 1) Pt–Au/C and (1 : 1) Pt–Au/C catalysts. DFT calculations reflect that, in DMFCs containing Pt–Au/C in 2 : 1 and 1 : 1 ratios, the probability of finding three neighboring Pt atoms on the surface is lower than in Pt and Pt–Au (3 : 1) making the former more methanol-tolerant in accordance with the experimental data.

Pt–Pd supported on carbon as a methanol tolerant cathode catalyst for DMFCs

Pt–Pd (40 wt%)/C catalyst was prepared by the formic acid method. In a typical synthesis, the required amount of Vulcan XC72R was dispersed in the aqueous solution of hexachloroplatinic acid, palladium chloride and formic acid accompanied with mild ultrasonication at room temperature (~25 °C) for 30 min. Subsequently, the solution was stored at room temperature for 72 h. After completion of the reduction reaction, the product was washed copiously with water and dried. Pt–Pd/C alloy catalysts containing Pt and Pd in varying atomic weight ratios, namely 1 : 1, 2 : 1 and 3 : 1, were prepared. Carbon-supported Pt was also prepared by the same procedure.

The diffraction peaks for Pt in Pt–Pd shifted to higher 2θ values compared to Pt in Pt/C as shown distinctly for (111) diffraction peaks for Pt–Pd/C and Pt/C samples. The shift, which is caused by the incorporation of Pd in the fcc structure of Pt, revealed the alloy formation between Pt and Pd.[156] The decrease in the lattice parameter and Pt–Pt inter-atomic distance in relation to Pt/C for all Pt–Pd catalyst further corroborated the alloy formation.

Electro-catalytic activities for Pt/C and Pt–Pd/C catalysts towards methanol oxidation have been studied by cyclic voltammetry as shown in Fig. 22(a) Pt/C, Pt–Pd(1 : 1)/C, Pt–Pd(2 : 1)/C and Pt–Pd(3 : 1)/C catalysts show two oxidation peaks to methanol oxidation and their intermediates produced during methanol oxidation. It is seen that the current densities due to methanol oxidation reaction on the Pt–Pd alloy catalysts are lower than that on the Pt/C catalyst, suggesting that the bimetallic catalysts for methanol oxidation are less active than pristine Pt catalyst. The current density due to the methanol oxidation reaction increases with Pt content within the Pt–Pd bimetallic catalysts. Methanol-oxidation activity for Pt/C catalyst is higher than Pt–Pd/C catalysts. This clearly establishes that the methanol oxidation activity for the Pt–Pd alloy catalyst is lower than the Pt/C catalyst. Methanol oxidation activity for various catalysts is found to be: Pt–Pd (1 : 1)/C < Pt–Pd (2 : 1)/C < Pt–Pd (3 : 1)/C < Pt/C. This indicates that on increasing Pd content with respect to Pt in Pt–Pd/C, methanol-oxidation activity of the catalyst tends to decrease. This is due to the inactivity of Pd towards MOR and, hence, the palladium-based catalyst shows a poor response towards methanol oxidation in acid solution.[157]

Most of the Pt–Pd/C catalysts have higher ORR activity than Pt/C, except for Pt–Pd (1 : 1)/C catalyst where Pd atoms preferentially occupy the surface sites hindering the access of oxygen molecules to adsorption on Pt sites. The onset potentials for ORR on Pt–Pd (1 : 1)/C, Pt/C, Pd (2 : 1)/C and Pt–Pd (3 : 1)/C are 0.899 V, 0.920 V, 0.948 V and 0.951 V *vs.* NHE, respectively. It is clearly seen that the onset potentials for ORR on Pt–Pd (2 : 1)/C and Pt–Pd (3 : 1)/C catalysts are more positive in relation to Pt–Pd (1 : 1)/C and Pt/C catalysts. The half-wave potentials for Pt–Pd (1 : 1)/C, Pt/C, Pd (2 : 1)/C and Pt–Pd (3 : 1)/C are 0.821 V, 0.841 V, 0.871 V and 0.881 V *vs.* NHE, respectively. The positive onset and half-wave potentials for Pt–Pd (2 : 1)/C and Pt–Pd (3 : 1)/C show improved catalytic activity. The improved ORR is due to the ready adsorption and easy dissociation of O_2 on the Pd-modified Pt surface in relation to pristine Pt.[158] Figure 22(c) shows ORR data on Pt/C and Pt–Pd/C alloy catalysts in the presence of 0.5 M CH_3OH as compared to that without methanol shown in Fig. 22(c). All the catalysts exhibit an increase in overpotential towards ORR in presence of methanol. The significant increase in ORR overpotential on Pt/C is due to the competition between oxygen reduction and methanol oxidation reactions. From the data in Fig. 22(c), it is clear that all Pt–Pd/C catalysts exhibit higher ORR activity than Pt/C catalyst. Among the various Pt–Pd/C catalysts, Pt–Pd (2 : 1)/C alloy catalyst exhibits the maximum ORR activity. As understood from the bi-functional mechanism for methanol oxidation on Pt surface, it would be desirable to block some or all the three Pt sites to impede methanol oxidation. However, the number of effective sites to be blocked may vary from catalyst to catalyst.

From cell polarization performance graph, it is clear that DMFCs comprising Pt–Pd/C cathode catalysts are superior in performance than the DMFC with Pt/C cathode catalyst. This is in agreement with the LSV data. The best fuel cell performance is observed for the Pt–Pd (2 : 1)/C cathode catalyst with a peak power density of 115 mW cm^{-2} in relation to the peak power density of 60 mW cm^{-2} observed for Pt/C. Since the anode catalyst and membrane for all the cases are kept identical, the enhanced performance obtained for the DMFC with the Pt–Pd/C catalyst is a clear manifestation of the improved ORR with accompanying methanol-tolerant behavior. The improvement in performance appears to be related to the specific structural configuration of the Pt–Pd catalyst. The addition of Pd atoms could induce an electronic modification in Pt by an alloying effect or surface-strain effect providing unique surface sites for the adsorption of O_2 molecules that help ameliorating electro-reduction of oxygen.[158] It is also noteworthy that for MOR to proceed, three Pt sites are initially required for the dissociative chemisorption of methanol. It appears that such a situation is hindered with Pt–Pd (2 : 1) catalyst.

Carbon-supported Pt encapsulated Pd nanostructure as methanol-tolerant oxygen reduction electrocatalyst

Pd/C nanoparticles were prepared by using different reducing agents. The average crystallite sizes calculated for Pd/C prepared using $NaBH_4$, ascorbic acid and formic acid from powder XRD patterns using Scherrer equation were 10.2, 17.4 and 14.2 nm, respectively. The results clearly showed that reducing agents play a crucial role on the crystallites size. ORR studies were performed to select the carbon-supported Pd nanostructure obtained using different reducing agents. Figure 23(a) shows comparative catalytic activities towards ORR for Pd/C prepared by different reducing agents. The onset potentials for ORR on Pd/C obtained using reducing agents $NaBH_4$, ascrobic acid and formic acid are 0.502, 0.521 and 0.597 V *vs.* SCE, respectively. It is clearly seen that the onset potential for ORR on Pd/C prepared by $NaBH_4$ reduction method is more positive in relation to Pd/C prepared using the other reducing agents revealing the higher catalytic activity of the former compared with the latter. Hence, we Pd/C obtained using $NaBH_4$ was selected, as a core for Pt. Pt modified Pd/C nanoparticles were prepared by a spontaneous displacement reaction following a procedure reported in the literature.[129]

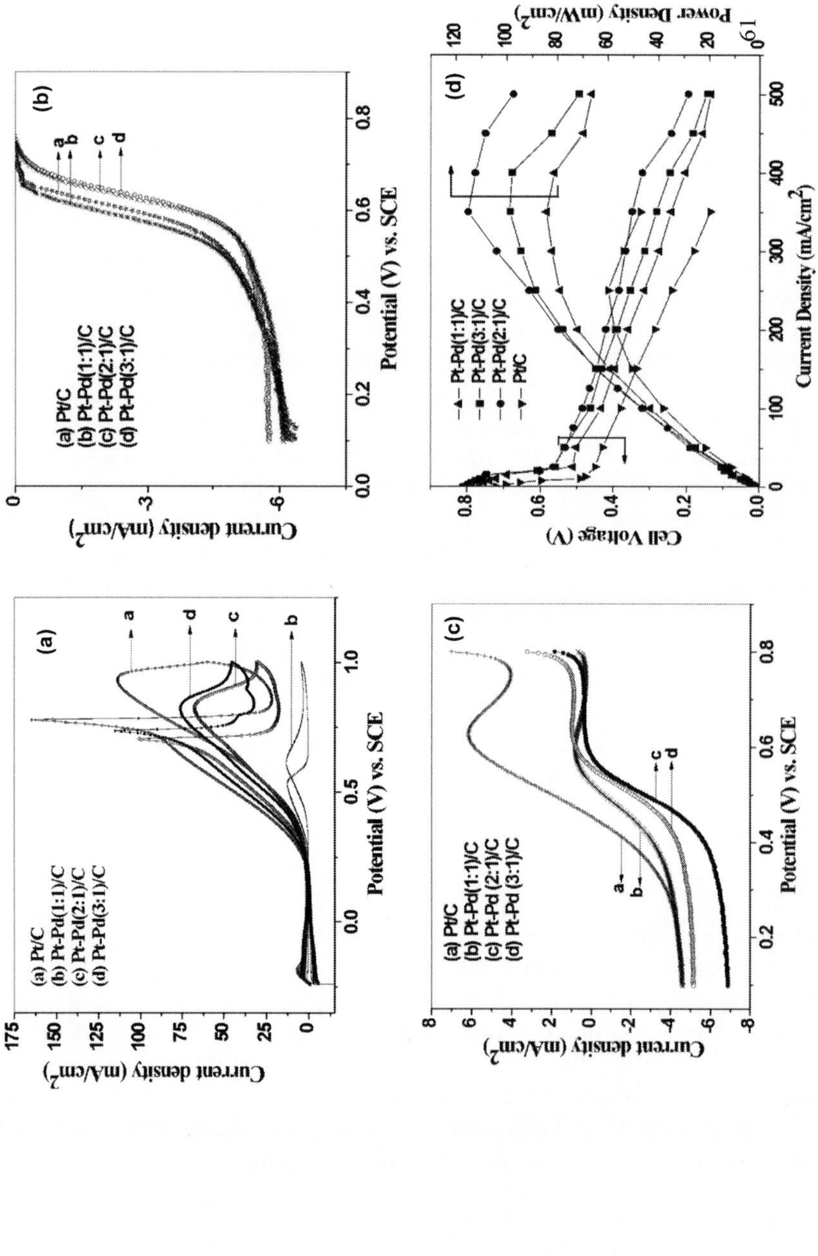

(a)
(a) Pt/C
(b) Pt-Pd(1:1)/C
(c) Pt-Pd(2:1)/C
(d) Pt-Pd(3:1)/C
Current density (mA/cm2)
Potential (V) vs. SCE
(b)
(a) Pt/C
(b) Pt-Pd(1:1)/C
(c) Pt-Pd(2:1)/C
(d) Pt-Pd(3:1)/C
Current density (mA/cm2)
Potential (V) vs. SCE
(c)
(a) Pt/C
(b) Pt-Pd(1:1)/C
(c) Pt-Pd (2:1)/C
(d) Pt-Pd (3:1)/C
Current density (mA/cm2)
Potential (V) vs. SCE
(d)
Pt-Pd(1:1)/C
Pt-Pd(3:1)/C
Pt-Pd(2:1)/C
Pt/C
Cell Voltage (V)
Power Density (mW/cm2)
Current Density (mA/cm2)

The diffraction peaks for Pt in Pt@Pd/C are shifted to higher 2θ values compared with Pt in Pt/C as shown distinctly for (200) diffraction peaks for Pt@Pd/C and Pt/C samples in Fig. 23(b). Encapsulation of Pd by Pt improves the electrocatalytic activity due to reduction in Pt–Pt inter-atomic distance and modified electronic and crystalline properties

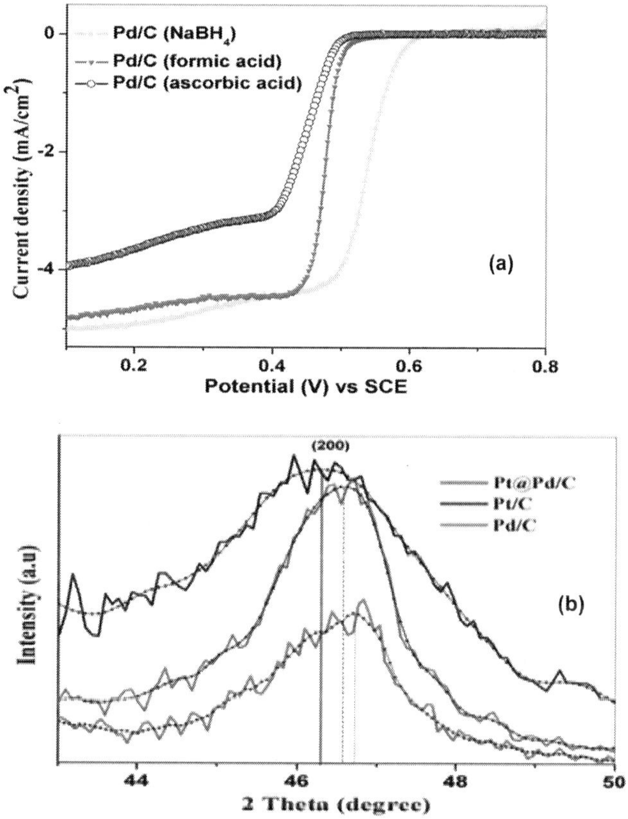

Fig. 23 (a) Voltammograms for Pd/C prepared using different reducing agents in 0.5 M $HClO_4$ solution saturated with pure oxygen at a scan rate of 1 mV s^{-1} (Electrode Rotation Rate: 1500 rpm); current densities are normalized to the geometric area. (b). Enlarged XRD(200) peaks of Pd/C, Pt/C and Pt@Pd/C. Reproduced from K. G. Nishanth, P. Sridhar and S. Pitchumani, Carbon-supported Pt encapsulated Pd nanostructure as methanol-tolerated oxygen reduction electro-catalyst, *Int. J. Hydrogen Energy*, **38**, 612–619. Copyright 2013 with permission from Elsevier B.V.

Fig. 22 (a) Cyclic voltammograms for Pt/C and Pt–Pd/C catalysts containing Pt and Pd in varying atomic ratios in N_2-saturated aqueous solution containing 0.5 M $HClO_4$ and 0.5 M CH_3OH at a scan rate of 50 mV s^{-1}. (b) LSVs of Pt/C and Pd–Pt/C alloy catalysts containing Pt and Pd in varying atomic ratios in 0.5 M $HClO_4$ saturated with pure oxygen at a scan rate of 1 mV s^{-1} (Electrode Rotation Rate: 1500 rpm); current densities are normalized to the geometric area. (c) LSVs for Pt/C and Pt–Pd/C alloy catalysts containing Pt and Pd in varying atomic ratios in aqueous solution containing 0.5 M $HClO_4$ and 0.5 M CH_3OH saturated with pure oxygen. The other conditions are identical to data presented in Fig. 22(b). (d) Steady-state performance data for DMFCs with Pt/C and Pt–Pd/C cathodes. Reproduced with permission from K. G. Nishanth, P. Sridhar, S. Pitchumani and A. K. Shukla., A DMFC with Methanol-Tolerant-Carbon-Supported-Pt-Pd-Alloy Cathode, *J. Electrochem. Soc.*, 2011, **158**(8), B871–B876, Copyright 2011 The Electrochemical Society.

of the Pt skin layer. The lattice parameter and the interatomic distance of Pt in Pt@Pd/C nanoparticle are 0.38885 nm and 0.27491 nm, respectively. They are smaller than the measured lattice parameter (0.3917 nm) and interatomic distance (0.27796 nm) of Pt in Pt/C and higher than the lattice parameter (0.38801 nm) and interatomic distance (0.27438 nm) of Pd in Pd/C.

The charge of a monolayer of adsorbed hydrogen is difficult to determine for palladium in contrast with platinum, due to its ability to absorb hydrogen. Hence, the electrochemical surface area (ECSA) is calculated from the charge of the monolayer of chemisorbed oxygen following the procedure reported by Alvarez *et al.*[159] The charge corresponding to the coverage of the oxygen monolayer (Q_O) for a smooth palladium electrode is twice that of the hydrogen monolayer (Q_H) on platinum, $Q_H = 1/2\ Q_O$ ($Q_H = 210\ \mu C\ cm^{-2}$).[159] The ECSA are 82.5 $m^2\ g^{-1}$ Pt and 92 $m^2\ g^{-1}$ metal (367.7 $m^2 g^{-1}$ Pt) for Pt/C and Pt@Pd/C, respectively. The hydrogen adsorption/desorption features of Pt@Pd/C are noticeably different from that of Pt/C as seen in Fig. 24(a), which could be due to the changes in the adsorption site geometry.[160] In the cathodic scan in Fig. 24(b), the oxide reduction peak on Pt@Pd/C is at 0.539 V, which is higher than 0.503 V that on Pt/C. The shift of the oxide reduction peak by 36 mV on Pt@Pd/C suggests weaker binding energy of the OH_{ads} species on the surface of the Pt@Pd/C.[160,161]

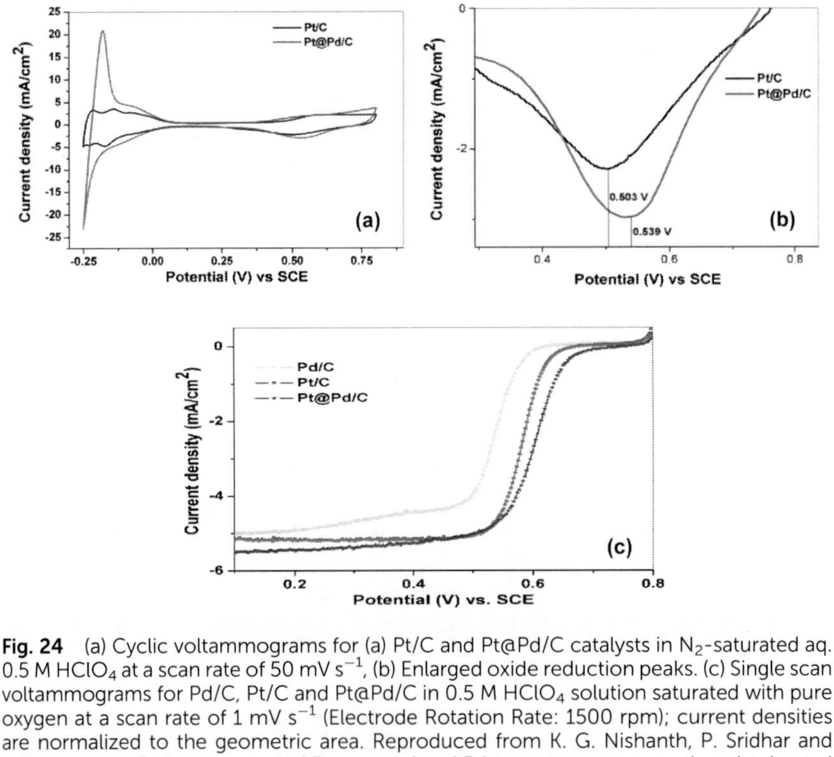

Fig. 24 (a) Cyclic voltammograms for (a) Pt/C and Pt@Pd/C catalysts in N_2-saturated aq. 0.5 M $HClO_4$ at a scan rate of 50 mV s^{-1}, (b) Enlarged oxide reduction peaks. (c) Single scan voltammograms for Pd/C, Pt/C and Pt@Pd/C in 0.5 M $HClO_4$ solution saturated with pure oxygen at a scan rate of 1 mV s^{-1} (Electrode Rotation Rate: 1500 rpm); current densities are normalized to the geometric area. Reproduced from K. G. Nishanth, P. Sridhar and S. Pitchumani, Carbon-supported Pt encapsulated Pd nanostructure as methanol-tolerated oxygen reduction electro-catalyst, *Int. J. Hydrogen Energy*, **38**, 612–619. Copyright 2013 with permission from Elsevier B.V.

Fig. 24(c) provides comparative ORR activities for Pd/C, Pt/C and Pt@Pd/C catalysts under similar experimental conditions. The onset potential for ORR on monometallic Pd/C and Pt/C catalyst are 0.597 V and 0.650 V, respectively. The onset potential of 0.685 V for Pt@Pd/C is positive in relation to Pd/C and Pt/C by 88 mV and 35 mV, respectively. The half-wave potentials for Pd/C, Pt/C and Pt@Pd/C are 0.528, 0.581 and 0.611 V *vs.* SCE, respectively. From the data, it is clear that the half-wave potential for Pt@Pd/C is positive than that of Pd/C and Pt/C by 83 mV and 30 mV, respectively. The positive onset and half-wave potentials for Pt@Pd/C compared with Pt/C and Pd/C reveal the improved catalytic activity. The mass activities for Pt/C and Pt@Pd/C at 0.605 V *vs.* SCE (0.85 V *vs.* NHE) are 11.17 and 48.68 Ag-1Pt, respectively. The highest mass activity for Pt@Pd/C in relation to conventional Pt/C is a clear manifestation of the more resourceful use of Pt. It has been reported that OH_{ads} is an intermediate of the ORR, and its strong adsorption on the catalyst surface has a negative impact on the performance of Pt-based catalysts.[160,162] The weaker binding energy of the OH_{ads} species on the surface of the Pt@Pd/C as discussed with reference to the results presented in Fig. 24(c) is definitely a contributing factor to the enhanced ORR activity of this catalyst. Furthermore, modification of the electronic structure could also affect the catalytic behavior of the materials. The electronic structure of Pt deposited on the Pd/C surface is different from that of bulk Pt because of the so-called strain and ligand effects of the core substrate. The weaker OH_{ads} binding could be attributed to the interaction between Pt and Pd resulting in favorable electronic and stress effects.[129,160,161] Pt (4f) XPS spectra for Pt@Pd/C and Pt/C catalysts are given in Fig. 25(a and b), respectively. The Pt (4f 7/2, 5/2) doublets in Pt@Pd/C (Fig. 25a) at 71.56 eV and 74.95 eV, and 72.68 eV and 76.96 eV have been assigned to Pt^0 and Pt^{2+}, respectively. For Pt/C (Fig. 25b), the Pt (4f 7/2, 5/2) peaks at 71.1 eV and 74.40 eV, and 72.56 eV and 76.52 eV have been assigned to Pt^0 and Pt^{2+}, respectively. In case of the Pt@Pd/C, 4f signals (Fig. 25a) have shifted positively in relation to Pt/C, which is due to the compressive strain introduced by depositing Pt with a larger lattice parameter over a Pd substrate with a smaller lattice parameter.

The XPS data provided the evidence for the electronic effects originating from Pt atoms withdrawing electrons from the neighboring Pd atoms in Pt@Pd/C.[160,161] The more complete filling of the Pt 5d band of Pt would repel OH_{ads}. The facilitated removal of OH_{ads} therefore benefited the ORR kinetics. Deposition of Pt with its larger lattice constant 0.3917 nm on Pd with a smaller lattice constant 0.38801 nm resulted in lattice mismatch with consequent strain in the Pt surface layer, which shifted the d-band center downward leading to a weaker interaction with the adsorbed species and hence a weaker chemisorptive bond.[160,161]

The I–V characteristics of ORR on Pt/C and Pt@Pd/C in pure $HClO_4$ solution and methanol-containing electrolyte. As shown in Fig. 26a, there is a peak at +0.556 V in which the ORR current is masked by the methanol oxidation current for Pt/C in presence of methanol. The observed behavior is caused by the competition between methanol oxidation and oxygen reduction reaction. However, the ORR activity of the Pt@Pd/C (Fig. 26b) catalyst is much less affected in the presence

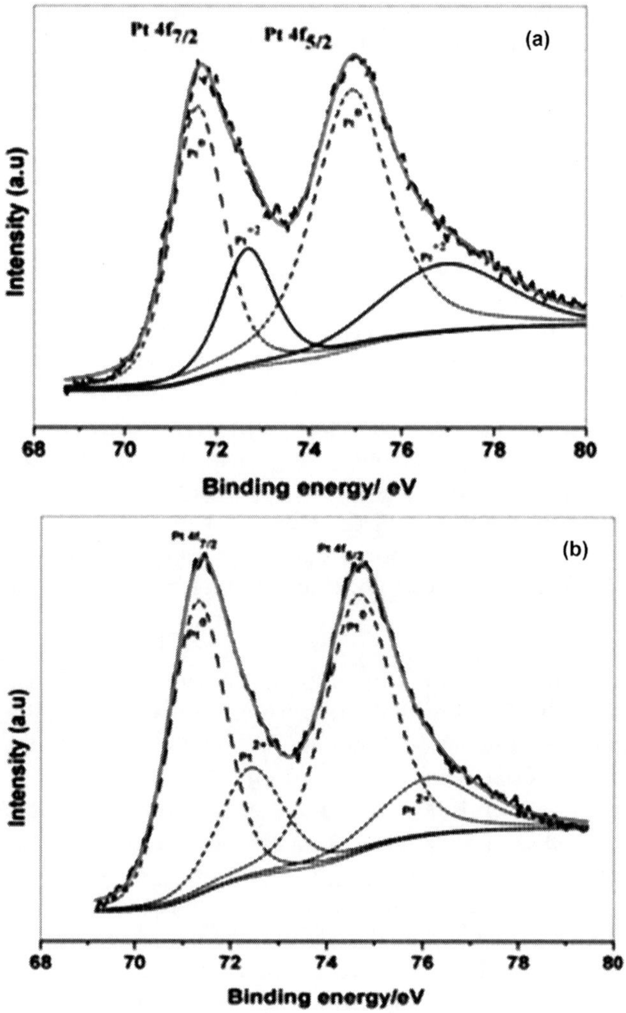

Fig. 25 Pt (4f) XPS spectra for (a) Pt@Pd/C and (b) Pt/C catalysts. Reproduced from K. G. Nishanth, P. Sridhar and S. Pitchumani, Carbon-supported Pt encapsulated Pd nanostructure as methanol-tolerated oxygen reduction electro-catalyst, *Int. J. Hydrogen Energy*, **38**, 612–619. Copyright 2013 with permission from Elsevier B.V.

of methanol, which clearly shows the potential use of this catalyst as a methanol tolerant oxygen reduction catalyst in DMFCs. The improved methanol tolerance is due to the presence of Pd, which is inactive for methanol electro-oxidation in acidic medium.

Durability of Pt@Pd/C catalysts is evaluated by cycling the electrode potential between −0.244 and 0.8 V *vs.* SCE for 1000 cycles. CV and LSV are recorded before and after the durability study. It can be seen from CV data presented in Fig. 26(c) that there is no noticeable change in the voltammogram before and after durability study, which clearly shows the stability of the electro-catalyst. The LSV data presented in Fig. 26(d), suggests that the ORR activity shows excellent reproducibility even after the durability study. The LSV results corroborate with CV study.

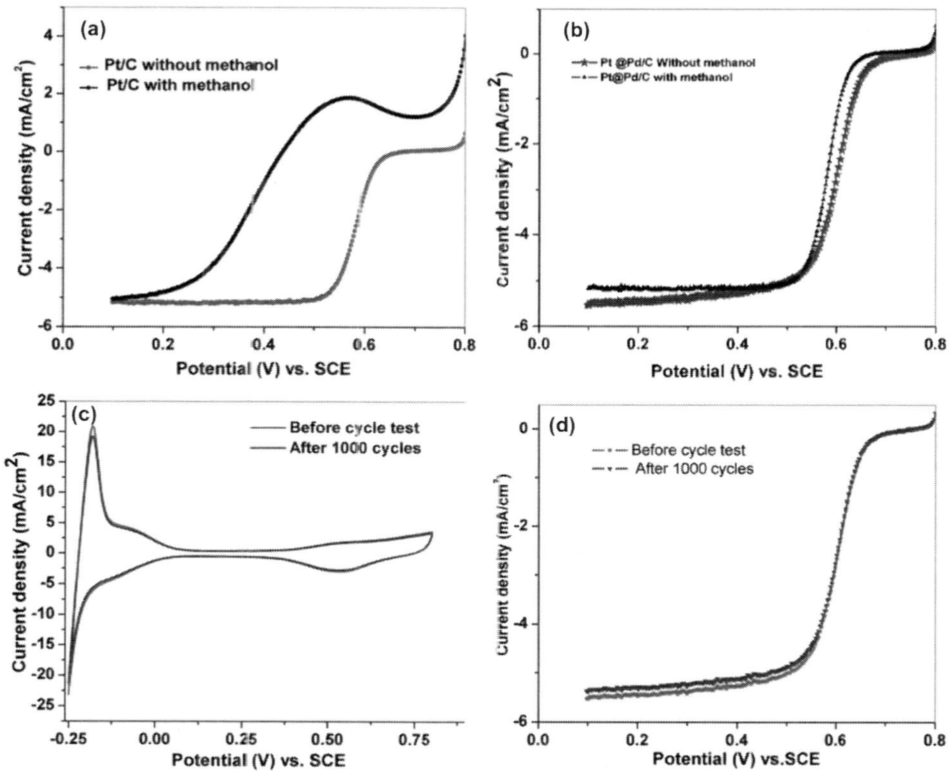

Fig. 26 Polarization curves for (a) Pt/C and (b) Pt@Pd/C catalysts in O_2-saturated 0 5 M $HClO_4$ with and without 0.5 M CH_3OH at a scan rate of 1 mV s^{-1} (Electrode Rotation Rate: 1500 rpm); current densities are normalized to the geometric area. (c) CVs of the Pt@Pd/C catalyst before and after 1000 potential cycles. (d) ORR polarization curves for the Pt @Pd/C catalyst before and after cycling. Reproduced from K. G. Nishanth, P. Sridhar and S. Pitchumani, Carbon-supported Pt encapsulated Pd nanostructure as methanol-tolerated oxygen reduction electro-catalyst, *Int. J. Hydrogen Energy*, **38**, 612–619. Copyright 2013 with permission from Elsevier B.V.

Carbon-supported Pt–TiO_2 as a methanol-tolerant oxygen-reduction catalyst for DMFCs

A sol–gel route was adopted to prepare Pt–TiO_2 nano-composite catalyst. In order to prepare titania sol, 10 μL of 0.5 M aq H_2SO_4 was added to titanium iso-propoxide solution and agitated for 1 h. Finally, the required quantity of carbon-supported Pt (40 w/o from Alfa Aesar) was added with continuous stirring. The resultant mixture was left for 24 h to form gel that was heated at 80 °C for 1 h to remove all volatile substances in it. The resulting cake was pulverized and subjected to heat treatment at 750 °C in a flowing mixture of 90% N_2-10% H_2 for 5 h and cooled to room temperature. For comparison, the commercial 40 w/o Pt/C (Alfa Aesar) was also subjected to heat treatment under similar conditions. In the following text, Pt–TiO_2/C samples heated at 750 °C is represented as 750 Pt–TiO_2/C. The carbon-supported platinum heat treated at 750 °C is represented as 750 Pt/C and as-received platinum catalyst is represented as Pt/C. Pt–TiO_2/C with varying atomic ratios of Pt to Ti, namely 3 : 1, 2 : 1 and 1 : 1, are denoted as Pt–TiO_2/C (3 : 1), Pt–TiO_2/C (2 : 1) and Pt–TiO_2/C (1 : 1), respectively, and the heat-treated samples are represented by the heat-treatment temperature followed by the sample representation, for example Pt–TiO_2/C (2 : 1) heat-treated at 750 °C is represented as 750 Pt–TiO_2/C (2 : 1).

Prior to evaluation of 750 Pt–TiO_2/C catalysts for DMFC cathode, the ability of the catalyst is assessed for methanol oxidation in half-cell mode. The methanol oxidation ability of Pt/C and 750 Pt–TiO_2/C with varying Pt to Ti atomic compositions, namely 3 : 1, 2 : 1 and 1 : 1, is investigated by CV in 0.1 M H_2SO_4 + 0.5 M CH_3OH solution and the corresponding data are shown in Fig. 27; the CV of bare GC electrode is also included. It is evident that the substrate has no catalytic activity for methanol oxidation. However, both Pt/C and 750 Pt–TiO_2/C with varying atomic ratios of Pt to Ti on glassy-carbon electrode show two oxidation peaks, which are related to the oxidation of methanol and the corresponding intermediates produced during the methanol oxidation. The methanol oxidation currents for Pt/C and 750 Pt/C are higher in relation to 750 Pt–TiO_2/C with varying atomic ratios of Pt to Ti. Among the 750 Pt–TiO_2/C catalysts, 750 Pt–TiO_2/C (2 : 1) and 750 Pt–TiO_2/C (1 : 1) show lesser methanol oxidation current in relation to 750 Pt–TiO_2/C (3 : 1).

The cell polarization data for 750 Pt–TiO_2/C cathodes with varying Pt to Ti atomic ratios in a methanol/O_2 DMFC are compared with the DMFCs employing Pt/C and 750 Pt/C cathodes in Fig. 27(b). It is seen that the DMFC with 750 Pt–TiO_2/C (2 : 1) cathode performs better in relation to the DMFCs comprising 750 Pt–TiO_2/C (3 : 1), 750 Pt–TiO_2/C (1 : 1) and Pt/C cathodes. DMFC comprising 750 Pt–TiO_2/C (2 : 1) cathode shows enhanced peak power-density of 180 mW cm^{-2} in relation to the peak power-density of 80 mW cm^{-2} for the DMFC with Pt/C cathode. Since the anode in all DMFCs is identical, the enhanced performance for the DMFC with 750 Pt–TiO_2/C (2 : 1) cathode is clearly due to enhanced ORR with decreased methanol oxidation on the catalyst. These data corroborate the CV data presented in Fig. 27(a).

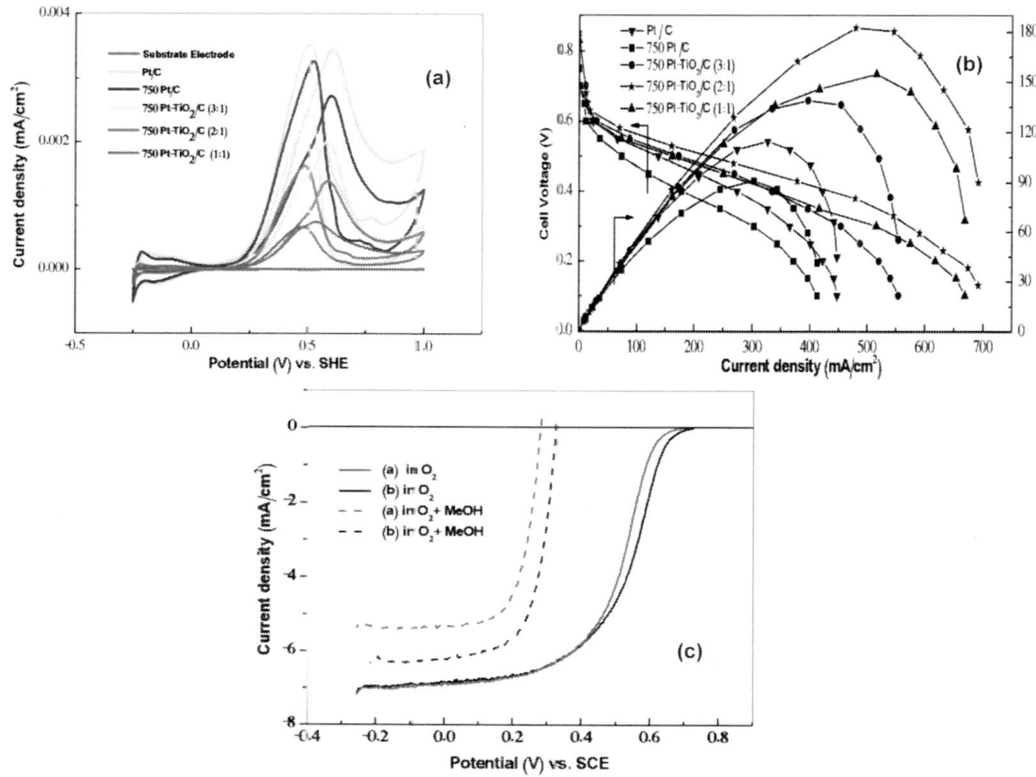

Fig. 27 (a) Cyclic voltammograms for Pt/C and 750 Pt–TiO_2/C catalysts with varying Pt to Ti atomic ratios in N_2-saturated aq 0.5 M $HClO_4$ + aq. 0.5 M CH_3OH solution with scan rate 50mV s^{-1}. (b). Steady-state performance data of DMFCs (CH_3OH & O_2) for Pt/C, 750 Pt/C and 750 Pt–TiO_2/C with varying Pt to Ti atomic ratios. (c) Linear sweep voltammetry data for ORR on (a) 750 Pt/C and (b) 750 Pt–TiO_2/C (2 : 1) catalysts in O_2 saturated aq 0.5 M $HClO_4$ in presence and absence of methanol at 1 mV s^{-1} scan rate (Electrode rotation rate: 1200 rpm) Reproduced with permission from G. Selvarani, S. Maheswari, P. Sridhar, S. Pitchumani and A. K. Shukla, Carbon-Supported Pt–TiO_2 as a Methanol-Tolerant Oxygen-Reduction Catalyst for DMFCs, *J. Electrochem. Soc.*, 2009, **156**(11), B1354–B1360. Copyright 2009, The Electrochemical Society.

Enhancement of electro-catalytic activity could be due to change in electronic structure of the catalyst arising due to the electronic interactions between Pt and TiO_2 due to the hypo d-electron character of TiO_2 and hyper d-electron character of Pt.[137]

The ORR activity data for Pt/C and 750 Pt–TiO_2/C (2 : 1) cathodes at 1200 rpm are shown in Fig. 27(c). The enhanced electro-catalytic activity for Pt–TiO_2/C cathode can be explained by an electronic factor, namely the change of the d-band vacancy in Pt, and by the geometric effect.[118] Both effects may enhance the reaction rate for oxygen adsorption and cleavage of O–O bond during ORR. Interestingly, the diffusion current (i_d) values for both Pt/C and 750 Pt–TiO_2/C (2 : 1) are almost similar indicating the ORR mechanism on both the catalysts to be same. The performance data for 750 Pt–TiO_2/C (2 : 1) and Pt/C cathodes towards ORR in presence of methanol are also presented in Fig. 27(c). 750 Pt–TiO_2/C (2 : 1) cathode exhibits higher ORR activity with minimum methanol oxidation. These data clearly indicate a higher overpotential for ORR at any current density in the presence of methanol for Pt/C cathode in relation to 750 Pt–TiO_2/C (2 : 1) cathode. The significant increase in the overpotential for ORR on Pt/C and 750 Pt–TiO_2/C (2 : 1) cathodes in presence of methanol reflects competing oxygen reduction and methanol oxidation reactions. These studies confirm that 750 Pt–TiO_2/C (2 : 1) cathode has higher ORR selectivity and better methanol tolerance in relation to Pt/C cathode.

Enhanced oxygen reduction reaction activity through spillover effect by Pt–$Y(OH)_3$/C catalyst in direct methanol fuel cells

Carbon-supported Pt–$Y(OH)_3$ (40 wt%) catalysts containing Pt and Y in varying atomic weight ratio, namely 1 : 1, 2 : 1 and 3 : 1 were prepared by employing the conventional $NaBH_4$ reduction method. In a typical synthesis, $H_2PtCl_6 \cdot 6H_2O$ and $Y(NO_3)_3 \cdot 6H_2O$ were used as the Pt and Y precursors, respectively.

In order to substantiate the composition effect toward methanol electro-oxidation for various Pt–$Y(OH)_3$/C and Pt/C, cyclic voltammetry in an aqueous solution containing 0.5 M $HClO_4$ and 0.5 M aq. CH_3OH at a scan rate of 50 mV s^{-1} was conducted as shown in Fig. 28(a). From the data it is clear that all the catalysts deposited on glassy carbon electrode show two oxidation peaks corresponding to methanol oxidation and their intermediates produced during methanol oxidation. It is seen that the current densities due to methanol oxidation reaction on the Pt–$Y(OH)_3$/C catalysts are lower than that on the Pt/C catalyst, suggesting that the hybrid catalysts for methanol oxidation are less active than pristine Pt catalyst. The current densities due to the methanol oxidation reaction increases with Pt content within the Pt–$Y(OH)_3$ catalysts. In addition to that the positive onset potential for all Y based catalyst in relation to Pt/C is emphasizing the less active behavior of the hybrid catalyst towards methanol electro-oxidation. This clearly establishes that the methanol oxidation activity for the Pt–$Y(OH)_3$ catalyst is lower than the Pt/C catalyst. It would be desirable to block some or all the three Pt sites to

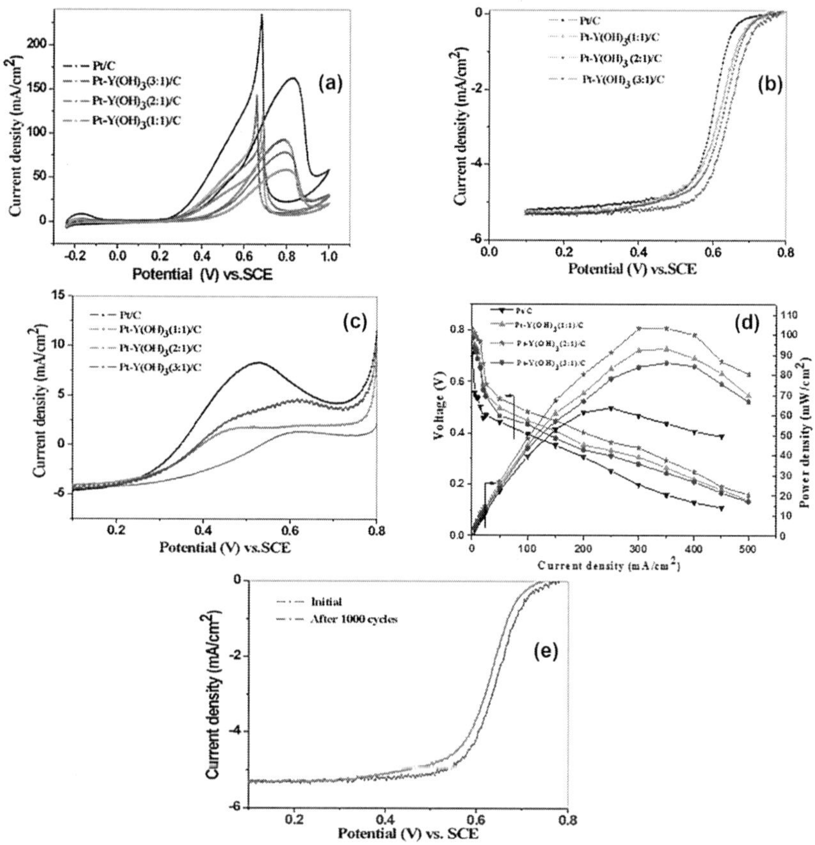

Fig. 28 (a) Cyclic voltammograms for Pt/C and Pt–Y$(OH)_3$ catalysts containing Pt and Y in varying atomic ratios in N_2-saturated aqueous solution containing 0.5 M $HClO_4$ and 0.5 M CH_3OH at a scan rate of 50 mV s^{-1} (b) LSVs of Pt/C and Pt–Y$(OH)_3$/C catalysts containing Pt and Y in varying atomic weight ratios in 0.5 M $HClO_4$ solution saturated with oxygen at a scan rate of 1 mV s^{-1}. (Electrode Rotation Rate: 1500 rpm) (c) LSVs in aqueous solution containing 0.5 M $HClO_4$ and 0.5 M CH_3OH saturated with pure oxygen (d) Steady-state performance data for DMFCs with Pt/C and Pt–Y $(OH)_3$/C cathodes. (e) LSVs for Pt–Y$(OH)_3$(3 : 1)/C catalyst before and after 1000 cycles. Reproduced from K. G. Nishanth, P. Sridhar and S. Pitchumani, Enhanced oxygen reduction reaction activity through spillover effect by Pt-Y(OH)3/C catalyst in direct methanol fuel cells, *Electrochem. Commun.*, **13**, 1465–1468. Copyright 2011 with permission from Elsevier B.V.

impede methanol oxidation. However, the number of effective sites to be blocked may vary from catalyst to catalyst.

From the Linear sweep voltammograms for commercial Pt/C and Pt–Y$(OH)_3$/C with Pt and Y in varying atomic ratio for ORR using an RDE at 1500 rpm presented in Fig. 28(b), it is clear that all the Pt–Y$(OH)_3$/C catalysts display much higher ORR activity than Pt/C, and maximum for Pt–Y$(OH)_3$(3 : 1)/C. The half-wave potential of all Pt–Y$(OH)_3$/C catalysts comprising Pt and Y in varying atomic ratios of 1 : 1, 2 : 1 and 3 : 1 show a positive shift of 0.023, 0.033 and 0.060 V, respectively compared to Pt/C. In addition to that, the onset potential for all the Pt–Y$(OH)_3$/C catalysts comprising Pt and Y in varying atomic ratios of 1 : 1, 2 : 1 and 3 : 1 show a positive shift of 0.035, 0.046 and 0.075 V, respectively. It is interesting to note that the positive half-wave potential shift for 3 : 1 is similar to that reported for bulk polycrystalline Pt–Y alloy prepared through ultra high vacuum sputtering.[163] The data show that all Pt–Y$(OH)_3$/C catalysts exhibit significant improvement in the ORR activity over Pt/C.

The improved ORR activity may be due to the dynamic spillover of primary oxides as described in the literature.[147–150] It has been shown that the ORR implies continuous source of the primary oxide to be catalytically promoted. In the case of $Y(OH)_3$ because of its altervalent character, OH-ions are easily transferable within the overall spillover mechanism as follows,

$$Y(OH)_3 + M \rightarrow Y(OH)_2{}^+ + M\text{-}OH + e^-$$

$$Y(OH)_2{}^+ + 2H_2O \rightarrow Y(OH)_3 + H_3O^+$$

$$\sum M + 2H_2O \rightarrow M\text{-}OH + H_3O^+ + e^-$$

The I–V characteristics of ORR on Pt/C and Pt–Y$(OH)_3$/C in methanol-containing electrolyte. As compared to the ORR in pure $HClO_4$ solution, all the catalysts showed an increase in overpotential for ORR in presence of methanol. The significant increase in ORR overpotential on Pt/C is due to the competitive reaction between oxygen reduction and methanol oxidation. From the Fig. 28(c), it is clear that all Pt–Y$(OH)_3$/C catalysts exhibit much higher ORR activity than Pt/C catalyst. Among the various Pt–Y$(OH)_3$/C catalysts, Pt–Y$(OH)_3$(2 : 1)/C hybrid catalyst exhibits the maximum ORR activity in presence of methanol. It is known that methanol adsorption-dehydrogenation process requires at least three neighboring Pt atoms with appropriate crystallographic arrangement.[118] It appears that such a situation is hindered with Pt–Y$(OH)_3$(2 : 1) catalyst.

The cell polarization data for DMFCs with Pt–Y$(OH)_3$(1 : 1)/C, Pt–Y$(OH)_3$(2 : 1)/C, Pt–Y$(OH)_3$(3 : 1)/C and Pt/C cathode are shown in Fig. 28(d). It is clear that DMFCs comprising Pt–Y$(OH)_3$/C cathode catalysts are superior in performance than the DMFC with Pt/C cathode catalyst. This is in agreement with the LSV data. A maximum peak power density of 105 mW cm^{-2} is obtained for the Pt–Y$(OH)_3$(2 : 1)/C cathode catalyst in relation to the peak power density of 64 mW cm^{-2} for Pt/C. Since anode catalyst and the membrane electrolyte are kept identical during the study, the enhanced performance obtained for the DMFC with the Pt–Y$(OH)_3$/C catalyst is a clear manifestation of the improved ORR with accompanying methanol-tolerant behaviour. performance data for DMFCs with Pt/C and Pt–Y $(OH)_3$/C cathodes.

The durability study of the high-performing catalyst, Pt–Y$(OH)_3$(3 : 1)/C, is presented in Fig. 28(e). The shift-in half-wave potential is only 16 mV compared to 85 mV reported for Pt/C.[164] It clearly shows that Pt–Y$(OH)_3$/C catalyst has better stability than Pt/C.

Methanol-tolerant ORR catalysts in ADMFCs

An alkaline fuel cell using an anion exchange membrane as an electrolyte is called alkaline polymer electrolyte membrane fuel cell (APEMFC)[165] and APEMFCs using methanol as fuel is referred as alkaline direct methanol fuel cells (ADMFCs).[166,167] Methanol crossover in ADMFCs may not be completely eliminated though it is lesser than that in DMFCs.[168] Conventional carbon-supported platinum-based materials so far have been the most active, efficient, relevant and successful electrocatalysts for electrochemical devices at the current technology stage in acidic medium. However, in alkaline electrolyte, a wide range of non-noble metals and their oxides are stable enough for practical applications due to lower corrosion than in acidic electrolyte. This allows the investigation of a large number of non-Pt (transition metal) catalysts, especially with regard to ORR and methanol tolerance (*i.e.*, an oxygen reduction reaction selective electrocatalyst) characteristics in alkaline medium.[169] Very high ORR catalytic activity of Pd in alkaline solution compared with Pt has been reported.[170–173] In recent years, Pd and Pd alloys have attracted greater attention as non-Pt catalysts because of the high proportion of surface to bulk atoms and about 50 times more abundance of Pd than Pt.[174] Recent developments in using Pd and Pd-based alloys as ORR catalysts have shown promise for their use in APEMFCs.[175] The main objective of Pd-based catalyst research is the development of synthesis methods resulting in high catalytic activity of Pd. Metal based alloys have been found to exhibit higher electrocatalytic activities than pure metal.[176] Bard *et al.* suggested that for Pd–M alloys, the metal M constitutes the site for breaking the O–O bonds.[177] Incidentally, the addition of a second metal such as Co, Fe to Pd decreased the metal particle size. So alloying of Pd with transition metals (Fe, Co, Cr, Ni, Ag, *etc.*) is of significance for improving the ORR activity due to the changes in Pd–Pd bond distance, modification of electron configuration and alteration of surface species and composition. According to Wang and Balbuena, if a catalyst consists of two metals, one with a low occupancy of d-orbitals (such as Co, Ni, Cr, or V) and the other with fully occupied d-orbitals (such as Pd, Au, and Ag), the d-orbital coupling effect between them can significantly decrease the Gibbs free energy of the electron transfer steps in ORR, resulting in an enhancement in ORR kinetics.[178] Fernandez explained that, thermodynamically, a PdCo alloy is the most promising binary system compared with AuCo and AgCo in regard to the catalytic activity.[177] In low pH media, dissolution of the transition-metal-alloyed catalysts is a major drawback because these transition metals are electrochemically soluble at a potential range of 0.3–1 V *vs.* NHE.[130] In acidic medium, it is found through XPS studies that Co dissolves as cobalt oxides, while in alkaline medium it is stable.[179] Zhang *et al.* have reported that the optimal heat-treatment temperature for the synthesis of Pd–Co alloys is 300 °C.[180] APEMFCs comprising PdCo(3 : 1)/C as cathode catalyst show superior performance in relation to PdCo(2 : 1)/C, PdCo(1 : 1)/C and Pd/C, which is in agreement with the kinetic data obtained by half-cell studies. The ORR studies with methanol show that PdCo(3 : 1)/C has higher methanol tolerant behavior and better selectivity compared with other catalysts.[181]

Catalysts based on metal–transition metal oxide are known to improve the ORR activity and methanol tolerance. The hypo d-electron transition metal oxides have high chemical-stability, good corrosion-resistance, and strong metal-support interaction, which can enhance the activity of the catalysts. They, as catalyst supports, are able to stabilize and disperse adequately a number of active phases, as well as to retain a high surface area.[182] Among different metal oxides, TiO_2 is attractive due to its promising electrochemical properties, chemical stability, non-toxic nature and when combined with carbon as catalyst support shows better catalytic activity than metal supported on TiO_2[183] due to poor conductivity of the latter compared with carbon. Further, Pt–TiO_2/C shows improved stability in PEMFCs compared with Pt/C.[184] Recently, F. Hu *et al.* have shown better ethanol oxidation and stability for Pd supported on carbonized TiO_2 nanotube in alkaline medium compared with Pd/C.[185] Nguyen *et al.* investigated the ethanol oxidation on Pd/TiO_2 and Pd/Ti_nO_{2n-1} in alkaline medium and found Pd/Ti_nO_{2n-1} to be suitable. But TinO_{2n-1} showed lower electronic conductivity than carbon[186] Mentus *et al.* have reported higher ORR activity for Pd/TiO_2 in alkaline medium than in acidic medium.[187] Pioneering work on ORR activity of Pt–TiO_2 in alkaline medium was reported by of Temmeveski *et al.*[188] Only a few studies have been performed on Pd–TiO_2/C in alkaline medium, however, they do not cover the methanol tolerance behavior and also the optimum requirement of Ti. Pd/TiO_2/C catalysts with varying atomic ratios (1 : 1, 2 : 1 and 3 : 1) of Pd to Ti were prepared and evaluated for ORR in the presence of methanol. The ORR studies with methanol shows that Pd–TiO_2(3 : 1)/C has higher methanol tolerance and better selectivity.[189]

There are three ways to obtain higher methanol tolerance (i) Methanol adsorption reduction that decreases both the mixed potential and CO poisoning, (ii) Lowering of CO oxidation, which reduces the mixed potential but does not reduce CO poisoning and (iii) CO oxidation enhancement that reduces CO poisoning but increases the mixed potential.[190] The most effective way is through reduced methanol adsorption, the so-called "ensemble effect", which is dilution of the active component with catalytically inert metals by alloying and affecting the distribution of active sites.

Non-noble metal catalysts for methanol tolerance reported in the literature are macrocyclic derivatives such as porphyrins and phthalocyanines of transition-metal compounds like Co and Fe and chalcogenide materials. Ruthenium based chalcogenides showed higher methanol tolerance compared with macrocyclic derivatives.[191] These electrocatalysts have shown nearly the same activity for ORR in the absence as well as in the presence of methanol but have shown poor electrocatalytic activity in relation to dispersed Pt in methanol free electrolytes.

A number of Ru-based chalcogenides (RuxZy; Z = S, Se, and Te) have been synthesized as cathode catalysts and tested for ORR in alkaline and acidic medium.[192,193] Among these, Ru-based chalcogenide catalysts, RuSex, have attracted much attention because of their activity for oxygen reduction combined with their complete inactivity for methanol oxidation,[191,193–195] which makes them interesting for use as cathode material in the ADMFC. The activity enhancement provided by the Se atoms is related to an interfacial effect resulting from the binary structure of the catalysts. In these systems, Ru atoms act as electrochemically active sites for ORR; Se atoms play a role in chemically stabilizing the Ru centres against oxidation, leading to improved catalytic activity and stability.[193] However, with and without methanol, these catalysts did not match the catalytic activity of carbon supported Palladium for ORR. Unlike in acid media, ORR and methanol oxidation activity of Pd in alkaline media is remarkably higher than that of Pt.[173,196,197]

Ternary electro-catalysts may be more active than their original mono and bimetallic counterparts, as the addition of a third component may modify the bifunctional, electronic or structural characteristics of the mono and bimetallic catalyst surface.[198,199]

Prakash and Joachin[200] reported that oxygen reduction on ruthenium electrode proceeds primarily by a direct 4-electron reduction pathway without producing significant amounts of solution phase peroxide in alkaline medium. Anastasijevic *et al.*[201] explained that the reaction kinetics and mechanism were found to depend on the oxidation state of the ruthenium surface and the kinetics was slower on a Ru surface with a thicker oxide layer. Ramaswamy *et al.*[192] first time investigated Ru based chalcogenides for ORR in alkaline medium and have shown Ru–Se/C as a better ORR catalyst compared with other Ru-based chalcogenides (Ru–S and Ru–Te/C). So far there is no report on RuSe/C catalyst for methanol tolerance in alkaline medium. Pd–HTRuSe(2 : 1)/C is a suitable methanol tolerant cathode material for oxygen reduction in ADMFCs as it shows good stability, oxygen reduction activity, and methanol tolerance.[202]

MWCNT has been oxidized by concentrated sulphuric acid and $KMNO_4$, and Pd is deposited on partially unzipped CNT by hydrazine reduction method. The CNT/graphene intermediate nanostructure exhibited high surface area, high aspect ratio and high dispersion properties, and is used as a support for the fuel cell catalyst. The ORR study with methanol shows that Pd/PU-CNT has higher methanol tolerance and better selectivity.[203]

Carbon supported Pd–Co alloy as a methanol tolerant cathode catalyst for ADMFCs

Procedure for the preparation of palladium-cobalt on carbon support by sodium borohydride reduction method reported in the literature[204] was followed. The structure, dispersion, electrochemical characterization and, surface area of PdCo/C were determined by X-ray diffraction (XRD), Transmission Electron Microscopy (TEM) and Cyclic Voltammetry (CV), respectively. The electrochemical activity for ORR was evaluated from Linear Sweep Voltammograms (LSV) obtained using rotating ring disk electrode.

Figure 29(a) shows the cyclic voltammograms for methanol oxidation in deaerated aqueous solution containing 0.1 M KOH and 0.1 M methanol. Pd/C shows a large anodic current for methanol oxidation above −0.5 V *vs.* MMO while PdCo(3 : 1)/C, PdCo(2 : 1)/C and PdCo(1 : 1)/C alloys exhibit less activity for methanol oxidation. The addition of Co to Pd has reduced the activity of Pd for methanol oxidation. In other words, Pd alloy electrocatalysts have ORR selectivity in presence of methanol.

The linear sweep voltammograms on the Pd and Pd alloys in aqueous solution containing 0.1 M KOH and 0.1 M methanol under an oxygen

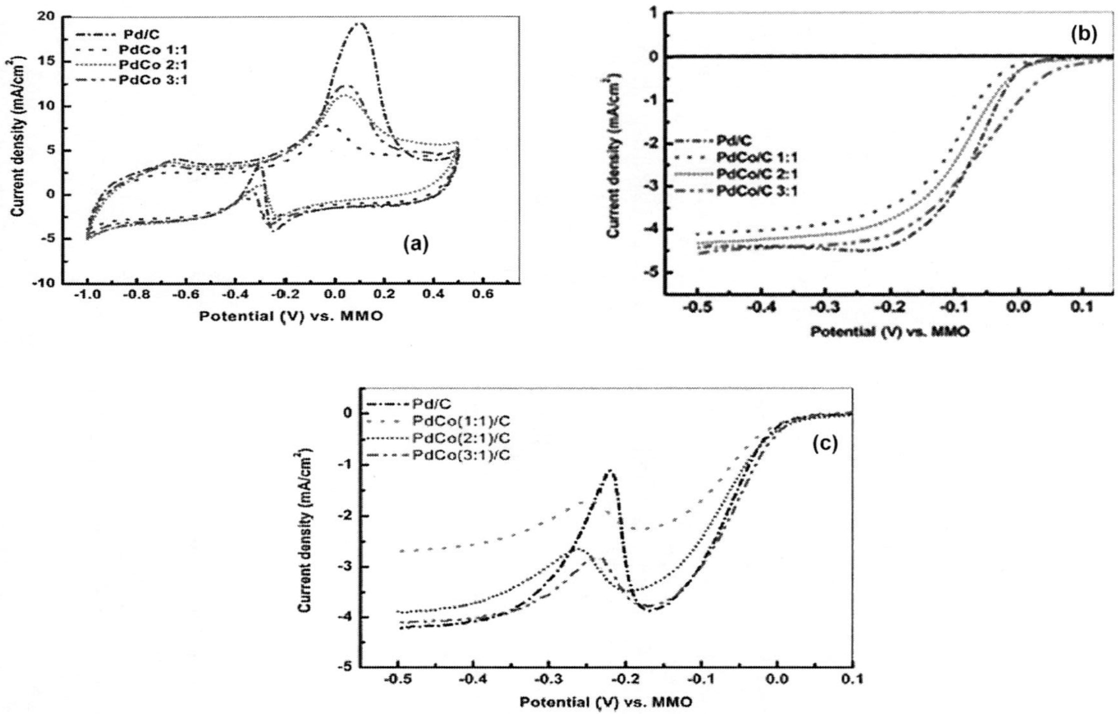

Fig. 29 (a) Cyclic Voltammograms of Pd/C, PdCo(1 : 1)/C, PdCo(2 : 1)/C and PdCo(3 : 1)/C in deaerated solution containing 0.1 M KOH and 0.1 M CH_3OH at a sweep rate of 50 mV s^{-1}. Linear sweep voltammograms of Pd/C, PdCo(1 : 1)/C, PdCo(2 : 1)/C and PdCo(3 : 1)/C in oxygen saturated 0.1 M KOH solution (b) without and (c) with methanol at 1600 rpm. Reproduced from ref. 181 with permission from the PCCP Owner Societies.

atmosphere are shown in Fig. 29(c). Pd exhibited a large anodic (reversal) current above −0.15 V *vs.* MMO. The current reversal is due to methanol oxidation reaction. Among all the catalysts, PdCo(3 : 1)/C shows the least oxidation current due to methanol. It is clearly seen from the comparative LSVs without and with methanol as shown in Fig. 29(b and c), respectively that the methanol oxidation caused the onset of cathodic limiting current to shift by ∼−200 mV in Fig. 29(c). This result is due to the formation of a mixed potential which is caused by the simultaneous reaction of the methanol oxidation and oxygen reduction on the surface of the Pd electrocatalyst. Such a mixed potential at the cathode would negatively affect the cathode performance of the DMFCs.

Pd–RuSe/C as ORR specific catalyst in ADMFCs

Carbon supported RuSe (RuSe/C) catalyst in varying atomic ratios of Ru to Se, namely, 1 : 1, 2 : 1, and 3 : 1 were prepared and their performances were compared with carbon supported Ru (Ru/C). Based on the performance, Palladium was incorporated into as prepared RuSe(2 : 1)/C and heat treated HTRuSe(2 : 1)/C. Ru/C, RuSe/C, and Pd–RuSe/C were characterized by X-ray diffraction (XRD) and transmission electron microscopy techniques. The XRD analyses of Ru/C, RuSe/C and Pd–HTRuSe/C show the formation of the hcp structure of Ru particles and the mean particle size was obtained from Ru(101) peak. The electrochemical characterizations of Ru/C, RuSe/C, Pd–HTRuSe(2 : 1)/C and Pd–RuSe(2 : 1)/C were conducted by cyclic voltammetry. Linear Sweep Voltammetric studies showed that incorporation of Pd in HTRu–Se(2 : 1)/C resulted in better catalytic activity toward oxygen reduction with resistance to methanol oxidation.

The linear sweep voltammograms on the Ru and RuSe/C catalysts in 0.1 M KOH containing 0.1 M methanol under oxygen atmosphere at a rotation rate of 1600 rpm and at a scan rate of 3 mV s^{-1} are shown in Fig. 30(a) LSV obtained for Ru/C matches well with that reported in the literature.[200,201] On modification of Ru surface by Se, the kinetic region shifted to more positive potential. RuSe (2 : 1)/C shows reduction in onset potential by 50 mV compared with Ru/C and exhibits higher catalytic activity than catalysts of other compositions. This positive shift indicates that Se is promoting the catalytic activity of Ru and stabilises the Ru surface toward adsorbed oxide/hydroxide formation, in other words, Se modification of Ru/C catalyst decreases the inhibiting effect of oxy-species adsorption on Ru.[191,205,206] According to Stolbov, Se atoms become negatively charged due to ionic Se–Ru bonding. Since O and OH also become negatively charged upon adsorption, Se and these intermediates repel each other electrostatically. This repulsion increases the total energy of the system and thus reduces the binding energies of oxy-species.[207]

However, RuSe(3 : 1)/C and RuSe(1 : 1)/C catalysts show lesser catalytic activity than RuSe (2 : 1)/C, which could be due to insufficient Se for inhibiting the formation of oxy-species on Ru in RuSe(3 : 1)/C and blocking of Ru active sites by Se in RuSe(1 : 1)/C. These catalysts show

Table 9 Onset and half-wave potentials.

Catalysts	HTRuSe(2 : 1)/C	Pd/C	Pd-HTRuSe(2 : 1)/C	Pd-RuSe(2 : 1)/C
On-set Potential (mV *vs.* MMO)	−50	16	16	−24
Half-wave potential (mV vs. MMO)	−197	−70	−75	−103

nearly the same catalytic activity for ORR in the presence and absence of 0.1 M methanol and no methanol oxidation peak is observed in 0.1 M KOH solution containing 0.1 M methanol.'Though Se modified Ru shows better ORR activity than Ru/C but is still lower than that on Pd/C.

The LSVs of RuSe(2 : 1)/C, Pd/C, Pd–HTRuSe(2 : 1)/C and Pd–RuSe(2 : 1)/C in 0.1 M KOH in presence of 0.1 M methanol under an oxygen atmosphere at a rotation rate of 1600 rpm and at a scan rate of 3 mV s^{-1}. The onset potential and $E_{1/2}$ values are given in Table 9 Pd–HTRuSe(2 : 1)/C shows enhanced ORR activity than HTRuSe(2 : 1)/C and nearly the same onset potential as that for Pd/C. But methanol oxidation current is lower by 200 mA cm^{-2} than Pd/C. It is very interesting to note that Pd–RuSe (2 : 1)/C shows more positive onset potential than HTRuSe(2 : 1)/C but less positive onset potential than that of Pd–HTRuSe(2 : 1)/C, while there is very little methanol oxidation current. Improved ORR activity of Pd–HTRuSe(2 : 1) compared with Pd–RuSe(2 : 1) could be due to the fact that without heat treatment Ru could be present predominantly in oxidized form as reported by Bron *et al.*[208]

ORR activity of Pd–HTRuSe(2 : 1)/C in absence of methanol is compared with that in presence of methanol as shown in Fig. 30(c). Onset and half-wave potentials for ORR in presence of methanol are more positive compared with that in absence of methanol while the trend is reversed in the case of Pd/C as can be seen in Fig. 30(d). The catalytic activity for ORR on Pd/C and Pd–HTRuSe(2 : 1)/C without methanol is nearly the same as revealed in Fig. 30(c and d). However, in case of Pd–HTRuSe(2 : 1)/C, influence of methanol oxidation current on ORR current is significantly less compared with that in case of Pd/C.

Steady state polarization studies are conducted for ADMFCs comprising 20% Pd/C, Pd–RuSe(2 : 1)/C and Pd–HTRuSe(2 : 1)/C as cathode catalyst and 40% Pt–Ru/C as anode catalyst at room temperature. Fig. 30(e) shows steady-state performance curves for ADMFCs with aqueous 6 M KOH containing 2 M methanol and oxygen as fuel and oxidant, respectively. Among the various ADMFCs, the cell comprising Pd–HTRuSe(2 : 1)/C shows the highest peak power density of 13.7 mW cm^{-2} in relation to Pd/C and Pd–RuSe(2 : 1)/C with a power density of 11.5 and 13.3 mW cm^{-2}, respectively matching the trend observed in the kinetic studies.

Pd–TiO_2/C as a methanol tolerant catalyst for oxygen reduction reaction in ADMFCs

TiO_2 on carbon was prepared by sol–gel method. Pd supported on TiO_2–C comprising Pd and Ti in varying atomic ratios namely, 1 : 1, 2 : 1 and 3 : 1

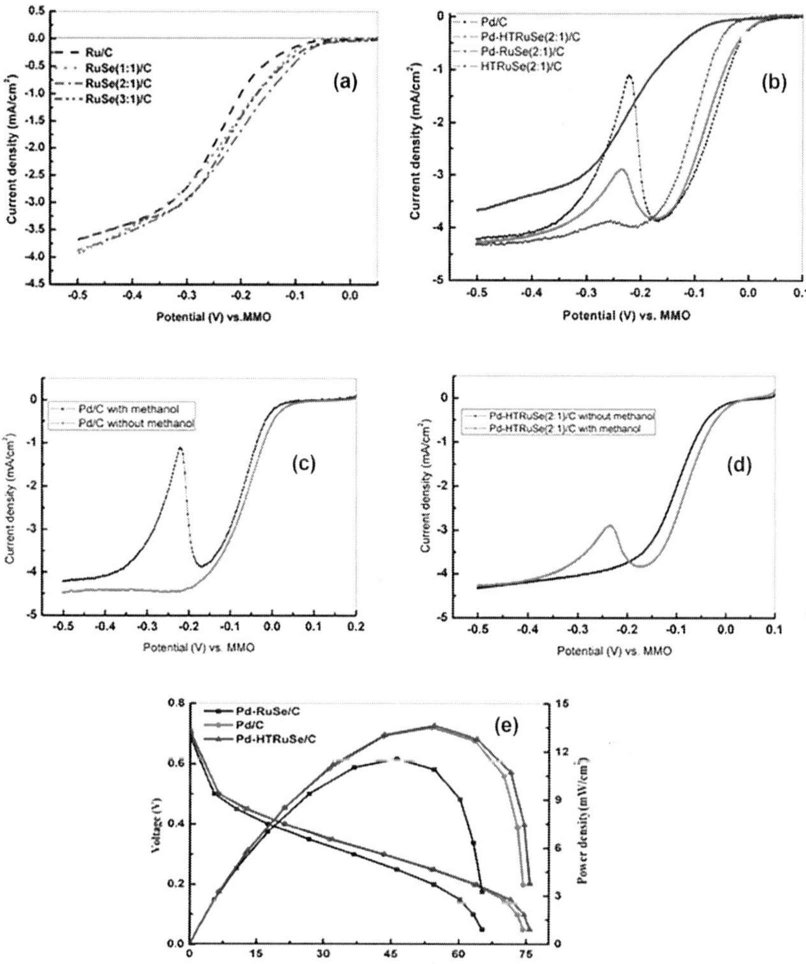

Fig. 30 (a) Linear sweep voltammograms of Ru/C, RuSe(1 : 1)/C, RuSe(2 : 1)/C and RuSe(3 : 1)/C in oxygen saturated 0.1 M KOH in presence of 0.1 M CH_3OH solution at 1600 rpm. (b) Linear sweep voltammograms of HTRuSe(2 : 1)/C, PdRuSe(2 : 1)/C, Pd/C and Pd–HTRuSe(2 : 1)/C in oxygen saturated 0.1 M KOH in presence of 0.1M CH_3OH solution at 1600 rpm. Linear sweep voltammograms of (c) Pd–HTRuSe(2 : 1)/C and (d) Pd/C in oxygen saturated 0.1 M KOH in presence and absence of 0.1 M CH_3OH solution at 1600 rpm. (e) Steady state performance curves for ADMFCs (methanol and oxygen) comprising Pd/C, Pd–RuSe(2 : 1)/C and Pd–HTRuSe(2 : 1)/C as cathode catalysts at room temperature. Reproduced with permission from S. Maheswari, P. Sridhar and S. Pitchumani, Pd–RuSe/C as ORR Specific Catalyst in Alkaline Solution Containing Methanol, *Fuel Cells*, 2012, **12**(6), 963–970. Copyright 2012 WILEY-VCH Verlag GmbH & Co. KGaA, Weinheim.

were prepared and characterised by X-ray diffraction and Transmission electron microscopy techniques. The electrocatalytic activity for oxygen reduction reaction (ORR) both in presence and absence of methanol in aqueous 0.1 M KOH solution was characterised by rotating disk electrode.

Figure 31(a) shows the CVs for methanol oxidation reaction (MOR) in deaerated aqueous solution containing 0.1 M KOH and 0.1 M methanol. Pd/C shows a larger anodic current for MOR above −0.4 V than that

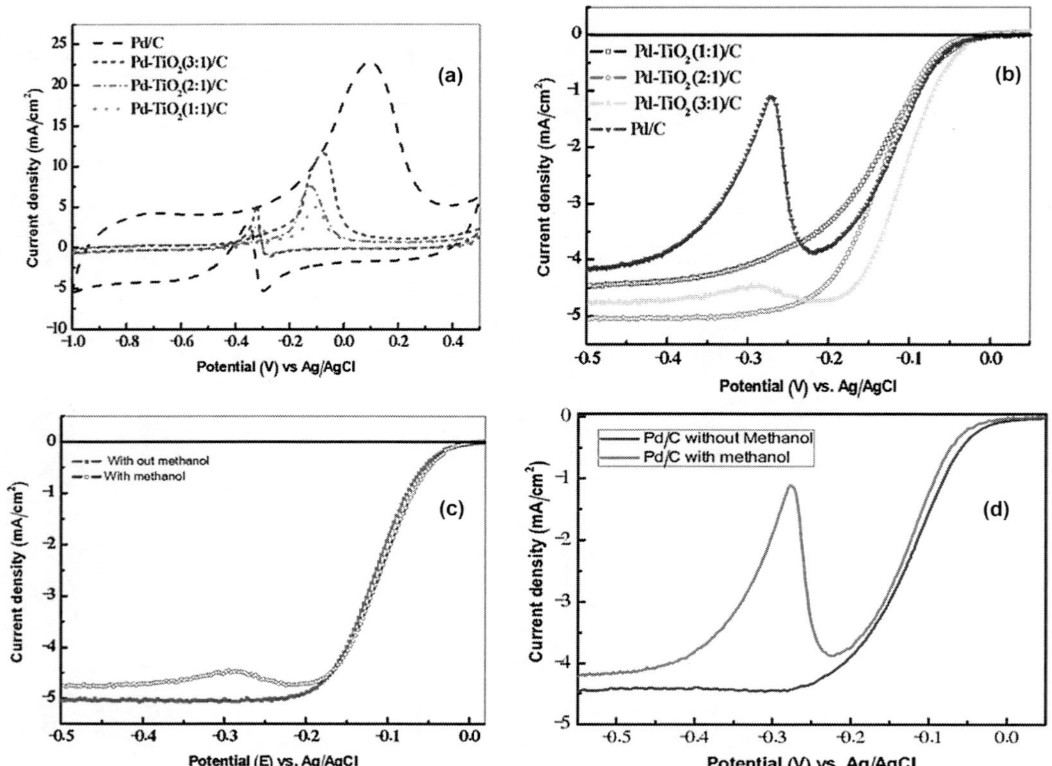

Fig. 31 (a) CV of Pd/C and Pd–TiO_2/C in aqueous 0.1 M KOH solution containing 0.1 M methanol. (b) LSVs of Pd/C and Pd–TiO_2/C in aqueous 0.1 M KOH solution containing 0.1 M methanol. LSVs of (c) Pd–TiO_2/C and (d) Pd/C in aqueous 0.1 M KOH solution in presence and absence of 0.1 M methanol at 1600 rpm. Reproduced with permission from S. Maheswari, P. Sridhar and S. Pitchumani, Pd-TiO2/C as a methanol tolerant catalyst for oxygen reduction reaction in alkaline medium, *Electrochem. Commun.*, **26**, 97–100. Copyright 2013 with permission from Elsevier B.V.

observed for $Pd-TiO_2(3:1)/C$, $Pd-TiO_2(2:1)/C$ and $Pd-TiO_2(1:1)/C$. The reduced MOR on $Pd-TiO_2/C$ reveals that TiO_2 addition possibly might not only block the active Pd sites for the oxidation of methanol but also inhibit methanol adsorption as revealed by Sanchez *et al.*[209]

The LSVs on the Pd and $Pd-TiO_2/C$ in aqueous solution containing 0.1 M KOH and 0.1 M methanol under an oxygen atmosphere at a rotation rate of 1600 rpm is shown in Fig. 31(b). Pd exhibited a large anodic (reversal) current above −0.2 V. The current reversal is due to MOR. Among all the catalysts, $Pd-TiO_2(3:1)/C$ shows the least oxidation current due to methanol. MOR caused the onset of cathodic limiting current to shift by ∼−200 mV. This is due to the formation of a mixed potential which is caused by MOR and ORR simultaneously on the surface of the Pd electrocatalyst. Such a mixed potential at the cathode would negatively affect the cathode performance of the ADMFCs. The enhanced catalytic activity of $Pd-TiO_2(3:1)/C$ is due to the hypo-hyper d-electron interaction between TiO_2 and Pd.[210]

ORR activity of $Pd-TiO_2(3:1)/C$ in absence of methanol is compared with that in presence of methanol as shown in Fig. 31(c). Onset and half-wave potentials for ORR in presence of methanol are more positive compared with that in absence of methanol while the trend is reversed in the case of Pd/C as shown in Fig. 31(d). The above trend is also reflected in the specific activities of $Pd-TiO_2/C$ catalysts containing Pd and Ti in varying atomic ratios of 1 : 1, 2 : 1 and 3 : 1 and Pd/C for ORR, which are found to be 0.28, 0.20, 0.50 and 0.22 $mA\,cm^{-2}$, respectively in presence of methanol, and 0.12, 0.18, 0.41 and 0.34 $mA\,cm^{-2}$, respectively in absence of methanol.

Two Tafel slope regions as shown in Fig. 32(a) accounted for an initial fast charge-transfer step followed by a chemical step, which is the rate-determining step.[211] The Tafel slope values are almost the same in presence and absence of methanol, suggesting that methanol does not affect the kinetics of oxygen reduction on $Pd-TiO_2(3:1)/C$. It is interesting to note that $Pd-TiO_2$ (3 : 1)/C shows much improved performance over Pd-IITRuSc(2 : 1)/C as revealed in Fig. 32(b).

Pd supported on partially unzipped carbon nanotube as a methanol tolerant oxygen reduction reaction catalyst for ADMFCs

In recent years, high-area carbon like reduced graphene oxide (RGO) and CNTs are predominantly used as supports for Pd nanoparticle catalysts. RGO can easily agglomerate and affect the dispersion of catalyst ink while MWCNTs being quite hydrophilic at the end of the tube and highly hydrophobic on the walls of the tube, the uniform dispersion of the catalyst is difficult in polar solvents.[212] However, RGO and MWCNTs have some advantages *i.e.* graphene has been widely considered as an ideal substrate for ultrathin coating of functional materials due to its unique structural and electrical properties, while the interconnected CNTs network is highly conductive and porous that is beneficial for electronic and ionic transport[213] besides it prevents face-to-face RGO agglomeration, simultaneously increasing the through-plane conductance.[214] From the literature survey, the main advantages of CNTs or graphene supports over activated carbon are: (i) the high purity of the material that

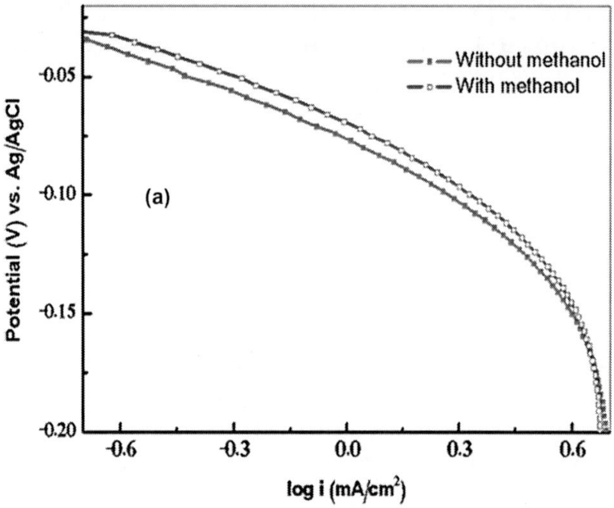

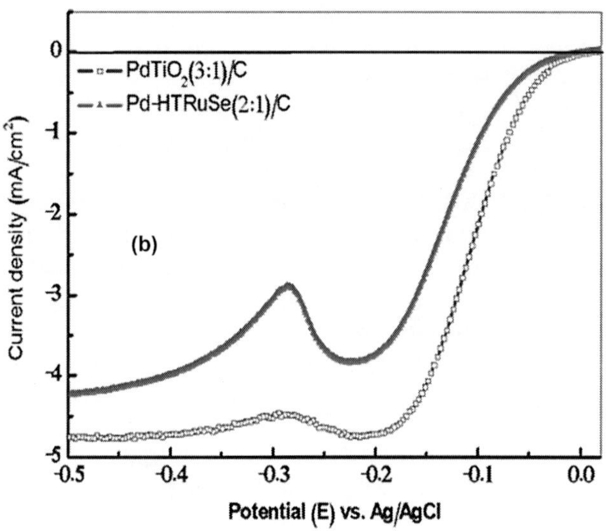

Fig. 32 (a) Tafel slopes for Pd–TiO_2 (3 : 1)/C in presence and absence of methanol. (b) Linear Sweep Voltammograms of Pd–TiO_2 (3 : 1)/C and Pd–HTRuSe (2 : 1)/C in aqueous 0.1 M KOH solution containing 0.1 M Methanol. Reproduced with permission from S. Maheswari, P. Sridhar and S. Pitchumani, Pd-TiO2/C as a methanol tolerant catalyst for oxygen reduction reaction in alkaline medium, *Electrochem. Commun.*, **26**, 97–100. Copyright 2013 with permission from Elsevier B.V.

can avoid self-poisoning, (ii) the mesoporous nature of these supports that can be of interest for liquid-phase reactions, thus limiting the mass transfer, and (iii) specific metal-support interactions that can directly affect the catalytic activity and the selectivity.[215]

In order to have the advantages of both RGO and CNTs, one successful approach is the recently reported longitudinal unzipping by Tour *et al.*[216] Partially unzipped CNTs has been reported as a superior catalyst support for PEM fuel cells.[217] The CNTs/graphene intermediate nanostructure exhibited high surface area, high aspect ratio and high dispersion properties, and is used as a support for the fuel cell catalyst.[218]

An intermediate material that contained a hybrid structure of 1D CNT and 2D graphene *via* partial opening of CNTs was developed. Palladium (Pd) nanoparticles were deposited on PU-CNTs by polyol method. Prepared Pd/PU-CNTs was characterized by X-ray diffraction (XRD), Transmission Electron Microscope (TEM), Scanning Electron Microscope (SEM) and Raman spectrometer. Electrochemical characterizations were performed by Cyclic Voltammetry (CV) and Linear Sweep Voltammetry (LSV).

Cyclic voltammograms of Pd/C, Pd/RGO and Pd/PU-CNTs catalysts in 0.1 M KOH in presence of 0.1 M methanol under deaerated atmosphere at a scan rate of 50 mV s^{-1} are shown in Fig. 33(a). Pd/RGO shows higher anodic current for MOR above −0.4 V than that observed for Pd/C and Pd/PU-CNTs. The reduced MOR on Pd/PU-CNTs reveals that PU-CNTs addition inhibits methanol adsorption.

The Linear Sweep Voltammograms of Pd/C, Pd/RGO and Pd/PU-CNTs in 0.1 M KOH in presence of 0.1 M methanol under an oxygen atmosphere at a rotation rate of 1600 rpm and at a scan rate of 3 mV s^{-1} are shown in Fig. 33(b). The onset potential and $E_{1/2}$ values are −30 mV, 30 mV, −48 mV and −119 mV, −76 mV, −130 mV for Pd/C, Pd/PU-CNT and Pd/RGO, respectively. Similarly, mass activities and kinetic current densities @ −0.05V are 3.5, 16, 1.3 A mg^{-1} and 0.2, 1.1, 0.09 mA cm^{-2} for Pd/C, Pd/PU-CNT and Pd/RGO, respectively. It is clear that Pd/PU-CNT

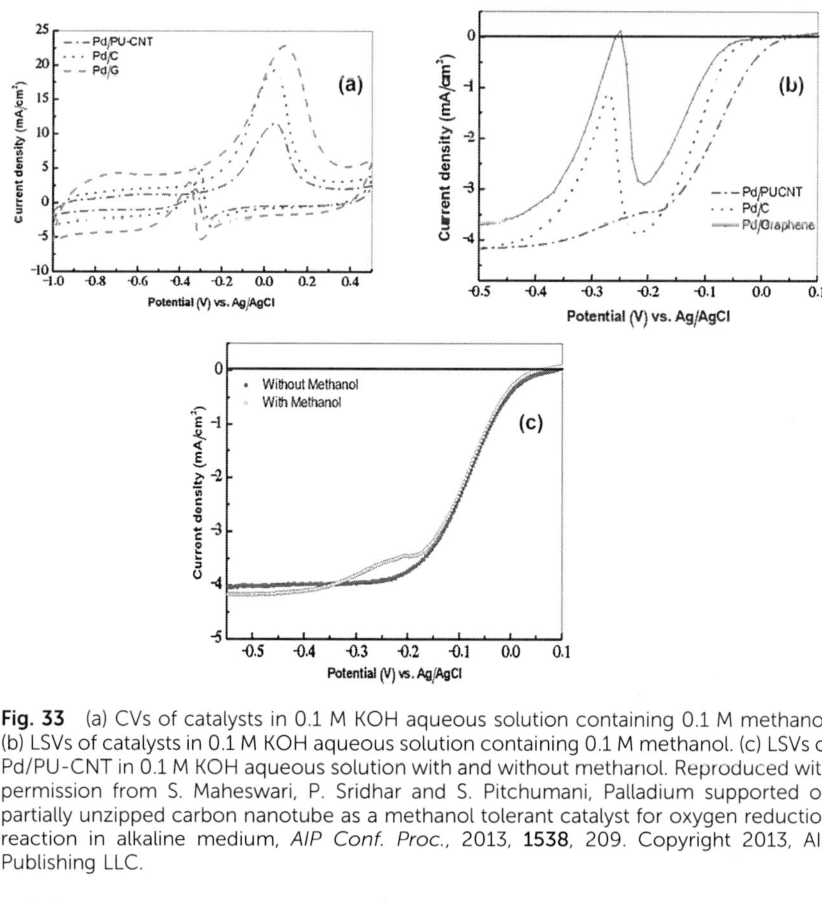

Fig. 33 (a) CVs of catalysts in 0.1 M KOH aqueous solution containing 0.1 M methanol. (b) LSVs of catalysts in 0.1 M KOH aqueous solution containing 0.1 M methanol. (c) LSVs of Pd/PU-CNT in 0.1 M KOH aqueous solution with and without methanol. Reproduced with permission from S. Maheswari, P. Sridhar and S. Pitchumani, Palladium supported on partially unzipped carbon nanotube as a methanol tolerant catalyst for oxygen reduction reaction in alkaline medium, *AIP Conf. Proc.*, 2013, **1538**, 209. Copyright 2013, AIP Publishing LLC.

has higher ORR activity in the presence of methanol than Pd/C and Pd/RGO. These results reveal that though Pd favors the methanol oxidation, PU-CNTs has better ORR selectivity due to reduced methanol adsorption.[219]

This is further validated by the similar catalytic activity for ORR observed on Pd/PU-CNTs in 0.1 M aqueous KOH solution containing with and without methanol as shown in Fig. 33(c).

The unique hybrid structure of 1D nanotube and 2D double-side graphene has resulted in higher methanol tolerance behavior. Density functional theory calculations suggest that the reactive nature of edge structures by itself introduces possibilities of ORR activity in carbon materials by lowering the oxygen adsorption barrier and the first electron transfer barrier significantly.[220]

Methanol oxidation catalysts. One of the main drawbacks of DMFCs is insufficient activities of the anode catalysts towards methanol oxidation reaction (MOR). It is known that platinum is the only single component catalyst that shows an excellent catalytic activity towards methanol electro-oxidation, especially at lower temperatures. However, pure platinum is susceptible to CO poisoning during the oxidation of methanol at low temperatures, resulting in a substantial decrease in its activity.[221] In order to improve the activity and CO-tolerance of the Pt catalysts, efforts on alloying Pt with other transition metals, viz. Ru, Mo, Sn, *etc.* have been expended. In general, CO-poisoned platinum can be regenerated *via* the reaction of surface CO with oxygen species associated with an element like Ru, Mo, *etc.* to yield CO_2. The addition of a metal to Pt-based catalysts significantly lowers the overpotential for methanol electrooxidation through the bi-functional mechanism. Pt–Ru is recognized as the most promising anode electro-catalyst for DMFCs because of its superior CO tolerance due to a bi-functional effect, where CO is oxidized by OH species generated by water discharge on Ru surface atoms.[222] But the CO tolerance of the Pt–Ru is still unsatisfactory for practical DMFC applications. To this end, it is necessary to further the catalytic activity of Pt–Ru catalyst.[223] It has been reported that the addition of transition metal oxides such as CeO_2, TiO_2, WO_3, MoO_3, *etc.* to Pt–Ru could improve CO tolerance and MOR activity.[224–226] Transition metal oxides are demonstrated to be good surface promoters for improving Pt activity towards methanol oxidation due to their special 'spillover' effect.[227] Among these metal oxides, TiO_2 seems to be a promising candidate owing to its natural abundance, cost, semiconductivity and stability in acidic environments.[228] The electrode using TiO_2 as the support for platinum changes the electronic property of the Pt surface such that the chemical adsorption of CO on the surface is weakened due to the interaction between TiO_2 and Pt.[228] It has been reported that there is some co-operative effect for platinum catalyst in the presence of TiO_2 that facilitates oxidation of CO_{ads}. Recent studies have shown an increase in the kinetics of MOR as well as CO tolerance for TiO_2 modified Pt–Ru/C and Pt–Ru/CNT.[227]

An effective synthetic method was developed to achieve homogenous nanocomposite catalyst without any post-pyrolysis heat treatment. Accordingly, titanium oxide–carbon nanocomposite with mesoporous structure is directly synthesized as a catalyst support *via* supramolecular self-assembly with an *in situ* crystallization process in order to achieve high surface area. Active sites on such a porous structure would be easily accessible to the reactant leading to reduced diffusion resistance/mass-transfer resistance. Methanol oxidation and accelerated durability study for Pt–Ru/TiO_2–C show enhanced catalytic activity and durability in relation to the carbon supported Pt–Ru.[229]

Pt–Ru decorated self-assembled TiO_2–carbon hybrid nanostructure for enhanced methanol electro-oxidation

Porous titanium oxide–carbon hybrid nanostructure (TiO_2–C) with a specific surface area of 350 $m^2\ g^{-1}$ and an average pore-radius of 21.8 Å is synthesized *via* supramolecular self-assembly with *in situ* crystallization process. Subsequently, TiO_2–C supported Pt–Ru electro-catalyst (Pt–Ru/TiO_2–C) is obtained and investigated as an anode catalyst for DMFCs.

Electrocatalytic activities for Pt–Ru/C and Pt–Ru/TiO_2–C catalysts for methanol oxidation reaction have been studied by cyclic voltammetry in 0.5 M aq. CH_3OH solution containing 0.5M $HClO_4$ at a scan rate of 50 mV s^{-1}. The resulting voltammograms are shown in Fig. 34(a). It is clear that TiO_2–C supported electrocatalyst shows higher catalytic activity in relation to Pt–Ru on carbon. The current density for methanol oxidation reaction on TiO_2–C supported catalyst is higher than the carbon supported electro-catalyst. The peak potentials for methanol adsorbate oxidation for Pt–Ru supported on TiO_2–C and carbon are 0.52 and 0.57 V, respectively revealing the oxidation of methanol occurring at a lower potential on TiO_2–C supported catalyst in relation to carbon-supported catalyst.

Methanol adsorbate electro-oxidation peak potential for TiO_2–C supported Pt is at 0.83 V while that for carbon supported Pt is at 0.93 V as shown in Fig. 34(b). Moreover, TiO_2–C supported Pt shows higher current density at lower potential. The negative shift in the methanol adsorbate electro-oxidation peak potential and higher current demonstrates the superior electro-catalytic activity for TiO_2–C supported Pt. The higher activity is likely due to enhanced electrochemical surface area in addition to the modification of electronic structure of Pt nano particles due to interaction with titanium oxide which leads to a change in the adsorption characteristic of methanol on Pt/TiO_2–C.[228] Furthermore, the hydroxylated Ti might facilitate the adsorption of methanol accelerating its diffusion. The concentration of methanol at the interface is increased due to enhanced chemisorption on TiO_2–C composite in comparison with that on Vulcan XC 72. It is reported that active oxygen containing surface species are necessary for the oxidative removal of CO-like poisoning intermediates on Pt surface.[230] It is reported that there is some co-operative effect for platinum catalyst in presence of TiO_2, which facilitates oxidation of adsorbed CO. In this catalytic system, TiO_2–C and Ru display anti-poisoning function, the carbon network benefits the electron transfer, and Pt works as the main dehydrogenation site.[228]

The electrochemical stability for Pt–Ru/TiO_2–C is evaluated by an accelerated durability test and compared with Pt–Ru/C. Figure 34(c and d) shows CVs for methanol oxidation on Pt–Ru/C and Pt–Ru/TiO_2–C

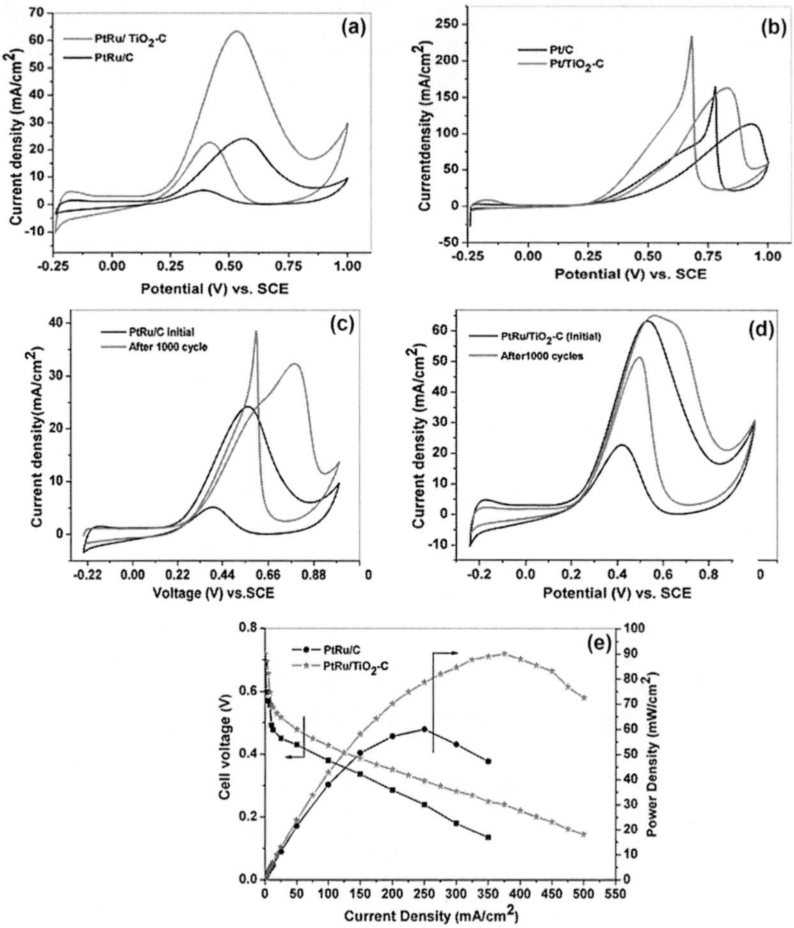

Fig. 34 (a) Cyclic voltammograms for methanol electro-oxidation on Pt–Ru/C and Pt–Ru/TiO_2–C in N_2-saturated aqueous solution containing 0.5 M $HClO_4$ and 0.5 M CH_3OH at a scan rate of 50 mV s^{-1}. (b) Cyclic voltammograms for methanol electro-oxidation on Pt/C and Pt/TiO_2/C. Cyclic voltammograms for methanol electro-oxidation on (c) Pt–Ru/C and (d) Pt–Ru/TiO_2–C catalysts before and after ADT. (e) Steady-state performance for DMFCs having Pt–Ru/TiO_2–C and Pt–Ru/C as anode catalysts at 65 °C. Reproduced with permission from K. G. Nishanth, Pt-Ru decorated self-assembled TiO_2-carbon hybrid nanostructure for enhanced methanol electrooxidation, *Bull. Mater. Sci.*, 2012, **36**(3), 353–359. Copyright Indian Academy of Sciences 2013. With permission of Springer.

before and after the durability study. It is seen that the methanol adsorbate electro-oxidation peak shifts from 0.57 V to 0.79 V after ADT for carbon supported catalyst, while, for TiO_2–C supported catalyst, the peak potential shifts from 0.52 V to 0.55 V after 1000 cycles. The potential shift for carbon supported catalyst after ADT is 220 mV while for TiO_2–C supported catalyst it is only 30 mV. The extent of increased positive potential shift in the case of carbon supported Pt–Ru, indicates decreased catalytic activity towards methanol electro-oxidation after ADT. Furthermore, the positive onset potential shift and higher current at peak potential after ADT suggest increased platinum character caused by a continuous Ru metal loss from Pt–Ru.[231,232] By contrast, TiO_2–C supported electro-catalyst shows only a small shift in onset potential and oxidation potential of methanol adsorbate after ADT. This is corroborated by ICP-MS analysis, which shows the concentration of leached out Ru to be: Pt–Ru/C (46 ppb) > Pt–Ru/TiO_2–C (14 ppb). These data suggest that the activity retention rate for TiO_2–C supported catalysts is superior than carbon supported catalyst. These data confirm the enhanced electrochemical stability of the Pt–Ru/TiO_2–C catalyst towards methanol electro-oxidation reaction. The stability may be due to the excellent corrosion tolerance of titanium oxide-carbon composite in comparison to carbon black in acidic medium. The improved durability for TiO_2–C supported electro-catalysts may be due to the strong interaction between Pt–Ru nano particles and TiO_2 support; it would be less likely for Pt–Ru atoms on TiO_2–C to agglomerate to produce larger particles that would cause gradual decrease in electrochemical activity or metal dissolution from the support surface.[233]

The catalysts have also been performance tested in DMFCs. The cell polarization data for DMFCs comprising Pt–Ru/TiO_2–C and Pt–Ru/C anode catalysts are given in Fig. 34(e). It is clear that the DMFC comprising TiO_2–C supported anode catalyst is superior in performance as compared to DMFC with carbon-supported Pt–Ru. The peak power-densities are 91 and 60 mW cm^{-2} for DMFCs with Pt–Ru/TiO_2–C and Pt–Ru/C, respectively. These data are in conformity with CV data. Since cathode catalyst and membrane during the study are kept identical, the enhanced performance for the DMFC with TiO_2–C supported catalyst is a clear manifestation of the improved methanol electro-oxidation behavior.

Ehtanol tolerant ORR catalysts for DEFCs. The most familiar oxygen reduction reaction (ORR) is based on noble metals, particularly platinum. Carbon supported platinum based materials are the most active, efficient and successful catalysts for electrochemical devices at the current technology stage. But the problem associated with Pt catalyst is the slowness of ORR which is due to the formation of –OH species at +0.8 V, which inhibits further reduction of oxygen and hence, results in loss of performance.[234] Research on alternative cathode catalysts is still necessary in order to find a material with an improved ORR activity and a higher ethanol tolerance than those of Pt. So researchers focused on alloys or composite catalysts such as that prevent the adsorption of –OH species, reduce oxygen selectively to water without being affected by the contaminants and tolerant to ethanol. The enhancement in the ORR activity observed when using supported Pt–M alloy electro-catalyst was ascribed to both geometric (decrease of the Pt–Pt bond distance) and electronic factors (increase of Pt d-electron vacancy).[235–237] Pt-based bimetallic catalyst (*e.g.* Pt–Fe, Pt–Ni, Pt–Co, *etc.*) for ORR has been reported.[238] Pt-based alloys comprising Pt and other transition metals, namely Co and Pd, have been extensively investigated owing to their enhanced ORR activity. Pd, Pd alloys, Pt–Co, NiCoFe/C and Pt–Fe/C cathode catalyst have been studied in the presence of alcohols demonstrating high performances for

ORR.[239–245] Recently, much of the research work is focused on the system of combining metal oxides, carbon and Pt. Strong d–d-Metal-Support Interaction (SMSI) of hyper-d-electronic metal with mostly hypo-d-oxide (TiO_2, ZrO_2, HfO_2) or various hypo-d-hypo-f-oxide supporting composites (TiO_2, CeO_2). SMSI implies that the d–d metal-oxide interaction (M/TiO_2) for heterogeneous catalysis, in accordance with the bonding strength, results in substantial weakening and even suppression of intermediate chemisorptive bonds (M–H, M–CO). A new concept additionally implies the spillover of interactive primary oxides (M–OH) and their decisive interference in the overall catalytic process. However, while the dynamic spillover effect of the primary oxide (M–OH) dipoles repulsion takes place all over the available metallic surface regardless its nano-sized magnitude, the stronger the hypo-hyper-d-inter ionic metal-oxide support interaction the higher the overall catalytic activity.[147–150,246–248] Pt–TiO_2/C as ethanol-tolerant cathode catalyst for DAFCs is discussed.[249]

Spillover effect induced Pt–TiO_2/C as ethanol tolerant oxygen reduction reaction catalyst for DEFCs

Hypo-hyper-d-electronic interactive nature is used to develop a new carbon supported HT–Pt–TiO_2 composite catalyst comprising Pt and Ti in varying atomic ratio, namely 1 : 1, 2 : 1 and 3 : 1. The electro-catalysts are characterized by XRD, TEM, SEM-EDAX, Cyclic Voltammetry (CV) and Linear sweep voltammetry (LSV) techniques.

Cyclic voltammograms of HT–Pt–TiO_2(1 : 1)/C, HT–Pt–TiO_2(2 : 1)/C, HT–Pt–TiO_2(3 : 1)/C, HT–Pt/C and Pt/C electro-catalysts in aq. 0.5 M $HClO_4$ at a scan rate of 50 mV s^{-1} between 0 and 1.0 V (*vs.* NHE) are presented in Fig. 35(a). All the catalysts show peaks associated with hydrogen adsorption/desorption between 0 and 0.3 V (*vs.* NHE) followed by the "double-layer" potential region; above 0.7 V (*vs.* NHE) oxide formation/reduction regions are observed. The electrochemical active surface area (ESA) is obtained using the charge corresponding to the reduction of oxide, which is formed due to adsorption of a monolayer of chemisorbed oxygen instead of using hydrogen adsorption/desorption peaks that may not give the exact charge as there is interference from hydrogen evolution reaction. ESA was calculated from the following equation.[159,181]

$$\text{ESA}\ (\text{cm}^2\ \text{g}^{-1}{}_{\text{Pt}}) = Q_{\text{O}}\ (\mu\text{C cm}^{-2})/\{420\ \mu\text{C cm}^{-2} \times \text{electrode loading}\ (\text{g}_{\text{Pt}}\ \text{cm}^{-2})\}$$

The calculated ESA values of all the catalysts are presented in Table 10. HT–Pt–TiO_2(2 : 1)/C catalyst shows higher ESA compared to all the other electro-catalysts. Increase in ESA is due to the formation of new sites at the interface between Pt and TiO_2.[140,141]

Linear scan voltammograms of Pt/C and HT–Pt–TiO_2/C catalysts containing Pt and Ti in varying atomic ratios in aq. 0.5 M $HClO_4$ saturated with pure oxygen at a scan rate of 5 mV s^{-1} are presented in Fig. 35(b).

The onset potentials for ORR on Pt/C, HT–Pt–TiO_2(1 : 1)/C, HT–Pt–TiO_2(2 : 1) and HT–Pt–TiO_2(3 : 1)/C catalysts are 0.89, 0.91, 0.99 and 0.98 V, respectively confirming that HT–Pt–TiO_2(2 : 1)/C and HT–Pt–TiO_2(3 : 1)/C

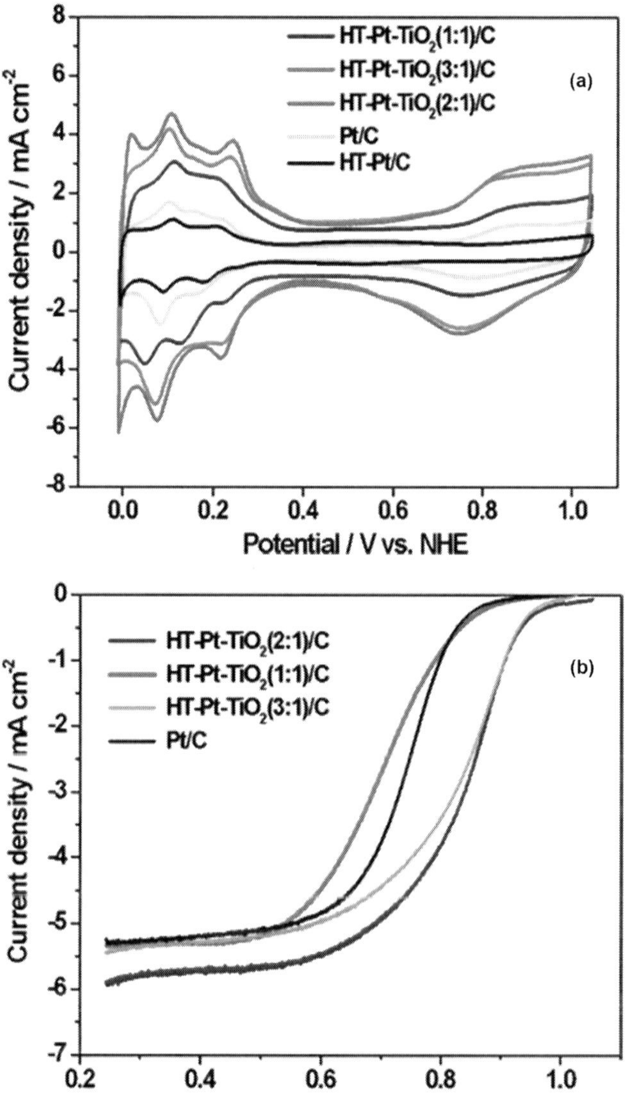

Fig. 35 (a) Cyclic voltammograms for Pt/C, HT–Pt/C and HT–Pt–TiO_2/C catalysts containing Pt and Ti in varying atomic ratios in N_2-saturated aq. 0.5 M $HClO_4$ at a scan rate of 50 mV s^{-1}. (b). Linear scan voltammograms of Pt/C and HT–Pt–TiO_2/C catalysts containing Pt and Ti in varying atomic ratios in aq. 0.5 M $HClO_4$ saturated with pure oxygen at a scan rate of 5 mV s^{-1} (Electrode rotation rate: 1600 rpm). Reproduced with permission from S. Meenakshi, K. G. Nishanth, P. Sridhar and S. Pitchumani, Spillover effect induced Pt-TiO2/C as ethanol tolerant oxygen reduction reaction catalyst for direct ethanol fuel cells, *Electrochim. Acta*, **135**, 52–59. Copyright 2013 with permission from Elsevier B.V.

have higher activities than HT–Pt $TiO_2(1:1)$/C and Pt/C. The onset potential for oxygen reduction is shifted to more positive potential by about 90 mV compared to Pt/C. The improved ORR is due to the ready adsorption and easy dissociation of O_2 on the TiO_2-modified Pt surface in relation to Pt/C.

The ORR curves of Pt/C and HT–Pt–TiO_2/C catalysts recorded in O_2-saturated solution containing aq. 0.5 M $HClO_4$ and 1 M ethanol at a

Table 10 Electrohemical Surface area of the catalysts.

Catalysts	Electrochemical surface area ($m^2\ g^{-1}$)
Pt/C	22
HT–Pt/C	06
HT–Pt–TiO_2(1 : 1)/C	36
HT–Pt–TiO_2(2 : 1)/C	73
HT–Pt–TiO_2(3 : 1)/C	68

scan rate of 5 mV s^{-1} are presented in Fig. 36(a–d). Anodic currents of 17 and 14.8 mA cm^{-2} are observed for Pt/C and HT–Pt–TiO_2(3 : 1)/C catalysts at 0.8 V (*vs.* NHE) while lower anodic currents of 0.8 and 1.84 mA cm^{-2} are obtained for HT–Pt–TiO_2(2 : 1)/C and HT–Pt–TiO_2(1 : 1)/C, respectively. The anodic current is due to the electrooxidation of ethanol. Therefore, it is evident that the HT–Pt–TiO_2(2 : 1)/C catalyst possesses a lower catalytic activity towards the ethanol oxidation compared to all the other catalysts. However, HT–Pt–TiO_2(2 : 1)/C exhibits a higher ORR activity with minimum ethanol oxidation. The onset potential for ORR in presence of ethanol on HT–Pt–TiO_2(2 : 1)/C is shifted to more positive potential by about 271 mV compared with that on Pt/C.

Ethanol adsorption and oxygen adsorption are competing with each other for the surface sites. These studies confirm that the HT–Pt–TiO_2(2 : 1)/C has a higher ORR selectivity and better ethanol tolerance in relation to Pt/C.

The performance of the DEFCs for Pt/C, HT–Pt/C and HT–Pt–TiO_2(2 : 1)/C cathode catalysts at 70 °C are shown in Fig. 36(e). The polarization curve for DEFC cathode containing HT–Pt–TiO_2(2 : 1)/C clearly shows superior performance than that containing Pt/C, reflecting its higher ethanol tolerance. It is possibly due to the presence of TiO_2 around Pt active sites that could block ethanol adsorption on Pt sites due to the dilution effect.

Ethanol oxidation catalysts for DEFCs. Many studies have investigated the carbon supported Pt as an anode catalyst for low temperature fuel cells. But Pt itself is rapidly poisoned on its surface by ethanol oxidation. Therefore, most of the researchers concentrated on binary and ternary metal based catalyst. For the binary system Pt–Ru/C,[250] Pt–Sn/C,[241,251–257] Pt–W/C,[258] Pt–Pd/C,[259] Pt–Re/C,[260] Pt–Rh/C,[261] Pt–Mo/C,[262] Pt–CeO_2/C,[263] Pt–ZrO_2/C,[264] Pt–RuO_2/C,[265] TNT/Pt/C[142] and Pt_x–WO_3/C[266] were used as anode catalysts. Among these, Pt–Sn based binary system is the most active electro-catalyst for ethanol electro-oxidation. Pt–Sn alloy and Pt–SnO_2/C,[267,268] carbon supported $Pt_{75}Sn_{25}$,[269] decoration of carbon supported Pt with Sn,[270] alloy Pt_7Sn_3 and bi-phase Pt–SnO_x nano-catalyst[271] were some of the catalysts that have been studied. Higuchi *et al.*[272] reported highly dispersed Pt–SnO_2 nanoparticle on carbon black. Zhu *et al.*[273] have reported on the effect of alloying degree in PtSn catalyst, SnO_2@Pt/C5 and also the enhanced ethanol oxidation reaction (EOR) activity of carbon supported $Pt_{50}Sn_{50}$ alloy catalyst.[274] There are two different schools of thought on the

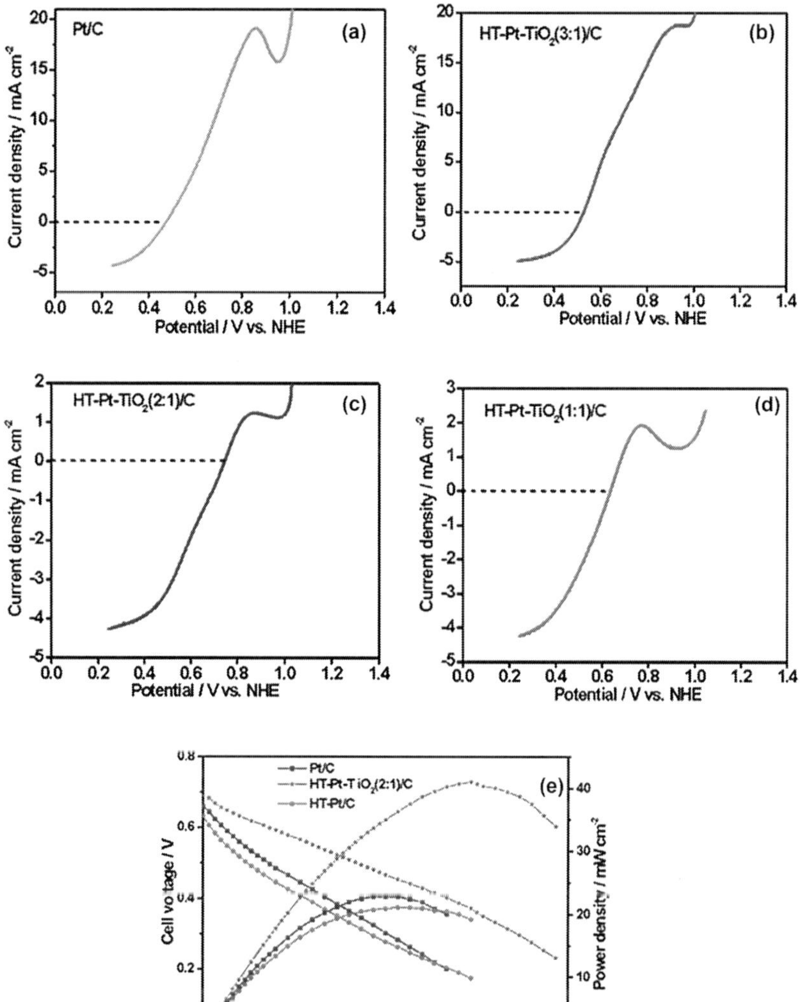

Fig. 36 (a–d) Linear scan voltammograms for Pt/C and HT–Pt–TiO_2/C catalysts containing Pt and Ti in varying atomic ratios in aq. solution containing 0.5 M $HClO_4$ and 1 M CH_3CH_2OH saturated with pure oxygen at a scan rate of 5 mV s^{-1} (Electrode rotation rate: 1600 rpm). (e) Steady-state performance data of DEFCs (CH_3CH_2OH and O_2) comprising Pt/C, HT–Pt/C and HT–Pt–TiO_2(2 : 1)/C cathode catalysts at 70 °C. Reproduced with permission from S. Meenakshi, K. G. Nishanth, P. Sridhar and S. Pitchumani, Spillover effect induced Pt-TiO2/C as ethanol tolerant oxygen reduction reaction catalyst for direct ethanol fuel cells, *Electrochim. Acta*, **135**, 52–59. Copyright 2013 with permission from Elsevier B.V.

enhanced activity for EOR while using Pt–Sn as an anode catalyst. One group attributes this to Pt–Sn in the alloy form while another group favours Pt–Sn as a binary mixture or Sn present in the form of SnO_2. Antolini *et al.*[275] investigated the effect of the alloy phase characteristics on carbon supported (PtSn)alloy/SnO_2 and (PtSnPd)alloy/SnO_2 catalyst. In order to enhance the fuel cell performance further they had introduced a third metal to the catalyst but not with success. The combined effect of weakened chemisorption of oxygen due to the interactive nature of Pt–Sn alloy and the ability of SnO_2 in transforming CO-like poisoning species on Pt into CO_2 makes Pt–Sn/SnO_2–C catalyst as a superior anode catalyst for ethanol electro-oxidation in direct ethanol fuel cells.[276]

Carbon supported Pt–Sn/SnO_2 anode catalyst for direct ethanol fuel cells

Binary Pt–Sn/SnO_2–C electro-catalysts comprising Pt and Sn in varying weight ratio, namely 31 : 9, 33 : 7 and 35 : 5, were synthesized by an alcohol-reduction process using ethylene glycol as solvent and reducing agent. The electro-catalysts were characterized by XRD, XPS, TEM, SEM-EDAX, ICP-OES, Cyclic Voltammetry (CV), Chronoamperometry and CO stripping techniques.

The diffraction peaks for Pt in Pt–Sn are shifted to 2θ values of 66.43° compared to 67.58° for Pt in Pt–C as shown distinctly in Fig. 37(a) for (220) crystalline plane confirming the alloy formation in the catalyst. In the diffraction pattern, the peak observed at 33.8° represents crystalline plane (101) of SnO_2 indicating that Sn is also present in the form of SnO_2 in the catalyst.

Pt(4f) spectra for Pt–C and Pt–Sn/SnO_2(33 : 7)-C catalysts are shown in Fig. 37(b and c), respectively. Spectra exhibited intense doublets that are assigned to Pt($4f_{7/2}$) and Pt($4f_{5/2}$) at 71.62 and 74.84 eV for Pt–C catalyst and at 71.94 and 75.12 eV for Pt–Sn/SnO_2(33 : 7)-C catalyst, respectively. It is clear that peak binding energies of Pt($4f_{7/2}$) and Pt($4f_{5/2}$) slightly shifted to higher binding energies (0.32 and 0.28 eV) for Pt–Sn/SnO_2(33 : 7)-C catalyst,[277] which may be due to the charge transfer from Sn to Pt because of the lower electronegativity of Sn (1.8) in relation to Pt (2.2).[278] The charge transfer from Sn atoms to Pt atoms increases the electron density around the Pt sites, which leads to weakened chemisorption energy with oxygen containing species. This modification of the electronic environment around Pt-sites enhances the electrocatalytic activity of the catalyst. Figure 37(d) indicates that Sn(3d) spectra exhibited intense doublets assigned to $3d_{3/2}$ and $3d_{5/2}$ at 495.7 and 487.3 eV, respectively corresponding to the binding energies of Sn(IV) in SnO_2[270,279] and also confirms that Sn is present in the elemental Sn(0) form based on the binding energy of 485.70 eV.[273] Sn in the oxide form will give an additional advantage of synergistic effect arising from the combination of Pt and SnO_2 through the bifunctional mechanism as suggested by Higuchi *et al.*[272] SnO_2 in the Pt–Sn/SnO_2–C catalyst helps in enhancing the ethanol oxidation by lowering the oxidation potential of Pt adsorbed CO *via* a bifunctional mechanism.[272,273]

Cyclic voltammograms of Pt–Sn/SnO_2(31 : 9)-C, Pt–Sn/SnO_2(33 : 7)-C, Pt–Sn/SnO_2(35 : 5)-C and Pt–C electro-catalyst in 0.5 M aq. $HClO_4$ at a scan rate of 50 mV s^{-1} are presented in Fig. 38(a). H_2 adsorption/desorption between 0 and 0.3 V (*vs.* NHE) followed by the "double-layer" potential region and above 0.7 V (*vs.* NHE) oxide formation/reduction

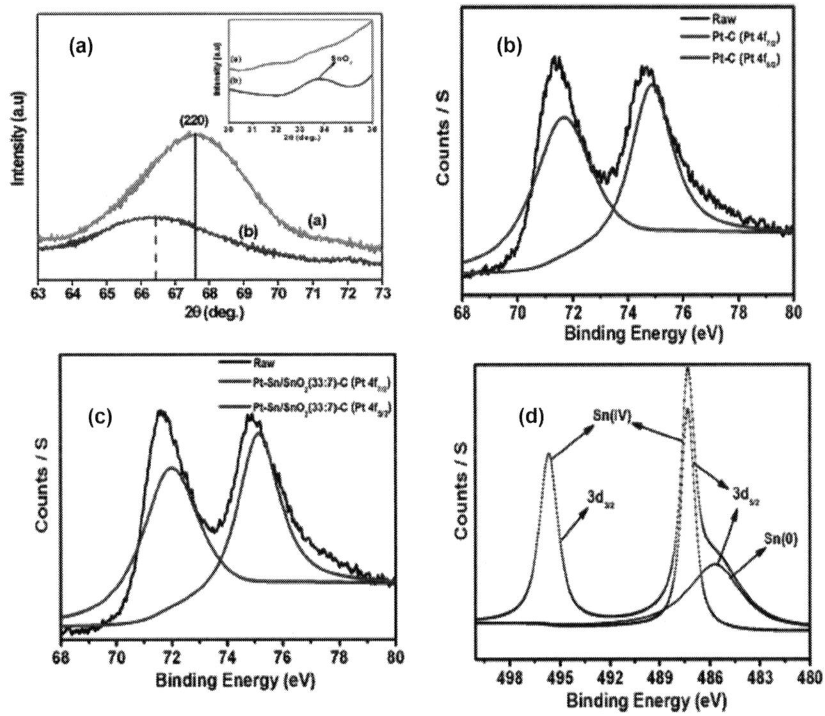

Fig. 37 (a) Powder XRD patterns and inset for (red) Pt–C and (blue) Pt–Sn/SnO$_2$(33 : 7)-C catalysts. X-ray photoelectron spectra for (b) Pt–C, (c) and (d) Pt–Sn/SnO2(33 : 7)-C catalysts. Reproduced from ref. 276 with permission from The Royal Society of Chemistry.

regions are observed. The voltammograms demonstrate the H_2 adsorption/desorption between 0 and 0.3 V (*vs.* NHE) for all the catalysts. It clearly shows that Pt–Sn based catalysts have a higher distinct H_2 adsorption/desorption region and the peaks are also shifted to lower potential in relation to Pt–C. Larger capacitive current observed in the double layer region for Pt–Sn/SnO$_2$(33 : 7)-C catalyst may be due to higher segregation of SnO_2 species in comparison to Pt–Sn/SnO$_2$(35 : 5)-C and Pt–Sn/SnO$_2$(31 : 9)-C catalysts.[272,280] Inset of Fig. 38(a) shows the voltammograms of oxide formation/reduction regions between 0.5 and 1.0 V for Pt–C and Pt–Sn/SnO2(33 : 7)-C catalysts, therein small peaks appear at around 0.59 and 0.71 V, which may be assigned to the adsorption and desorption of oxygen-containing species resulting from the dissociation of water on SnO_2.[256] The electrochemical active surface area (ESA) was calculated using the H_2 desorption peak area of the CV curves. The ESA can be calculated from the following equation:[273]

$$\text{ESA}\ (\text{cm}^2\ \text{g}^{-1}{}_{\text{Pt}}) = Q_{\text{H}}\ (\mu\text{C cm}^{-2})/\{210\ \mu\text{C cm}^{-2} \times \text{electrode loading}\ (\text{g}_{\text{Pt}}\,\text{cm}^{-2})\}$$

where Q_H represents the charge of hydrogen desorption and 210 μC cm^{-2} is the charge required to oxidize a monolayer of H_2 on smooth platinum surface. The estimated ESA values of all the catalysts are given in

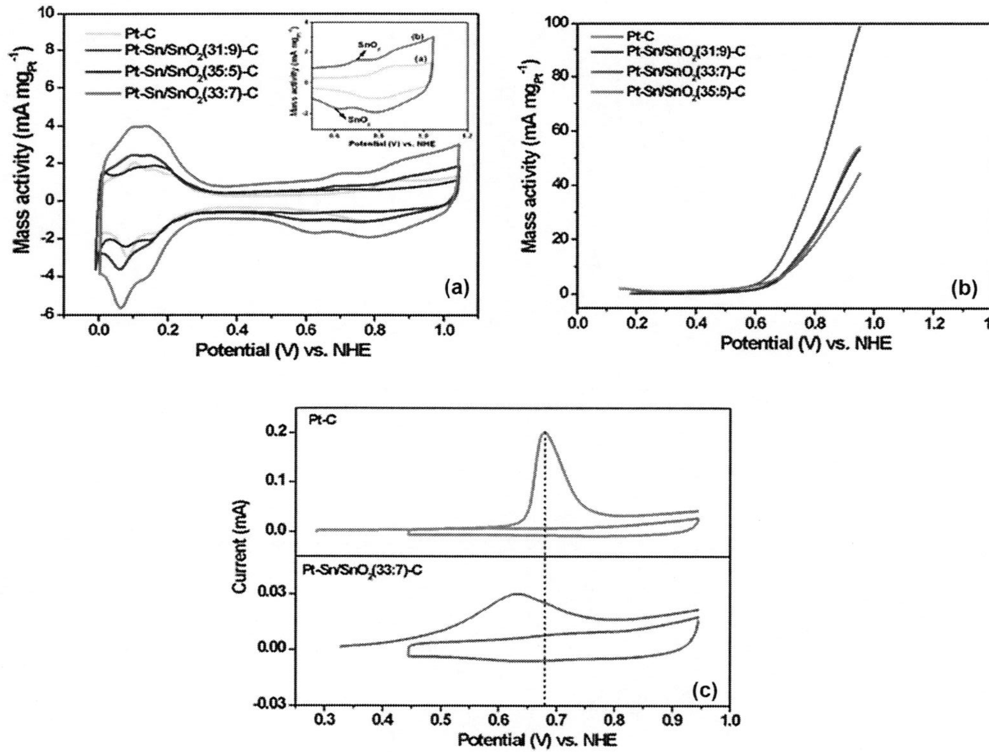

Fig. 38 (a) Cyclic voltammograms for Pt–Sn/SnO_2(31 : 9)-C, Pt–Sn/SnO_2(33 : 7)-C, Pt–Sn/SnO_2(35 : 5)-C and Pt–C catalysts containing Pt and Sn in varying weight ratios and inset shows the voltammogram of oxide formation/reduction regions of (green) Pt–C and (pink) Pt–Sn/SnO_2(33 : 7)-C catalysts in N_2-saturated aq. 0.5 M $HClO_4$ at a scan rate of 50 mV s^{-1}, (b) Cyclic voltammograms for Pt–Sn/SnO_2(31 : 9)-C, Pt-Sn/SnO_2(33 : 7)-C, Pt–Sn/SnO_2(35 : 5)-C and Pt–C catalysts containing Pt and Sn in varying weight ratios in aq. solution containing 0.5 M $HClO_4$ and 1 M CH_3CH_2OH saturated with N_2 at a scan rate of 50 mV s^{-1}, (c) Stripping voltammograms of adsorbed CO on Pt–C and Pt–Sn/SnO_2 (33 : 7)-C electrocatalysts recorded at 10 mV s^{-1}. Reproduced from ref. 276 with permission from The Royal Society of Chemistry.

Table 11. Pt–Sn/SnO_2(33 : 7)-C shows higher ESA compared with all the other catalysts. The improved ESA is due to the quick adsorption and easy desorption of H_2 on the Sn modified Pt surface in relation to carbon supported pristine Pt in $HClO_4$.

Cyclic voltammograms for electro-oxidation of ethanol on Pt–Sn/SnO_2(31 : 9)-C, Pt–Sn/SnO_2(33 : 7)-C, Pt–Sn/SnO_2(35 : 5)-C and Pt–C electro-catalysts in 0.5 M aq. $HClO_4$ and 1 M aq. CH_3CH_2OH at a scan rate of 50 mV s^{-1} are given in Fig. 38(b). The onset potential shifted towards negative for Pt–Sn/SnO_2(33 : 7)-C catalyst compared to all the other catalysts, the lowest onset potential obtained for the oxidation of ethanol is due to an electronic effect and bifunctional effect in the Sn modified Pt-based materials. Mass activities are calculated for the above referred catalysts at 0.8 V and the results are given in Table 11. Pt–Sn/SnO_2(33 : 7)-C catalyst gives higher mass activity as compared to Pt–Sn/SnO_2(31 : 9)-C, Pt–Sn/SnO_2(35 : 5)-C and Pt–C catalysts.

CO species are the main poisoning intermediates during ethanol electro-oxidation.[276] An effective catalyst should have excellent CO electro-oxidation ability, which can be ascertained by CO stripping experiment. Figure 38(c) shows CO stripping curves for Pt–C and Pt–Sn/SnO_2(33 : 7)-C catalysts obtained at a sweep rate of 10 mV s^{-1} in $HClO_4$ electrolyte at room temperature. The onset potentials for CO oxidation are around 620 and 330 mV while the peak potentials are at 680 and 635 mV (*vs.* NHE) for Pt–C and Pt–Sn/SnO_2(33 : 7)-C catalysts, respectively. The potential shifted negative for Pt–Sn catalyst in relation with Pt–C, which is in agreement with literature.[274,275] The negative shift in potential in the stripping voltammogram for the catalyst is due to the presence of oxygenated species on Sn sites formed at lower potentials in comparison with Pt. According to bifunctional mechanism, these oxygenated species allow the oxidation of CO to CO_2 at lower potentials. Vigier *et al.*[260] accounted that SnO_2 could supply the oxygen species for the oxidation of CO adsorbed on the Pt sites, thereby enhancing the oxidation of ethanol at lower potentials. Furthermore, the shape of the CO stripping peak depends on the nature of the catalyst, the range of potential allowing for oxidation of CO on the Pt–Sn/SnO_2(33 : 7)-C catalyst is wider than Pt–C. This confirms that SnO_2 enhances the catalytic activity of Pt–Sn/SnO_2(33 : 7)-C catalyst.

Single cell DEFC performances are examined for Pt–Sn/SnO_2–C anode catalyst. Figure 39 shows the performance for commercial Pt–C, prepared Pt–C and Pt–Sn/SnO_2(33 : 7)-C catalyst (chosen based on half-cell studies)

Table 11 Electrochemical surface area and mass activities of the catalysts.

Catalyst	Electrochemical surface area (m^2 g^{-1})	Mass activity $I_{0.8\ V}$ (mA $mg^{-1}{}_{Pt}$)
Pt–C	56	21
Pt–Sn/SnO_2(31 : 9)-C	72	23
Pt–Sn/SnO_2(33 : 7)-C	115	42
Pt–Sn/SnO_2(35:5)-C	62	19

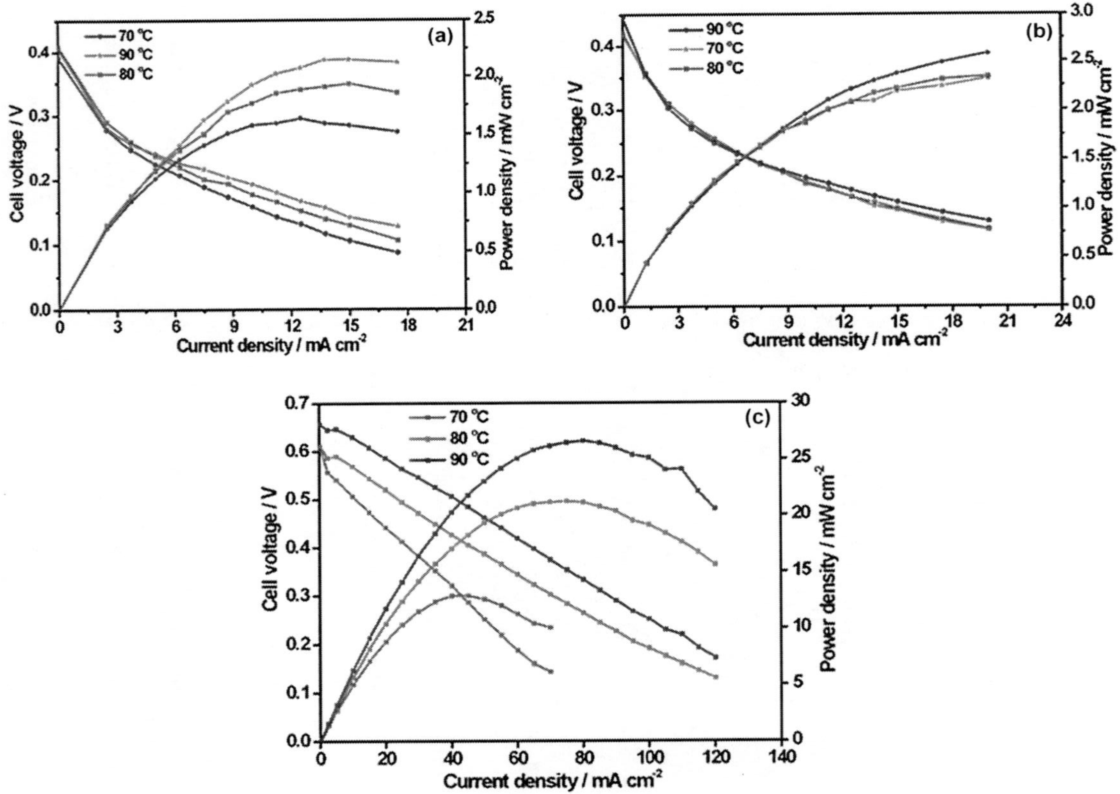

Fig. 39 Steady-state performance data of DEFCs (CH_3CH_2OH and O_2) comprising MEA prepared using decal transfer method and (a) commercial Pt–C (b) prepared Pt–C and (c) Pt–Sn/SnO_2(33 : 7)-C anode (0.75 mg cm^{-2}) catalyst, cathode (Pt/C)-1.25 mg cm^{-2}, Nafion 117, ethanol fed (2 M) – 2 mL min^{-1} and oxygen – 300 mL min^{-1} under atmospheric pressure. Reproduced from ref. 276 with permission from The Royal Society of Chemistry.

at 70, 80 and 90 °C. Peak power densities of 2.6, 2.2 and 27.0 mW cm^{-2} are obtained for MEAs comprising commercial Pt–C, prepared Pt–C and Pt–Sn/SnO_2(33 : 7)-C catalysts respectively, at 90 °C under ambient pressure. Pt–Sn/SnO_2–C as anode catalyst exhibited enhanced peak power density with reduced loading of 0.75 mg cm^{-2} compared with Pt–C and also that reported in the literature with a higher loading of 2.0 mg cm^{-2}.[86,88,268,281]

4 Conclusions

The above reviewed membrane electrolytes has shown tremendous potential for its application in DAFCs. Inorganic fillers such as SiO_2, MZP, MTP, porous zeolites along with graphene has significant contribution towards mitigation of methanol crossover and in improving the DMFC performances. Among all, Nafion-Al-MCM-41 hybrid membrane deliver peak-power densities of 246 mW cm^{-2} which is highest among all other hybrid membrane studied here. Excellent DMFC performance of Nafion-aluminosilicate is attributed to the strong Brønsted acidity and quasi-solid electrolytes properties of the aluminosilicate in which protons move along the hydrogen bonds. Further, attractive textural and surface properties of porous aluminosilicate facilitate specific ionic-interactions with the host Nafion matrix which contributes in enhancing DMFC performance. sPEEK-POP blends have shown restricted methanol cross-over due to the hydrophobic nature of POP partially blocking the ionic channels with slight compromise in the ionic conduction, whereas sPEEK composite by incorporating zeolites and clays have shown better ionic conductivity and DAFC performance with compromise in long term stability. Carbon nanostructures incorporation like PSSA-CNT and Sulfonated Fullerene to sPEEK has shown improvement in both the aspects in terms of long term stability and DAFC performance due to specific intermolecular interaction between sPEEK and CNT/Fullerene. Anion exchange membrane of polyphenylene oxide (APPO) has also shown restricted alcohol cross-over property in alkaline DAFCs due to the optimized degree of quaternization in the host matrix.

It would be desirable to block some or all the three Pt sites to impede methanol oxidation. The enhanced electrocatalytic activity and methanol tolerance for Pt–Au/C alloy is explained by the electronic factor, namely, the change of the d-band vacancy in Pt upon alloying and/or by a geometric effect. Both the effects may enhance the reaction rate for oxygen adsorption and cleavage of O–O bond during the reduction reaction. DFT calculations reflect that, in DMFCs containing Pt–Au/C in 2 : 1 and 1 : 1 ratios, the probability of finding three neighboring Pt atoms on the surface is lower than in Pt and Pt–Au (3 : 1) making the former more methanol-tolerant in accordance with the experimental data. This has been validated in the case of Pt–Pd (2 : 1)/C alloy catalyst, which exhibits the maximum ORR activity in DMFCs.

750 Pt–TiO_2/C (2 : 1) cathode catalyst has higher ORR selectivity and better methanol tolerance in relation to Pt/C cathode. Enhancement of

electro-catalytic activity could be due to change in electronic structure of the catalyst arising due to the electronic interactions between Pt and TiO_2 due to the hypo d-electron character of TiO_2 and hyper d-electron character of Pt. Pt–$Y(OH)_3$(2 : 1)/C hybrid catalyst exhibits the maximum ORR activity in presence of methanol. DEFC employing HT–Pt–TiO_2(2 : 1)/C as cathode catalyst exhibits better performance than that with Pt/C. The improved ORR activity in these catalysts may also be due to the dynamic spillover of primary oxides.

The maximum electrocatalytic activity for ORR in presence of methanol was observed for PdCo(3 : 1)/C in ADMFCs. The DFT calculations established that PdCo(3 : 1)/C has the better catalytic activity than the other catalysts based on band gap, bond lengths etc. Pd–TiO2(3 : 1)/C shows much improved performance over Pd–HTRuSe(2 : 1)/C as cathode catalyst in ADMFCs.

Methanol oxidation and accelerated durability studies for Pt–Ru/TiO_2–C show enhanced catalytic activity and durability in relation to the carbon supported Pt–Ru. There is some co-operative effect for platinum catalyst in presence of TiO_2, which facilitates oxidation of adsorbed CO. In this catalytic system, TiO_2–C and Ru display anti-poisoning function, the carbon network benefits the electron transfer, and Pt works as the main dehydrogenation site The improved durability for TiO_2–C supported electro-catalysts may be due to the strong interaction between Pt–Ru nano particles and TiO_2 support; it would be less likely for Pt–Ru atoms on TiO_2–C to agglomerate to produce larger particles that would cause gradual decrease in electrochemical activity or metal dissolution from the support surface.

The mass activity is higher for Pt–Sn/SnO_2(33 : 7)-C cathode catalyst than the prepared and commercial Pt–C and also than that reported for Pt–Sn/C catalyst in DEFCs. The higher activity obtained is due to the fact that SnO_2 can transform CO-like poisoning species on Pt into CO_2, leaving the active sites on Pt for further adsorption and oxidation of ethanol by the bifunctional mechanism.

Acknowledgements

Authors thank CSIR for the funding under 12th Five Year Plan Programme of HYDEN (CSC-0122), D-NEED (PSC0109) and NMITLI (TLP-01/13). Authors thank research scholars Mr. Rambabu Gutru, Mrs. S. Sasikala, Mr. K Hari Gopi, Mr. V. Parthiban and Mr. A. Arunchander for their timely help and support. Authors also thank Director, CSIR-CECRI and Scientist-In-Charge, CECRI Madras Unit for the support. One of the authors (P. Sridhar) thanks the management M/s Mesha Energy Solutions Pvt. Ltd. for permitting to spare time in the preparation of the manuscript.

References

1 W. J. Zhou, B. Zhou, W. Z. Li, Z. H. Zhou, S. Q. Song, G. Q. Sun, Q. Xin, S. Douvartzides, M. Goula and P. Tsiakaras, *J. Power Sources*, 2004, **126**, 160.

2 A. Brouzgou, A. Podias and P. Tsiakaras, *J. Appl. Electrochem.*, 2013, **43**, 119.

3 E. Antolini, J. R. C. Salgado and E. R. Gonzalez, *Appl. Catal., B*, 2006, **63**, 137.
4 U. Bossel, The Physics of the Hydrogen Economy, *European Fuel Cell News*, 2003. p. 16.
5 H. R. Corti and E. R. Gonzalez, *Direct Alcohol Fuel Cells - Materials, Performance, Durability and Applications*, Springer, ISBN 978-94-007-7707-1, 2014, ch. 6, p. 121.
6 S. Song, V. Maragou and P. Tsiakaras, *J. Fuel Cell Sci. Technol.*, 2006, **4**, 203.
7 S. Song and P. Tsiakaras, *Appl. Catal., B*, 2006, **63**, 187.
8 E. Antolini, *J. Power Sources*, 2007, **170**, 1.
9 V. Neburchilov, J. Martin, H. Wang and J. Zhang, *J. Power Sources*, 2007, **169**, 221.
10 H. K. Kim and H. Chang, *J. Membr. Sci.*, 2007, **288**, 188.
11 H. Liu, C. Song, L. Zhang, J. Zhang, H. Wang and D. P. Wilkinson, *J. Power Sources*, 2006, **155**, 95.
12 A. K. Sahu, G. Selvarani, S. Pitchumani, P. Sridhar and A. K. Shukla, *J. Electrochem. Soc.*, 2007, **154**, 123.
13 A. K. Sahu, G. Selvarani, S. Pitchumani, P. Sridhar, A. K. Shukla, N. Narayanan, A. Banerjee and N. Chandrakumar, *J. Electrochem. Soc.*, 2008, **155**, B686.
14 F. Langenbucher, *J. Pharm. Pharmacol.*, 1972, **24**, 979.
15 K. Kosmidis, P. Argyrakis and P. Macheras, *Pharm. Res.*, 2003, **20**, 988.
16 V. Papadopoulou, K. Kosmidis, M. Vlachou and P. Macheras, *Int. J. Pharm.*, 2006, **309**, 44.
17 A. K. Sahu, S. D. Bhat, S. Pitchumani, P. Sridhar, V. Vimalan, C. George, N. Chandrakumar and A. K. Shukla, *J. Membr. Sci.*, 2009, **345**, 305.
18 Y. Jin, S. Qiao, L. Zhang, Z. P. Xu, S. Smart, J. C. Diniz da costa and G. Q. Lu, *J. Power Sources*, 2008, **185**, 664.
19 A. K. Sahu, S. Meenakshi, S. D. Bhat, A. Shahid, P. Sridhar, S. Pitchumani and A. K. Shukla, *J. Electrochem. Soc.*, 2012, **159**, F702.
20 S. Meenakshi, A. K. Sahu, S. D. Bhat, P. Sridhar, S. Pitchumani and A. K. Shukla, *Electrochim. Acta*, 2013, **89**, 35.
21 G. Alberti and M. Casciola, in *Proton Conductors*, ed. P. Colomban, Cambridge University Press, Cambridge, 1992, p. 238.
22 W. Fan, M. A. Snyder, S. Kumar, P. Lee, W. C. Yoo, A. V. Mccormick, R. L. Penn, A. Stein and M. Tsapatsis, *Nat. Mater.*, 2008, **7**, 984.
23 C. W. Lin and Y. S. Lu, *J. Power Sources*, 2013, **237**, 187.
24 Y. S. Ye, C. Y. Tseng, W. C. Shen, J. S. Wang, K. J. Chen, M. Y. Cheng, J. Rick, Y. J. Huang, F. C. Chang and B. J. Hwang, *J. Mater. Chem.*, 2011, **21**, 10448.
25 B. G. Choi, J. Hong, Y. C. Park, D. H. Jung, W. H. Hong, P. T. Hammond and H. S. Park, *ACS Nano*, 2011, **5**, 5167.
26 S. Ansari, A. Kelarakis, L. Estevez and E. P. Giannelis, *Small*, 2010, **6**, 205.
27 J. J. Liang, Y. Huang, L. Zhang, Y. Wang, Y. F. Ma, T. Y. Guo and Y. S. Chen, *Adv. Funct. Mater.*, 2009, **19**, 2297.
28 H. C. Chien, L. D. Tsai, C. P. Huang, C. Kang, J. N. Lin and F. C. Chang, *Int. J. Hydrogen Energy*, 2013, **38**, 13792.
29 V. Parthiban, S. Akula, S. G. Peera, N. Islam and A. K. Sahu, *Energy Fuels*, 2016, **30**, 725.
30 R. Kumar, C. Xu and K. Scott, *RSC Adv.*, 2012, **2**, 8777.
31 Z. Liu, F. Wang and T. L. Chen, *Macromol. Rapid Commun.*, 2001, **22**, 579.
32 K. D. Kreuer, *J. Membr. Sci.*, 2001, **185**, 29.
33 S. P. Nunes, B. Ruffmann, E. Rikowski, S. Vetter and K. Richau, *J. Membr. Sci.*, 2002, **203**, 215.

34 P. Xing, G. P. Robertson, M. D. Guiver, S. D. Mikhailenko and S. Kaliagune, *J. Polym. Sci., Part A: Polym. Chem.*, 2004, **42**, 2866.
35 S. Xue and G. P. Yin, *Polymer*, 2006, **47**, 5044.
36 C. J. Zhao, Z. Wang, D. W. Bi, H. D. Lin, K. Shao, T. Z. Fu, S. L. Zhong and H. Na, *Polymer*, 2007, **48**, 3090.
37 X. F. Li, C. P. Liu, D. Xu, C. J. Zhao, Z. Wang, G. Zhang, H. Na and W. Xing, *J. Power Sources*, 2006, **162**, 1.
38 X. F. Li, D. J. Chen, D. Xu, C. J. Zhao, Z. Wang, H. Lu and H. Na, *J. Membr. Sci.*, 2006, **275**, 134.
39 R. Carter, R. Wycisk, H. Yoo and P. N. Pintauro, *Electrochem. Solid-State Lett.*, 2002, **5**, A195.
40 H. L. Wu, C. C. M. Ma, H. C. Kuan, C. H. Wang, C. Y. Chen and C. L. Chiang, *J. Polym. Sci., Part B: Polym. Phys.*, 2006, **44**, 565.
41 P. Potin and R. DeJaeger, *Eur. Polym. J.*, 1991, **27**, 341.
42 H. R. Allcock, R. J. Fitzpatrick and L. Salvati, *Chem. Mater.*, 1991, **3**, 1120.
43 H. R. Allcock and M. W. Richard, *J. Polym. Sci., Part B: Polym. Phys.*, 2006, **44**, 2358.
44 S. G. Peera, S. Meenakshi, K. H. Gopi, S. D. Bhat, P. Sridhar and S. Pitchumani, *RSC Adv.*, 2013, **3**, 14048.
45 W. Han-Lang, M. M. Chen-Chi, L. Chia-Hsun, L. Tzong- Ming, C. Chih-Yuan, C. Chin-Lung and W. Chen, *J. Membr. Sci.*, 2006, **280**, 501.
46 M. L. Di Vona, E. Sgreccia, A. Donnadio, M. Casciola, J. F. Chailan, G. Auer and P. Knauthe, *J. Membr. Sci.*, 2011, **369**, 536.
47 M. J. Sumner, W. L. Harrison, R. M. Weyers, Y. S. Kim, J. E. McGrath, J. S. Riffle, A. Brink and M. H. Brink, *J. Membr. Sci.*, 2004, **239**, 199.
48 N. Carretta, V. Tricoli and F. Picchioni, *J. Membr. Sci.*, 2000, **166**, 189.
49 Y. Gao, G. P. Robertson, M. D. Guiver, X. G. Jian, S. D. Mikhailenko, K. P. Wang and K. Serge, *J. Polym. Sci., Part A: Polym. Chem.*, 2003, **41**, 2731.
50 N. P. Chen and L. Hong, *Polymer*, 2004, **45**, 2403.
51 S. Meenakshi, S. D. Bhat, A. K. Sahu, P. Sridhar and S. Pitchumani, *Fuel Cells*, 2013, **13**, 851.
52 C. P. Zhang, S. M. Sharkh, X. Li, F. C. Walsh, C. N. Zhang and J. C. Jiang, *Energy Convers. Manage.*, 2011, **52**, 3391.
53 D. W. Breck, *Zeolite Molecular Sieves, Structure, Chemistry and Use*, John Wiley & Sons, New York, USA, 1974.
54 Z. Huang, Y. Shi, R. Wen, Y. Guo, J. Su and T. Matsuura, *Sep. Purif. Technol.*, 2006, **51**, 126.
55 H. V. Bekkum, E. R. Geus and H. W. Kouwenhoven, *Stud. Surf. Sci. Catal.*, 1994, **85**, 509.
56 H. Chon, S. I. Woo and S. E. Park, *Recent Advances and New Horizons in Zeolite Science and Technology*, Elsevier, NewYork, USA, 1996.
57 C. Zhao, X. Li, Z. Wang, Z. Dou, S. Zhong and H. Na, *J. Membr. Sci.*, 2006, **280**, 643.
58 B. Budiyono, T. D. Kusworo, A. F. Ismail, I. N. Widiasa, S. Johari and S. Sunarso, *Int. J. Basic Appl. Sci.*, 2010, **10**, 1.
59 S. D. Bhat, A. K. Sahu, C. George, S. Pitchumani, P. Sridhar, N. Chandrakumar, K. K. Singh, N. Krishna and A. K. Shukla, *J. Membr. Sci.*, 2009, **340**, 73.
60 D. Shah, K. Kissick, A. Ghorpade, R. Hannah and D. Bhattacharyya, *J. Membr. Sci.*, 2000, **179**, 185.
61 Z. Gaowen and Z. Zhentao, *J. Membr. Sci.*, 2005, **261**, 107.
62 Y. F. Lin, C. Y. Yen, C. H. Hung, Y. H. Hsiao and C. C. M. Ma, *J. Power Sources*, 2007, **168**, 162.

63 D. J. Kim, H. Y. Hwang, S. B. Jung and S. Y. Nam, *J. Ind. Eng. Chem.*, 2012, **18**, 556.
64 S. Sasikala, S. Meenakshi, S. D. Bhat and A. K. Sahu, *Electrochim. Acta*, 2014, **135**, 232.
65 N. Stankovic, M. Logar, J. Lukovic, J. Pantic, M. Miljevic, B. Babic and A. R. Mihajlovic, *Process. Appl. Ceram.*, 2011, **5**, 97.
66 Q. H. Hu, S. Z. Qiao, F. Haghseresht, M. A. Wilson and G. Q. Lu, *Ind. Eng. Chem. Res.*, 2006, **45**, 733.
67 K. G. Bhattacharyya and S. S. Gupta, *Appl. Clay Sci.*, 2008, **41**, 1.
68 S. M. I. Sajidu, I. Persson, W. R. L. Masamba and E. M. T. Henry, *J. Hazard. Mater.*, 2008, **158**, 401.
69 M. A. Akl, A. M. Youssef and M. M. Al-Awadhi, *J. Anal. Bioanal. Tech.*, 2013, **4**, 1.
70 E. Eren, B. Afsin and Y. Onal, *J. Hazard. Mater.*, 2009, **161**, 677.
71 C. H. Rhee, H. K. Kim, H. Chang and J. S. Lee, *Chem. Mater.*, 2005, **17**, 1691.
72 A. Yahiaoui, M. Belbachir and A. Hachemaoui, *Int. J. Mol. Sci.*, 2003, **4**, 548.
73 J. T. Wang, S. Wasmus and R. F. Savinell, *J. Electrochem. Soc.*, 1996, **143**, 1239.
74 W. Zhang and S. R. P. Silva, *Rev. Adv. Mater. Sci.*, 2011, **29**, 1.
75 R. Kannan, B. A. Kakade and V. K. Pillai, *Angew. Chem., Int. Ed.*, 2008, **47**, 2653.
76 N. P. Cele, S. S. Ray, S. K. Pillai, M. Ndwandwe, S. Nonjola, L. Sikhwivhilu and M. K. Mathe, *Fuel Cells*, 2010, **10**, 64.
77 S. H. Joo, C. Pak, E. A. Kim, Y. H. Lee, H. Chang, D. Seung, Y. S. Choi, J. B. Park and T. K. Kim, *J. Power Sources*, 2008, **180**, 63.
78 W. Zhou, J. Xiao, Y. Chen, R. Zeng, S. Xiao, H. Nie, F. Li and C. Song, *Polym. Adv. Technnol.*, 2011, **22**, 1747.
79 G. Rambabu and S. D. Bhat, *Chem. Eng. J.*, 2014, **243**, 517.
80 X. Zhang, Q. Tang, D. Yang, W. Hua, Y. Yue, B. Wang, X. Zhang and J. Hu, *Mater. Chem. Phys.*, 2011, **126**, 310.
81 K. Tasaki, J. Gasa, H. Wang and R. DeSousa, *Polymer*, 2007, **48**, 4438.
82 S. Saga, H. Matsumoto, K. Saito, M. Minagawa and A. Tanioka, *J. Power Sources*, 2008, **176**, 16.
83 G. Rambabu and S. D. Bhat, *Electrochim. Acta*, 2015, **176**, 657.
84 Y. Kanbur and Z. Küçükyavuz, *Nanostructures*, 2012, **20**, 119.
85 J. Ji, G. Zhang, H. Chen, S. Wang, G. Zhang, F. Zhang and X. Fan, *Chem. Sci.*, 2011, **2**, 484.
86 C. Lamy, S. Rousseau, E. M. Belgsir, C. Coutanceau and J. M. Leger, *Electrochim. Acta*, 2004, **49**, 3901.
87 L. Geng, Y. He, D. Liu, X. Dia and C. Lu, *Microporous Mesoporous Mater.*, 2012, **148**, 8.
88 S. Rousseau, C. Coutanceau, C. Lamy and J. M. Leger, *J. Power Sources*, 2006, **158**, 18.
89 A. D. Blasi, V. Baglio, A. Stassi, C. D. Urso, V. Antonucci and A. S. Arico, *ECS Trans.*, 2006, **3**, 1317.
90 L. Barbora, S. Acharya and A. Verma, *Macromol. Symp.*, 2009, **277**, 177.
91 T. Mathuraiveeran, K. Roelofs, D. Senftleben and T. Schiestel, *Desalination*, 2006, **200**, 662.
92 K. S. Roelofs, A. Kampa, T. Hirth and T. Schiestel, *J. Appl. Polym. Sci.*, 2009, **111**, 2998.
93 K. S. Roelofs, T. Hirth and T. Sciestel, *J. Membr. Sci.*, 2010, **346**, 215.
94 A. R. Tan, L. M. Carvalho, F. G. Ramos Filho and A. S. Gomes, *Macromol. Symp.*, 2006, **245–246**, 470.

95 S. Xue, G. Yin, K. Cai and Y. Shao, *J. Membr. Sci.*, 2007, **289**, 51.
96 S. Meenakshi, A. Manokaran, S. D. Bhat, A. K. Sahu, P. Sridhar and S. Pitchumani, *Fuel Cells*, 2014, **14**, 842.
97 R. Jiang, H. R. Kunz and J. M. Fenton, *J. Membr. Sci.*, 2006, **272**, 116.
98 J. R. Varcoe and R. C. T. Slade, *Fuel Cells*, 2005, **5**, 187.
99 S. Lu, J. Pan, A. Huang, L. Zhuang and J. Lu, *Proc. Natl. Acad. Sci.*, 2008, **105**, 20611.
100 G. Couture, A. Alaaeddine, F. Boschet and B. Ameduri, *Prog. Polym. Sci.*, 2011, **36**, 1521.
101 R. Janarthanan, S. K. Pilli, J. L. Horan, D. A. Gamarra, M. R. Hibbs and A. M. Herring, *J. Electrochem. Soc.*, 2014, **161**, F944.
102 A. Katzfuß, V. Gogel, L. Jörissen and J. Kerres, *J. Membr. Sci.*, 2013, **425–426**, 131.
103 K. HariGopi, S. D. Bhat, A. K. Sahu and P. Sridhar, *J. Appl. Polym. Sci.*, 2016, **133**, 43693.
104 P. Hernández-Fernández, S. Rojas, P. Ocón, J. L. Gómez de la Fuente, J. S. Fabián, Sanza, M. A. Peňa, F. J. Garcìa-Garcìa, P. Terreros and J. L. G. Fierro, *J. Phys. Chem. C*, 2007, **111**, 2913.
105 J. Mathiyarasu and K. L. N. Phani, *J. Electrochem. Soc.*, 2007, **154**, B1100.
106 H. Li, Q. Xin, W. Li, Z. Zhou, L. Jiang, S. Yang and G. Sun, *Chem. Commun.*, 2004, **23**, 2776.
107 M. H. Shao, T. Huang, P. Liu, J. Zhang, K. Sasaki, M. B. Vukmirovic and R. R. Adzic, *Langmuir*, 2006, **22**, 10409.
108 G. Selvarani, S. Vinod Selvaganesh, S. Krishnamurthy, G. V. M. Kiruthika, P. Sridhar, S. Pitchumani and A. K. Shukla, *J. Phys. Chem. C*, 2009, **113**, 7461.
109 Y. Shimodaira, T. Tanaka, T. Miura, A. Kudo and H. Kobayashi, *J. Phys. Chem. C*, 2007, **111**, 272.
110 M. Haruta, *Nature*, 2005, **437**, 1098.
111 M. Haruta, *Catal. Today*, 1997, **36**, 153.
112 M. S. El-Deab and T. Ohsaka, *Electrochem. Commun.*, 2002, **4**, 288.
113 M. S. El-Deab and T. Ohsaka, *J. Electroanal. Chem.*, 2003, **553**, 107.
114 M. O. Pedersen, S. Helveg, A. Ruban, I. Stensgaard, E. Laegsgaard, J. K. Nørskov and F. Besenbacher, *Surf. Sci.*, 1996, **426**, 395.
115 H. Moller and P. C. Pistorius, *J. Electroanal. Chem.*, 2004, **570**, 243.
116 K. Tekaia-Elhsissen, F. Bonet, P. Y. Silvert and R. J. Herrera-Urbina, *J. Alloys Compd.*, 1999, **292**, 96.
117 J. Luo, P. N. Njoki, Y. Lin, L. Wang and C. Zhong, *Electrochem. Commun.*, 2006, **8**, 581.
118 P. Hernández-Fernández, S. Rojas, P. Ocǒn, A. de Frutos, J. M. Figueroa, P. Terreros, M. A. Peňa and J. L. G. Fierro, *J. Power Sources*, 2008, **177**, 9.
119 J. Wang, G. Yin, G. Wang, Z. Wang and Y. Gao, *Electrochem. Commun.*, 2008, **10**, 831.
120 J. Zhang, K. Sasaki, E. Sutter and R. R. Adzic, *Science*, 2007, **315**, 220.
121 O. Savadogo, K. Lee and K. Oishi, *Electrochem. Commun.*, 2004, **6**, 105.
122 R. K. Pandey and V. Laksminarayanan, *J. Phys. Chem. C*, 2009, **113**, 21596.
123 J. Zhang, Y. Mo, M. B. Vukmirovic, Y. Xu, M. Mavrikakis and R. R. Adzic, *Angew. Chem.*, 2005, **44**, 2132.
124 A. H. Creus, Y. Gimeno, P. Diaz, L. Vazquez, S. Gonzalez, R. C. Salvarezza and J. Arvia, *J. Phys. Chem. B*, 2004, **108**, 10785.
125 K. G. Nishanth, P. Sridhar, S. Pitchumani and A. K. Shukla, *J. Electrochem. Soc.*, 2011, **158**, B871.
126 A. Sarkar and A. Manthiram, *J. Phys. Chem. C*, 2010, **114**, 4725.

127 Y. Jinhua, Z. Weijiang, H. C. Chin, Y. L. Jim and L. Zhaolin, *ACS Appl. Mater. Interfaces*, 2010, **2**, 119.
128 W. Rongfang, L. Hao, J. Shan, W. Hui and L. Ziqiang, *Electrochim. Acta*, 2010, **55**, 1519.
129 W. Deli, L. X. Huolin, Y. Yingchao, W. Hongsen, R. Eric, A. M. David and H. D. Abruña, *J. Am. Chem. Soc.*, 2010, **132**, 17664.
130 E. Antolini, J. R. C. Salgado and E. R. Gonzalez, *J. Power Sources*, 2006, **160**, 957–968.
131 K. G. Nishanth, P. Sridhar and S. Pitchumani, *Int. J. Hydrogen Energy*, 2013, **38**, 612.
132 C. Wang and A. J. Appleby, *J. Electrochem. Soc.*, 2003, **150**, A493.
133 S. R. D'Souza, J. Ma and C. Wang, *J. Electrochem. Soc.*, 2006, **153**, A1795.
134 G. Selvarani, A. K. Sahu, G. V. M. Kiruthika, P. Sridhar, S. Pitchumani and A. K. Shukla, *J. Electrochem. Soc.*, 2009, **156**, B118.
135 T. Ioroi, Z. Siroma, N. Fujiwara, S. I. Yamazaki and K. Yasuda, *Electrochem. Commun.*, 2005, **7**, 183.
136 J. M. Chen, L. S. Sarma, C. H. Chen, M. Y. Cheng, S. C. Shih, G. R. Wang, D. G. Liu, J. F. Lee, M. T. Tang and B. J. Hwang, *J. Power Sources*, 2006, **159**, 29.
137 J. Shim, C. R. Lee, H. K. Lee, J. S. Lee and E. J. Cairns, *J. Power Sources*, 2001, **102**, 172.
138 N. Rajalakshmi, N. Lakshmi and K. S. Dhathathreyan, *Int. J. Hydrogen Energy*, 2008, **33**, 7521.
139 Y. Fu, Z. D. Wei, S. G. Chen, L. Li, Y. C. Feng, Y. Q. Wang and M. J. Liao, *J. Power Sources*, 2009, **189**, 982.
140 H. J. Kim, D. Y. Kim, H. Han and Y. G. Shul, *J. Power Sources*, 2006, **159**, 484.
141 L. Xiong and A. Manthiram, *Electrochim. Acta*, 2004, **49**, 4163.
142 H. Song, X. Qiu, X. Li, F. Li, W. Zhu and L. Chen, *J. Power Sources*, 2007, **170**, 50.
143 M. Gustavsson, H. Ekstrom, P. Hanarp, L. Eurenius, G. Lindbergh, E. Olsson and B. Kasemo, *J. Power Sources*, 2007, **163**, 671.
144 L. Brewer and P. R. Wengert, *Metall. Trans.*, 1973, **4**, 83.
145 B. Hammer and J. K. Norskov, *Adv. Catal.*, 2000, **45**, 71.
146 G. Selvarani, S. Maheswari, P. Sridhar, S. Pitchumani and A. K. Shukla, *J. Electrochem. Soc.*, 2009, **156**, B1354.
147 F. Rosalbinoa, S. Delsante, G. Borzone and E. Angelini, *Int. J. Hydrogen Energy*, 2008, **33**, 6696.
148 M. M. Jaksic, *Electrochim. Acta*, 2000, **45**, 4085.
149 S. G. Neophytides, K. Murase, S. Zafeiratos, G. D. Papakonstantinou, F. E. Paloukis, N. V. Krstajic and M. M. Jaksic, *J. Phys. Chem. B*, 2006, **110**, 3030.
150 J. M. Jaksic, D. Labou, G. D. Papakonstantinou, A. Siokou and M. M. Jaksic, *J. Phys. Chem. C*, 2010, **114**, 18298.
151 K. G. Nishanth, P. Sridhar and S. Pitchumani, *Electrochem. Comm.*, 2011, **13**, 1465.
152 J. B. Goodenough, R. Manoharan, A. K. Shukla and K. V. Ramesh, *Chem. Mater.*, 1989, **1**, 391.
153 A. K. Shukla, A. S. Aricò, K. M. El-Khatib, H. Kim, P. L. Antonucci and V. Antonucci, *Appl. Surf. Sci.*, 1999, **137**, 20.
154 S. J. Trasatti, *Electroanal. Chem.*, 1971, **33**, 351.
155 J. Li, H. J. Zhai and L. S. Wang, *Science*, 2003, **299**, 864.
156 T. Lopes, E. Antolini and E. R. Gonzalez, *Int. J. Hydrogen Energy*, 2008, **33**, 5563.
157 W. Wenming, H. Qinghong, L. Juanyin, Z. Zhiqing, L. Zhilin and Y. Hui, *Electrochem. Commun.*, 2008, **10**, 1396.

158 L. Huanqiao, S. Gongquan, L. Na, S. Shiguo, S. Dangsheng and X. Qin, *J. Phys. Chem. C*, 2007, **111**, 5605.
159 G. F. Alvarez, M. Mamlouk and K. Scott, *Int. J. Electrochem.*, 2011, 684535.
160 Y. Jinhua, Y. L. Jim, Z. Qingbo, Z. Weijiang and L. Zhaolin, *J. Electrochem. Soc.*, 2008, **155**, B776.
161 J. Simon, T. Karaked, S. Montree, S. F. John, F. Janet, T. David and S. C. E. Tsang, *ChemCatChem*, 2010, **2**, 1089.
162 H. A. Gasteiger, S. S. Kocha, B. Sompalli and F. T. Wagner, *Appl. Catal., B*, 2005, **56**, 9.
163 J. Greeley, I. E. L. Stephens, A. S. Bondarenko, T. P. Johansson, H. A. Hansen, T. F. Jaramillo, J. Rossmeisl, I. Chorkendorff and J. K. Nørskov, *Nat. Chem.*, 2009, **1**, 552.
164 Q. Zhu, S. Zhou, X. Wang and S. Dai, *J. Power Sources*, 2009, **193**, 495.
165 M. Cifrain and K. V. Kordesch, *J. Power Sources*, 2004, **127**, 234.
166 H. F. Hunger, Proceeding of Annual Power Sources Conference, 1960, **14**, 55.
167 C. Bianchini and P. K. Shen, *Chem. Rev.*, 2009, **109**, 4183.
168 E. H. Yu and K. Scott, *J. Power Sources*, 2004, **137**, 248.
169 H. Liu and J. Zhang in *Electrocatalysis of Direct Methanol Fuel Cells*, ed. H. Liu and J. Zhang, Wiley-VCH, Weinheim, Germany, 2009, p. 504.
170 L. Jiang, A. Hsu, D. Chu and R. Chen, *J. Electrochem. Soc.*, 2009, **156**, B643.
171 F. H. B. Lima, J. Zhang, M. H. Shao, K. Sasaki, M. B. Vukmirovic, E. A. Ticianelli and R. R. Adzic, *J. Phys. Chem. C*, 2007, **111**, 404.
172 K. Q. Ding, *Int. J. Electrochem. Sci.*, 2010, **5**, 668.
173 S. Maheswari, G. Selvarani, P. Sridhar, S. Pitchumani and A. K. Shukla, *ECS Trans.*, 2010, **33**, 1795.
174 E. Antolini, *Energy Environ. Sci.*, 2009, **2**, 915.
175 M. H. Aho, K. Sasaki, P. Liu and R. R. Adzic, *Z. Phys. Chem.*, 2007, **221**, 1175.
176 M. Shao, P. Liu, J. Zhang and R. Adzic, *J. Phys. Chem. B*, 2007, **111**, 6772.
177 J. L. Fernandez, D. A. Walsh and A. J. Bard, *J. Am. Chem. Soc.*, 2005, **127**, 357.
178 Y. X. Wang and P. B. Balbuena, *J. Phys. Chem. B*, 2005, **109**, 18902.
179 H. T. Duong, M. A. Rigsby, W. P. Zhou and A. Wieckowski, *J. Phys. Chem. C*, 2007, **111**, 13460.
180 L. Zhang, K. Lee and J. Zhang, *Electrochim. Acta*, 2007, **52**, 3088.
181 S. Maheswari, S. Karthikeyan, P. Murugan, P. Sridhar and S. Pitchumani, *Phys. Chem. Chem. Phys.*, 2012, **14**, 9683.
182 E. N. Muhamad, T. Takeguchi, G. Wang, Y. Anzai and W. Ueda, *J. Electrochem. Soc.*, 2009, **156**, B32.
183 X. Guo, D. J. Guoa, X. P. Qiu, L. Q. Chen and W. T. Zhu, *J. Power Sources*, 2009, **194**, 281.
184 S. Vinod Selvaganesh, G. Selvarani, P. Sridhar, S. Pitchumani and A. K. Shukla, *J. Electrochem. Soc.*, 2010, **157**, B1000.
185 F. Hu, F. Ding, S. Song and P. K. Shen, *J. Power Sources*, 2006, **163**, 415.
186 S. T. Nguyen, J. M. Lee, Y. Yang and X. Wang, *Ind. Eng. Chem. Res.*, 2012, **51**, 9966.
187 S. Mentus, A. Abu-Rabi and D. Jašin, *Electrochim. Acta*, 2012, **69**, 174.
188 K. Tammeveski, T. Tenno, A. Rosental, P. Talonen, L. S. Johansson and L. Niinistöd, *J. Electrochem. Soc.*, 1999, **146**, 669.
189 S. Maheswari, P. Sridhar and S. Pitchumani, *Electrochem. Commun.*, 2013, **26**, 97.
190 E. Antolini, T. Lopes and E. R. Gonzalez, *J. Alloys compd.*, 2008, **461**, 253.
191 L. Colmenares, Z. Jusys and R. J. Behm, *J. Phys. Chem. C*, 2007, **111**, 1273.
192 N. Ramaswamy, R. J. Allen and S. Mukerjee, *J. Phys. Chem. C*, 2011, **115**, 12650.

193 D. Cao, A. Wieckowski, J. Inukai and N. Alonso-Vante, *J. Electrochem Soc.*, 2006, **153**, A869.
194 G. Vázquez and O. S. Feria, *J. New Mater Electrochem. Syst.*, 2009, **12**, 17.
195 H. Cheng, W. Yuan and K. Scott, *Fuel cells*, 2007, **1**, 16.
196 L. Jiang, A. Hsu, D. Chu and R. Chen, *J. Electrochem. Soc.*, 2009, **156**, B370.
197 C. Xu, L. Cheng, P. Shen and Y. Lu, *Electrochem. Commun.*, 2009, **9**, 997.
198 V. Stamenkovic, B. S. Moon, K. J. Mayerhofer, P. N. Ross, N. Markovic, J. Rossmeisl, J. Greeley and J. K. Norskov, *Angew. Chem., Int. Ed.*, 2006, **45**, 2897.
199 A. Kowal, M. Li, M. Shao, K. Sasaki, M. B. Vukmirovic, J. Zhang, N. S. Marinkovic, P. Liu, A. I. Frenkel and R. R. Adzic, *Nat. Mater.*, 2009, **8**, 325.
200 J. Prakash and H. Joachin, *Electrochim. Acta*, 2000, **44**, 2289.
201 N. A. Anastasijevic, Z. M. Dimitrijevic and R. R. Adzic, *J. Electroanal. Chem. Interfacial Electrochem.*, 1986, **199**, 351.
202 S. Maheswari, P. Sridhar and S. Pitchumani, *Fuel Cells*, 2012, **12**, 963.
203 S. Maheswari, P. Sridhar and S. Pitchumani, *AIP Conf. Proc.*, 2013, **1538**, 209.
204 R. Adzic and T. Huang, *US Pat.*, 7232601 B2, 2009.
205 V. Rao and B. Viswanathan, *J. Phys. Chem. C*, 2007, **111**, 16538.
206 L. Colmenares, Z. Jusys and R. J. Behm, *Langmuir*, 2006, **22**, 10437.
207 S. Stolbov, *J. Phys. Chem. C*, 2012, **116**, 7173.
208 M. Bron, P. Bpgdanoff, S. Fiechter, M. Hilgendorff, J. Radnik, I. Dorbandt, H. Schulenburg and H. Tributsch, *J. Electoanal. Chem.*, 2001, **517**, 85.
209 V. M. Sanchez, E. D. Llave and D. A. Scherlis, *Langmuir*, 2011, **27**, 2411.
210 J. M. Jakšić, C. M. Lačnjeva, N. V. Krstajič and M. M. Jakšić, *Chem. Ind. Chem. Eng. Q.*, 08, **14**, 119.
211 N. Alexeyeva, A. Sarapuu, K. Tammeveski, F. J. Vidal-Iglesias, J. Solla-Gullón and J. M Feliu, *Electrochim. Acta*, 2011, **56**, 6702.
212 G. Jürmann and K. Tammesveski, *J. Electroanal. Chem.*, 2006, **597**, 119.
213 Y. Cheng, S. Lu, H. Zhang, C. V. Varanasi and S. Liu, *Nano Lett.*, 2012, **12**, 1206.
214 J. Hou, Y. Shao, M. W. Ellis, B. R. Moore and B. Yi, *Phys. Chem. Chem. Phys.*, 2011, **13**, 15384.
215 P. Serp, M. Corrias and P. S. Kalck, *Appl. Catal., A*, 2003, **253**, 337.
216 D. V. Kosynkin, A. L. Higginbotham, A. Sinitskii, J. R. Lomeda, A. Dimiev, B. K. Price and J. M. Tour, *Nature*, 2009, **458**, 872.
217 D. Long, W. Li, W. Qiao, J. Miyawaki, S. H. Yoon, I. Mochidab and L. Ling, *Chem. Commun.*, 2011, **47**, 9429.
218 V. C. Tung, L. M. Chen, M. J. Allen, J. K. Wassei, K. Nelson, R. B. Kaner and Y. Yang, *Nano Lett.*, 2009, **9**, 1949.
219 Y. Li, W. Zhou, H. Wang, L. Xie, Y. Liang, F. Wei, J. C. Idrobo, S. J. Pennycook and H. Dail, *Nat. Nanotechnol.*, 2012, **7**, 394.
220 H. Kim, K. Lee, S. L. Woo and Y. Jung, *Phys. Chem. Chem. Phys.*, 2011, **13**, 17505.
221 Z. Wang, G. Chen, D. Xia and L. Zhang, *J. Alloys Compd.*, 2008, **450**, 148.
222 D. He, L. Yang, S. Kuang and Q. Cai, *Electrochem. Commun.*, 2007, **9**, 2467.
223 J. Tian, G. Sun, L. Jiang, S. Yan, Q. Mao and Q. Xin, *Electrochem. Commun.*, 2007, **9**, 563.
224 K. W. Park, J. H. Choi, K. S. Ahn and Y. E. Sung, *J. Phys. Chem. B*, 2004, **108**, 5989.
225 C. Song, M. Kanfar and P. G. Pickup, *J. Appl. Electrochem.*, 2006, **36**, 339.
226 Z. Sun, X. Wang, Z. Liu, H. Zhang, P. Yu and L. Mao, *Langmuir*, 2010, **26**, 12383.

227 M. Hepel, I. Dela, T. Hepel, J. Luo and C. J. Zhong, *Electrochim. Acta*, 2007, **52**, 5529.

228 S. Shanmugam and A. Gedanken, *Small*, 2007, **3**, 1189.

229 K. G. Nishanth, P. Sridhar, S. Pitchumani and A. K. Shukla, *Bull. Mater. Sci.*, 2013, **36**, 353.

230 T. Saida, N. Ogiwara, Y. Takasu and W. Sugimoto, *J. Phys. Chem. C*, 2010, **114**, 13390.

231 Z. Jusys, J. Kaiser and R. J. Behm, *Electrochim. Acta*, 2002, **47**, 3693.

232 K. G. Nishanth, P. Sridhar, S. Pitchumani and A. K. Shukla, *ECS Trans.*, 2010, **33**, 2027.

233 F. F. Cao, X. L. Wu, S. Xin, Y. G. Guo and L. J. Wan, *J. Phys. Chem. C*, 2010, **114**, 10308.

234 C. C. Liang and A. L. Juliard, *J. Electroanal. Chem.*, 1965, **9**, 390.

235 S. Mukerjee and S. Srinivasan, *J. Electroanal. Chem.*, 1993, **357**, 201.

236 Q. Huang, H. Yang, Y. Tang, T. Lu and D. L. Akins, *Electrochem. Commun.*, 2006, **8**, 1220.

237 E. Antolini, R. R. Passos and E. A. Ticianelli, *Electrochim. Acta*, 2002, **48**, 263.

238 W. Li, W. Zhou, H. Li, Z. Zhou, B. Zhou, G. Sun and Q. Xin, *Electrochim. Acta*, 2004, **49**, 1045.

239 O. Savadogo and F. J. R. Valera, *ECS Trans.*, 2006, **1**, 247.

240 F. J. R. Varela and O. Savadogo, *ECS Trans.*, 2006, **3**, 181.

241 L. Jiang, G. Sun, S. Sun, J. Liu, S. Tang, H. Li, B. Zhou and Q. Xin, *Electrochim. Acta*, 2005, **50**, 5384.

242 E. Antolini, J. R. C. Salgado, M. J. Giz and E. R. Gonzalez, *Int. J. Hydrogen Energy*, 2005, **30**, 1213.

243 T. Lopes, E. Antolini, F. Colmati and E. R. Gonzalez, *J. Power Sources*, 2007, **164**, 111.

244 N. Park, T. Shiraishi, K. Kamisugi, Y. Hara, K. Iizuka, T. Kado and S. Hayase, *J. Appl. Electrochem.*, 2008, **38**, 371.

245 A. M. Castro Luna, A. Bonesi, W. E. Triaca, A. D. Blasi, A. Stassi, V. Baglio, V. Antonucci and A. S. Aricò, *J. Nanopart. Res.*, 2010, **12**, 357.

246 J. M. Jaksic, N. V. Krstajic, L. M. Vracar, S. G. Neophytides, D. Labou, P. Falaras and M. M. Jaksic, *Electrochim. Acta*, 2007, **53**, 349.

247 M. M. Jaksic, *Int. J. Hydrogen Energy*, 2001, **26**, 559.

248 X. Wang, W. Xu, X. Zhou, T. Lu, W. Xing, C. Liu and J. Liao, *J. Solid State Electrochem.*, 2009, **13**, 1449.

249 S. Meenakshi, K. G. Nishanth, P. Sridhar and S. Pitchumani, *Electrochim. Acta*, 2014, **135**, 52.

250 H. Pramanik and S. Basu, *Can. J. Chem. Eng.*, 2007, **85**, 781.

251 M. Brandalise, M. M. Tusi, R. M. S. Rodrigues, E. V. Spinacé and A. O. Neto, *Int. J. Electrochem. Sci.*, 2010, **5**, 1879.

252 S. C. Zignani, E. R. Gonzalez, V. Baglio, S. Siracusano and A. S. Aricò, *Int. J. Electrochem. Sci.*, 2012, **7**, 3155.

253 B. Su, K. Wang, C. Tseng, C. Wang and Y. Hsueh, *Int. J. Electrochem. Sci.*, 2012, 7, 5246.

254 L. Jiang, G. Sun, Z. Zhou, W. Zhou and Q. Xin, *Catal. Today*, 2004, **93–95**, 665.

255 A. O. Neto, R. W. R. Verjulio-Silva, M. Linardi and E. V. Spinacé, *Ionics*, 2010, **16**, 85.

256 D. H. Lim, D. H. Choi, W. D. Lee, D. R. Park and H. I. Lee, *Electrochem. Solid-State Lett.*, 2007, **10**, B87.

257 B. Liu, Z. W. Chia, Z. Y. Lee, C. H. Cheng, J. Y. Lee and Z. L. Liu, *Fuel Cells*, 2012, **12**, 670.
258 S. Tanaka, M. Umeda, H. Ojima, Y. Usui, O. Kimura and I. Uchida, *J. Power Sources*, 2005, **152**, 34.
259 W. J. Zhou, W. Z. Li, S. Q. Song, Z. H. Zhou, L. H. Jiang, G. Q. Sun, Q. Xin, K. Poulianitis, S. Kontou and P. Tsiakaras, *J. Power Sources*, 2004, **131**, 217.
260 F. Vigier, C. Coutanceau, A. Perrard, E. M. Belgsir and C. Lamy, *J. Appl. Electrochem.*, 2004, **34**, 439.
261 S. S. Gupta and J. Datta, *J. Electroanal. Chem.*, 2006, **594**, 65.
262 D. M. dos Anjos, K. B. Kokoh, J. M. Léger, A. R. De Andrade, P. Olivi and G. Tremiliosi-Filho, *J. Appl. Electrochem.*, 2006, **36**, 1391.
263 Y. Bai, J. Wu, X. Qiu, J. Xi, J. Wang, J. Li, W. Zhu and L. Chen, *Appl. Catal., B*, 2007, **73**, 144.
264 Y. Bai, J. Wu, J. Xi, J. Wang, W. Zhu, L. Chen and X. Qiu, *Electrochem. Commun.*, 2005, 7, 1087.
265 M. L. Calegaro, H. B. Suffredini, S. A. S. Machado and L. A. Avaca, *J. Power Sources*, 2006, **156**, 300.
266 D. Zhang, Z. Ma, G. Wang, K. Konstantinov, X. Yuan and H. Liu, *Electrochem. Solid-State Lett.*, 2006, **9**, A423.
267 J. C. M. Silva, L. S. Parreira, R. F. B. De Souza, M. L. Calegaro, E. V. Spinacé, A. O. Neto and M. C. Santos, *Appl. Catal. B*, 2011, **110**, 141.
268 L. Jiang, L. Colmenares, Z. Jusys, G. Q. Sun and R. J. Behm, *Electrochim. Acta*, 2007, **53**, 377.
269 F. Colmati, E. Antolini and E. R. Gonzalez, *Appl. Catal. B*, 2007, **73**, 106.
270 G. Li and P. G. Pickup, *J. Power Sources*, 2007, **173**, 121.
271 E. A. Baranova, T. Amir, P. H. J. Mercier, B. Patarachao, D. Wang and Y. L. Page, *J. Appl. Electrochem.*, 2010, **40**, 1767.
272 E. Higuchi, K. Miyata, T. Takase and H. Inoue, *J. Power Sources*, 2011, **196**, 1730.
273 M. Zhu, G. Sun and Q. Xin, *Electrochim. Acta*, 2009, **54**, 1511.
274 F. Colmati, E. Antolini and E. R. Gonzalez, *J. Solid State Electrochem.*, 2008, **12**, 591.
275 E. Antolini, F. Colmati and E. R. Gonzalez, *J. Power Sources*, 2009, **193**, 555.
276 S. Meenakshi, P. Sridhar and S. Pitchumani, *RSC Adv.*, 2014, **4**, 44386.
277 N. Murata, T. Suzuki, M. Kobayashi, F. Togoh and K. Asakura, *Phys. Chem. Chem. Phys.*, 2013, **15**, 17938.
278 A. S. Aricò, V. Antonucci, N. Giordano, A. K. Shukla, M. K. Ravikumar, A. Roy, S. R. Barman and D. D. Sarma, *J. Power Sources*, 1994, **50**, 295.
279 X. Li, B. Lin, B. Xu, Z. Chen, Q. Wang, J. Kuang and H. Zhu, *J. Mater. Chem.*, 2010, **20**, 3924.
280 J. C. M. Silva, R. F. B. De Souza, L. S. Parreira, E. Teixeira Neto, M. L. Calegaro and M. C. Santos, *Appl. Catal., B*, 2010, **99**, 265.
281 C. Lamy, C. Coutanceau and J.-M. Leger in *Catalysis for Sustainable Energy Production*, ed. P. Barbaro and C. Bianchini, Wiley-VCH Verlag Gmbh and Co. KGaA, Weinheim, 2009, ch. 1–2, p. 1.

Application of Metal Organic Framework (MOF) in the electrocatalytic process

Y. Luo and N. Alonso-Vante*

DOI: 10.1039/9781782622727-00194

1 Introduction

1.1 Electrocatalysis for energy conversion

Electrocatalytic processes play a significant role in the conversion of chemical energy to electrical energy in, *e.g.*, fuel cell systems, and electrolysers. Electrochemical processes such as the oxygen reduction reaction (ORR), oxygen evolution reaction (OER), hydrogen evolution reaction (HER), hydrogen oxidation reaction (HOR), carbon dioxide reduction, and oxidation of fuels, are key to improve the performance of energy-converting devices, such as fuel cells and electrolysers for fuel production.[1,2] A unitized regenerative fuel cell (URFC) can be a promising device for a sustainable fuel-energy recycling system. This kind of device works in two-modes, *viz.*, electrolyser and fuel cell,[3–5] Fig. 1. In the electrolyser mode,[3–5] water molecules are oxidized involving the HER (eqn (1)) at the cathode, and OER (eqn (2)) at the anode. This device consumes energy to generate hydrogen as a fuel.

$$2H^+ + 2e^- \rightarrow H_2 \qquad E^0 = 0 \text{ V/SHE} \tag{1}$$

$$2H_2O \rightarrow O_2 + 4H^+ + 4e^- \quad E^0 = 1.23 \text{ V/SHE} \tag{2}$$

$$2H_2O \rightarrow 2H_2 + O_2 \qquad \Delta E = 1.23 \text{ V} \tag{3}$$

The fuel cell mode is based on H_2/O_2 or H_2/air. This low-temperature fuel cell (so-called proton-exchange or polymer electrolyte membrane-PEMFC) consumes hydrogen as fuel to produce electrical energy yielding water (eqn (6)).[6–9] Thus the HOR (eqn (4)) occurs at the anode, and ORR (eqn (5)) at the cathode. As mentioned above, the energy conversion is reversible.

$$H_2 \rightarrow 2H^+ + 2e^- \quad E^0 = 0 \text{ V/SHE} \tag{4}$$

$$O_2 + 4H^+ + 4e^- \qquad E^0 = 1.23 \text{ V/SHE} \tag{5}$$

$$2H_2 + O_2 \rightarrow 2H_2O \tag{6}$$

In order to achieve a rechargeable fuel cell, highly efficient and stable catalysts for the electrolysis mode and fuel cell mode are necessary.[10]

IC2MP, University of Poitiers, 4 rue Michel Brunet, 86022 Poitiers Cedex, France. E-mail: nicolas.alonso.vante@univ-poitiers.fr

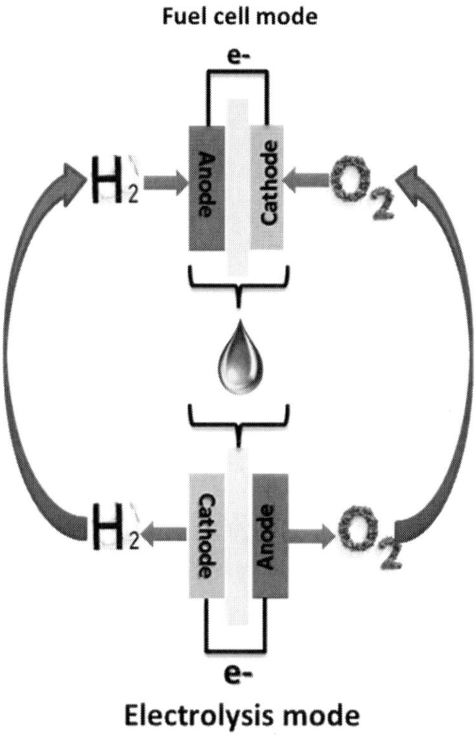

Fig. 1 The schematics of a unitized regenerative fuel cell.

Taken as a half-reaction, the HER for example, the electrocatalytic process improvement is expected to be obtained by an enhanced exchanged current density and larger real surface of the electrode.[11] The HER kinetics is essentially correlated to the performance of the electrode materials. Many materials can be used as electrocatalysts for HER, for example, mono- or bi-metallic bulk or nano-materials,[12–20] RANEY® type materials (RANEY® Ni, Zn and so on),[21–25] oxides,[26–28] chalcogenides,[27,29–32] *etc.* However, Pt group metals are more active than the others (Fig. 2a).[33] On the other hand, for the HOR, the most widely used and studied catalysts are Pt-based materials.[20,34–38] Pd-based catalysts are also good candidates for HOR.[39,40] However, Pt- and Pd-based catalysts are precious metals, therefore efforts have been devoted to explore non-precious metals for HOR.[41]

OER is efficient on various materials, *e.g.*, transition metal oxides,[42–46] among them, IrO_2 that shows an enhanced activity with respect to other oxides.[47]

The best and most widely used catalysts for the ORR are Pt-based materials (Fig. 2b).[48–55] Among them, rare earth metal alloyed Pt catalysts were predicted as the most active and stable materials for ORR, with respect to Pt alloyed with transition metals, *cf.* Fig. 2c.[48] In addition, Pd-based catalysts[39,56–63] showed a similar ORR kinetics to the former ones. Nonetheless, non-precious metals and their oxides/nitrides,[64–69] carbon-based materials,[70–73] molecular catalysts,[74–77] and chalcogenides[78–82] could be promising candidates.

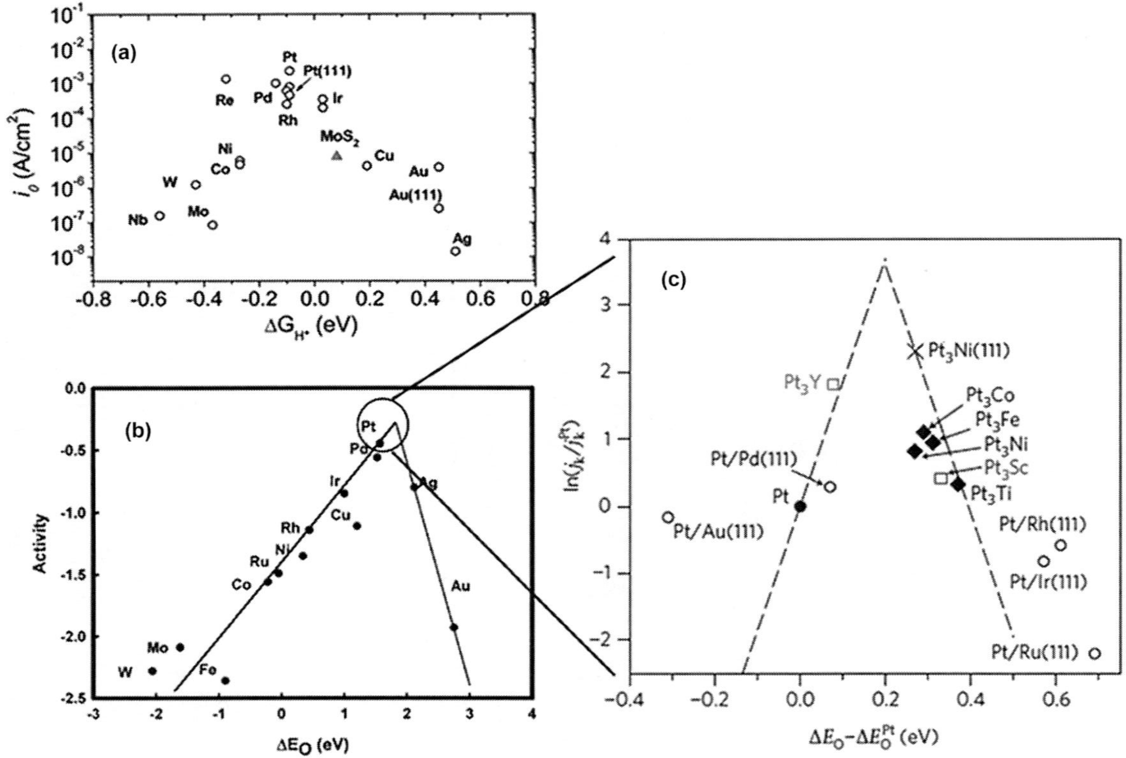

Fig. 2 Volcano plots for HER (a), reprinted with permission from T. F. Jaramillo, K. P. Jørgensen, J. Bonde, J. H. Nielsen, S. Horch and I. Chorkendorff, Identification of Active Edge Sites for Electrochemical H_2 Evolution from MoS_2 Nanocatalysts, *Science*, 2007, **317**, 100–102. Copyright (2007) The American Association for the Advancement of Science; ORR (b) reprinted with permission from Macmillian Publishers Ltd: J. Greeley, I. E. L. Stephens, A. S. Bondarenko, T. P. Johansson, H. A. Hansen, T. F. Jaramillo, J. Rossmeisl, I. Chorkendorff and J. K. Nørskov, Alloys of platinum and early transition metals as oxygen reduction electrocatalysts, *Nat. Chem.*, **1**, 552–556. Copyright (2009) Nature Publishing Group; and (c) reprinted with permission of J. K. Nørskov, J. Rossmeisl, A. Logadottir, L. Lindqvist, J. R. Kitchin, T. Bligaard and H. Jónsson, Origin of the Overpotential for Oxygen Reduction at a Fuel-Cell Cathode, *J. Phys. Chem. B*, 2004, **108**, 17886–17892. Copyright (2004) American Chemical Society. Figures based on density of functional theory (DFT)-calculated Gibbs free energy of absorbed atomic hydrogen and oxygen.

Recycling CO_2 is also a potential technology to realise sustainable energy conversion. For ecological purposes, CO_2 could be transformed into fuels. For example, CO_2 can be reduced to CO (eqn (7)), HCOOH (eqn (8)), HCHO (eqn (9)), CH_3OH (eqn (10)), and CH_4 (eqn (11)). The reduction is hard to obtain, since CO_2 is a highly stable molecule, and the formation of the anion radical $CO_2^{-\bullet}$, as an intermediate, is an energy demanding process (eqn (12)).

$$CO_2 + 2H^+ + 2e^- \rightarrow CO + 2H_2O \quad E^0 = -0.53\ \text{V/NHE, pH} = 7 \tag{7}$$

$$CO_2 + 2H^+ + 2e^- \rightarrow HCOOH \quad E^0 = -0.61\ \text{V/NHE, pH} = 7 \tag{8}$$

$$CO_2 + 4H^+ + 4e^- \rightarrow HCHO + H_2O \quad E^0 = -0.48\ \text{V/NHE, pH} = 7 \tag{9}$$

$$CO_2 + 6H^+ + 6e^- \rightarrow CH_3OH + H_2O \quad E^0 = -0.38\ \text{V/NHE, pH} = 7 \tag{10}$$

$$CO_2 + 8H^+ + 8e^- \rightarrow CH_4 + 2H_2O \quad E^0 = -0.24\ \text{V/NHE, pH} = 7 \tag{11}$$

$$CO_2 + e^- \rightarrow CO_2^{-\bullet} \quad E^0 = -1.90\ \text{V/NHE, pH} = 7 \tag{12}$$

The overpotential for the CO_2 reduction could be reduced to <0.4 V when current density of CO_2 reduction is higher than 1 $mA\,cm^{-2}$.[83] For example, the CO_2 reduction can take place on the surface of Pd-,[84,85] Cu-,[86–88] Au-,[83,89,90] Ag-,[91] and Mo-based[86,92] nanoparticles (NPs).[93] The CO_2 reduction can be performed on molecular catalysts, based on Fe-,[94–97] Mn-,[98–100] Ru-,[101–103] Os-,[104,105] Ir-,[106] and Re-[106–108] metallic centres coordinated by organic ligands, Fig. 3. The mechanism of CO_2 reduction on macrocyclic and polypyridine involving organometallic complexes can be divided into two steps: (1) reduction of catalyst, where the electron transfer takes place from ligand to metal; (2) the reduced metal centre ejects an electron to CO_2, forming $CO_2^{-\bullet}$. Let us take the Mn(I)(bpy)(CO)$_3$Br (bpy = 2,2′-pyridine),[98] and (tpp)Fe(I) (tpp = tetra-phenyl-porphyrin)[94] as examples. The mechanism for CO_2 reduction on these two catalysts is shown in Fig. 3. One can find that the CO_2 reduction on these catalysts is based on the formation and breaking of chemical bonds between the active redox metal centre and the carbonyl ligand (CO).

In addition to electrocatalysis, the photoelectrocatalytic-process for CO_2 reduction attracts lots of attention. In such a system, semiconducting materials are used as photocathodes, responsible for the conversion of light to electricity. Several semiconducting materials can be used in such a system, for example, TiO_2, Si wafer, Si nanowires (SiNWs), p-Gallium Phosphide, and p-Gallium Arsenide.[109–115] Electrocatalysts are usually involved in the CO_2 reduction. In Fig. 4,[110] it can be seen that the CO_2 onset potential is positively shifted on the molecular catalyst at the Si wafer, and Si NWs photocathode, due to the photovoltaic effect, indicating that the overpotential for CO_2 electro-reduction at the photocathode can be reduced.

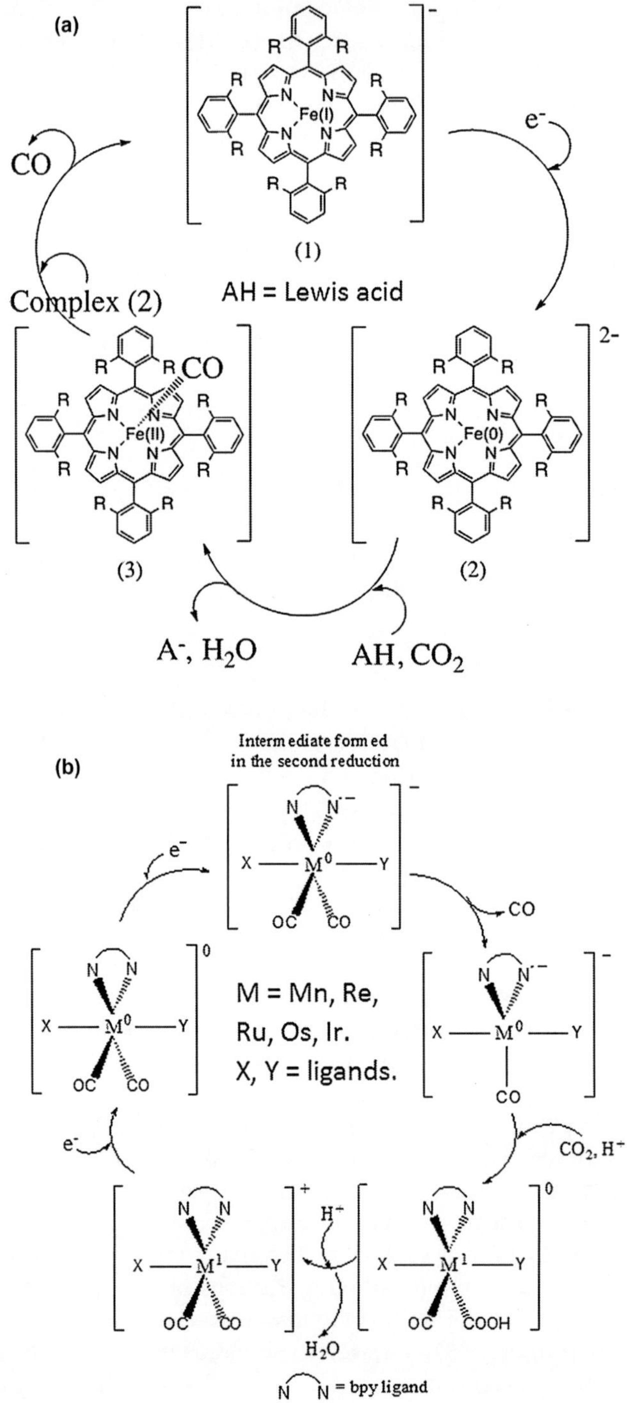

Fig. 3 Molecular catalysts for catalytic CO_2 reduction mechanism on Fe-,[94–97] Mn-,[98–100] Ru-,[101–103] Os-,[104,105] Ir-,[106] and Re-.[106–108]

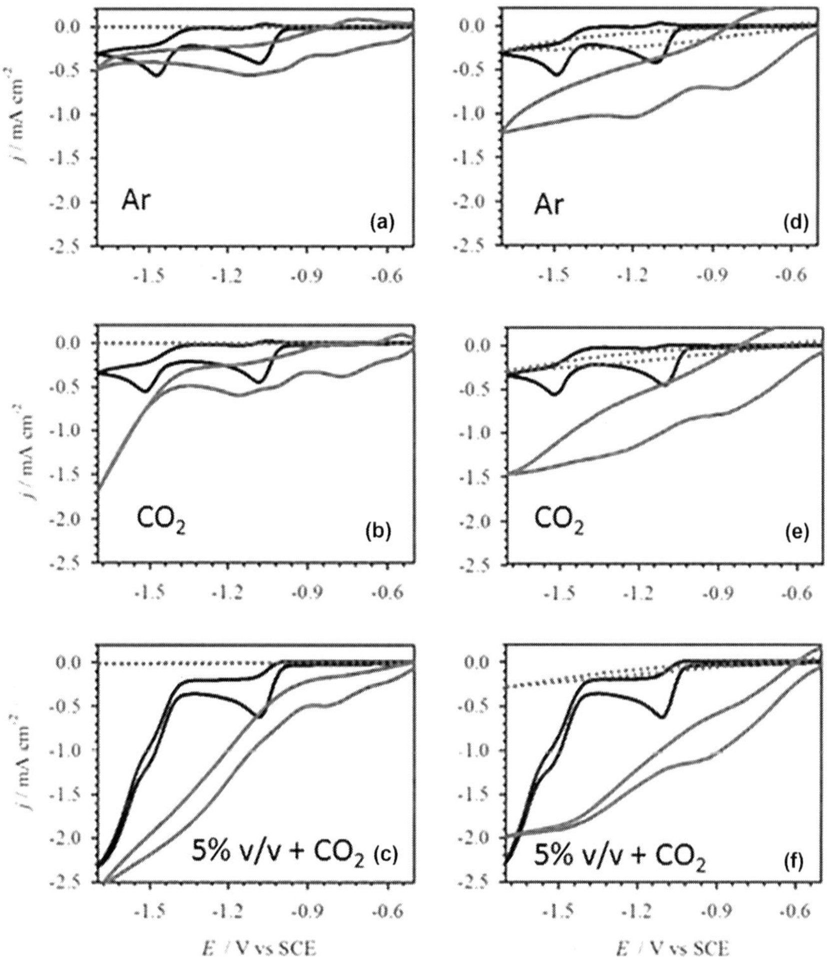

Fig. 4 Cyclic voltammograms (CVs) of CO_2 on $\{Mn(bpy)(CO)_3(CH_3CN)\}PF_6$ at silicon (a–c) and silicon nanowire (d–f) in $CH_3CN + 0.1$ M Bu_4NClO_4, at a scan rate of 0.1 V s^{-1}. Black lines are CVs at glassy carbon electrode. Red dash lines are CVs on dark Si electrodes. Red solid lines are CVs on illuminated Si electrodes (20 mW cm^{-2}). Reprinted with permission of E. Torralba-Peñalver, Y. Luo, J.-D. Compain, S. Chardon-Noblat and B. Fabre, Selective Catalytic Electroreduction of CO, *ACS Catal.*, 2015, **5**, 6138–6147. Copyright (2015) American Chemical Society.

1.2 The Metal Organic Framework-MOF

MOF also called coordination polymer, is an infinite one-, two- or three-dimensional network. It is comprised of metal centres connected by organic ligands. Generally, the MOF is formed when the ligands donate one pair of electrons to fill the orbitals of the metal cations (*via* the Lewis acid/base relationship). In order to form such a supramolecular framework, the ligands have to be at least bi-functional, which allows to form an infinite network. Over the last decades, a large amount of MOFs have been synthesized because of their fascinating network topologies and potential applications in gas storage/separation,[116–118] catalysis,[119,120] luminescent sensing,[121,122] and magnetism.[123] The formation of the MOF is based on a self-assembly of covalent bonds between metal ions and ligands *via* a soft-chemical route, *e.g.* hydro-, solvo-thermal route, and direct reactions.

The metal ions in the MOF structure are usually referred as "nodes", and ligands as "linkers". The construction of the polymeric architecture depends on the coordination mode of the nodes with the linkers. The coordination number of metal centres in MOFs varies from 2 to 14, providing large possibilities in coordination geometries of nodes, Fig. 5.[124] The transition metal centre, presenting partially filled d-orbitals, can hybridize with the orbitals of the ligands. The coordination number is usually from 4 to 8. Lanthanides (Ln) can also be used as metal nodes, presenting an unfilled f-orbital. The coordination number of lanthanide varies from 7 to 12. Thus, lanthanides have more possibilities of coordination geometry than other transition metals, Fig. 5a. Unfortunately, because of this feature, the architecture of Ln-based MOF is hard to control. However, it is still worth to be developed, due to its unique luminescent and magnetic properties. The organic ligand must coordinate with the metal cation, acting as the bridge between multiple metal centres. There are two types of ligands, namely, mono-dentate, bi-dentate and poly-dentate, according to the number of coordination bonds. The mono-dentate ligands have only one coordination bond, for example, H_2O, NH_3, OH^-, and so forth. The ligands with two coordination bonds are called bi-dentate, whereas those with more than two coordination bonds are poly-dentate. Some examples of bi- and poly-dentate ligands are shown in Fig. 5b. Among these ligands, multi-dentate carboxylates are extensively used in the domain of MOF. First, they are able to form various coordination modes, resulting in a large diversity of infinite topologies. Second, they are essential to stabilize the resulting neutral open framework and prevent the formation of interpenetrating nets.[125–128] The coordination modes of multi-dentate carboxylates are depicted in Fig. 5c.

1.3 Is MOF a promising material for catalysis?

The infinite matrix, composed by metal ions coordinated to organic ligand (MOF) may serve as a precursor in various fields of catalysis, *viz.* electrocatalysis: (1) a direct utilisation. Herein, the ligand acts as electron acceptor coming from the metal centre. This latter plays the role of an active site, thus responsible for electrocatalysis. (2) MOF as support for metallic nanoparticles (NPs). Highly porous MOF materials can be used as supports for catalytic materials, *e.g.*, Pt and Au NPs. (3) MOF-driving materials. Highly porous carbon and oxide-carbon composite can be obtained during the calcination of MOF materials at high temperature (>600 °C) under inert gas atmosphere.[129–131] These MOF-derived materials can be widely used as support or as electrocatalysts.[129–132] In fact, Morozan and Jaouen summarized the electrochemical applications of MOFs in 2012.[133] They listed the utilizations of coordination polymers in the domain of electrochemistry, such as energy conversion devices (batteries,

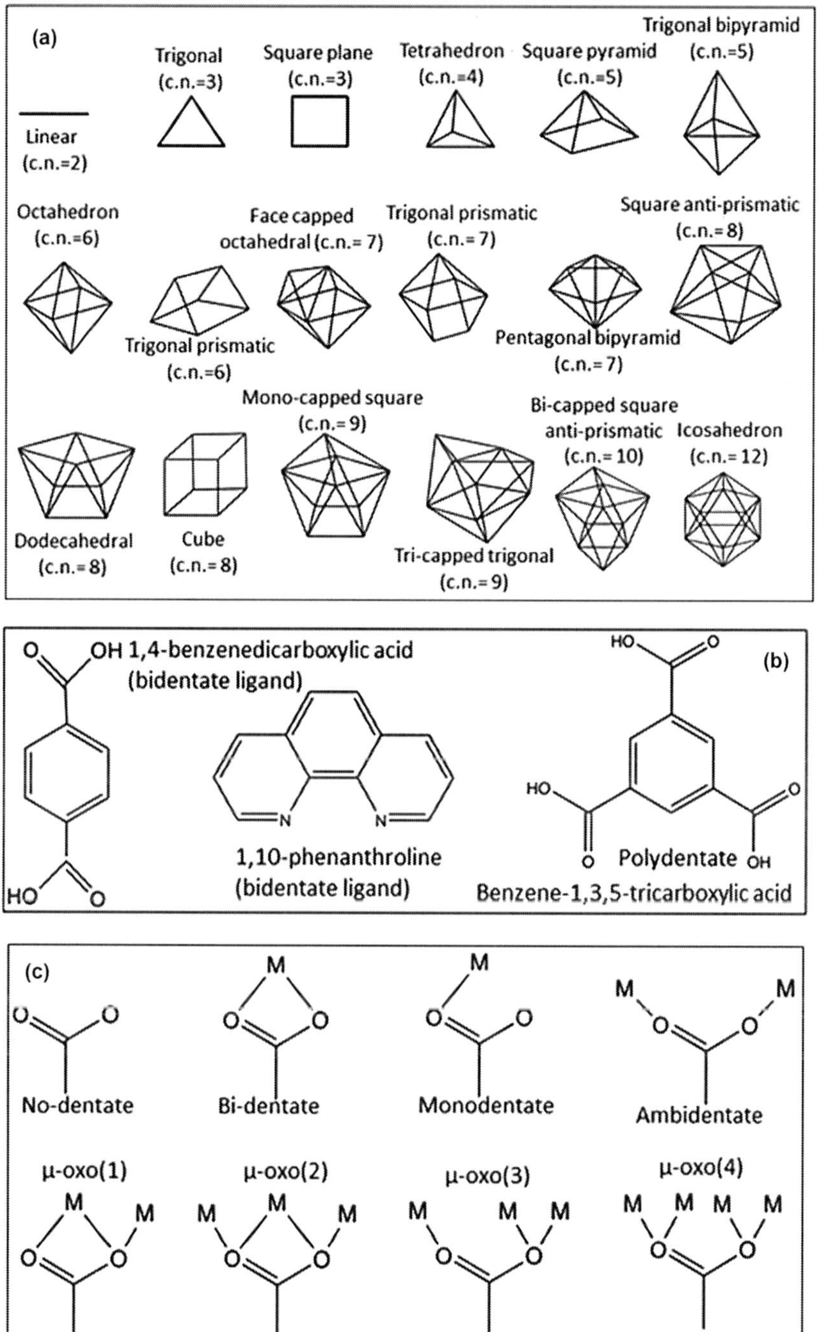

Fig. 5 (a) Coordination geometries of lanthanides based coordination polymers, (b) examples of bi- and poly-dentate ligands for MOF synthesis and (c) the coordination modes of multi-dentate carboxylates. Data extracted from ref. 124.

fuel cells, *etc.*), electrocatalysis, electrolytes, corrosions, *etc.*[133] However, from 2012 to 2016, the publications in this domain have largely increased. Some breakthroughs on, for instance, conducting MOF, MOF-based or MOF-driving electrocatalysts have been reported so far, largely promoting the development of electrochemistry. Therefore, it was necessary to reintroduce these interesting materials in the field of electrocatalytic energy conversion.

2 The Metal Organic Framework in catalysis

2.1 As molecular catalysts

For electrocatalytic purposes, MOF is a poor electrical conductor since the overlap between the π-orbital of the ligand and d-orbital of metal ions is poor. Though some approaches have been proposed to prepare electronic conducting MOFs, based on second or third-row transition metals as metal nodes, hetero-bimetallic metal centres and redox-active ligands,[134] few conducting MOFs have been so far obtained.[135] Therefore, the improvement of the conductivity of MOF is still an open challenge.

2.1.1 Conducting MOF structure. Commonly, Ni, Pd or Pt-based one-dimensional (1D) chains attract interest for their electrical transport behavior.[135] The metal centres in these materials are coordinated by organic ligands (for example, triethylenediamine, ethylenediamine, 1*R*,2*R*-diaminocyclohexane and so on) forming entities linked by halogens (for example, Cl, Br and I). The conductivity of Pt group metal based 1D chains was reported to vary from 2×10^{-14} to 1×10^{-1} S cm^{-1}, Fig. 6.[136–141] In addition, two-dimensional (2D) MOFs, based on inorganic linkers, nitrogen-containing ligands and organosulfur ligands, were reported for their good conductivity.[142–144] For example, $Rh_2(acam)_4(\mu_3\text{-}Cl)_2\cdot4nH_2O$ (acam = acetamidato) (Fig. 7a), formed by six

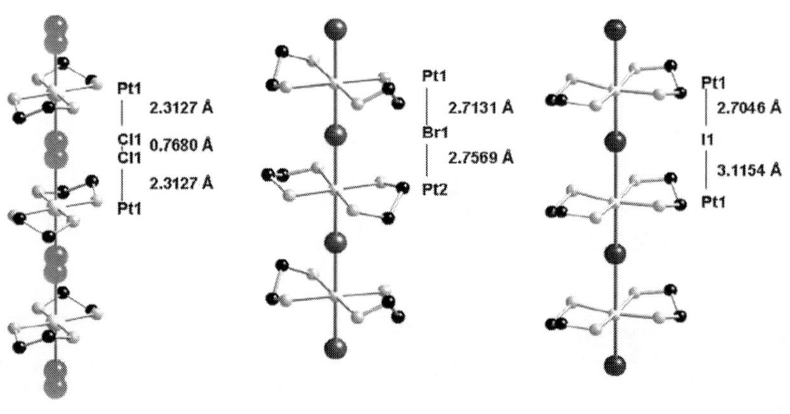

Fig. 6 Crystallographic structures of (a) $\{Pt(en)_2\}\{Pt(en)_2Cl_2\}(ClO_4)_4$ (en = ethylenediamine), (b) $\{Pt(en)_2\}\{Pt(en)_2Br_2\}(ClO_4)_4$ and (c) $\{Pt(en)_2\}\{Pt(en)_2I_2\}(ClO_4)_4$. Counter-ions and hydrogen atoms are omitted for clarity. Reproduced from ref. 135 with permission from the Royal Society of Chemistry.

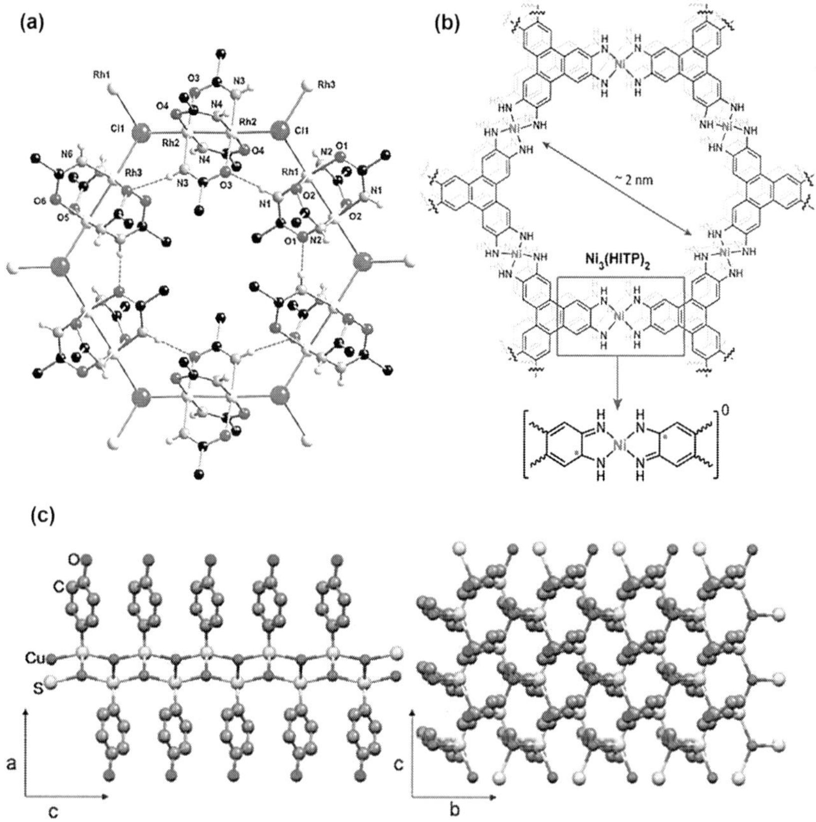

Fig. 7 (a) Crystallographic structures of $Rh_2(acam)_4(\mu_3\text{-}Cl)_2 \cdot 4nH_2O$ (acam = acetamidato). Reprinted from ref. 135 with permission from the Royal Society of Chemistry. (b) Structure of $Ni_3(HITP)_2$ (HITP = 2,3,6,7,10,11-hexaiminotriphenylene). Reprinted with permission from D. Sheberla, L. Sun, M. A. Blood-Forsythe, S. Er, C. R. Wade, C. K. Brozek, A. Aspuru-Guzik and M. Dincă, High Electrical Conductivity in Ni3(2,3,6,7,10,11-hexaiminotriphenylene)2, a Semiconducting Metal–Organic Graphene Analogue, *J. Am. Chem. Soc.*, 2014, **136**, 8859–8862. Copyright (2014) American Chemical Society. (c) Structure of CuHT (HT = 4-hydroxythiophenolate). Reproduced from ref. 149 with permission from the Royal Society of Chemistry.

$Rh_2(acam)_4$ dimers in sides and six Cl-hexagons in corners, consisted of a honeycomb-like layer.[145] This framework showed a conductivity of 2×10^{-7} S cm^{-1}. $Cu_2Br(IN)_2$ (IN = isonicotinate = py-COO$^-$) consisted of copper ions coordinated by oxygen atoms, nitrogen atoms from isonicotinate ligands and bromide ions.[146] It showed a semiconducting behavior with a conductivity of 1.2×10^{-5} S cm^{-1} at room temperature.[146] In addition, Dinca's group developed an excellent conducting MOF of $Ni_3(HITP)_2$ (HITP = 2,3,6,7,10,11-hexaiminotriphenylene) Fig. 7b.[147] The conductivity for black powder and violet thin film of $Ni_3(HITP)_2$ was 2 and 40 S cm^{-1} at room temperature, respectively.[147] The value for powder (or pellet) was lower than highly conducting organic polymers, *e.g.*, TTF-TCNQ of 10 S cm^{-1}.[148] But the value of the thin film was one of the best for conducting coordination polymers.[135] Another 2D MOF synthesized *via* self-assembly of Cu(I) and 4-hydroxythiophenol displayed an electrical conductivity of 120 S cm^{-1}, Fig. 7c.[149] Such a structure composed by Cu_3S_3 units, resembling the graphene structure, exhibited an excellent stability.[149] Its electrical conductivity can be tuned. It turned from conductor to semiconductor after treatment with *O*-acetylation.[149]

As to the three-dimensional (3D) MOFs, it is more attractive if the macro- or micro-porous frameworks possess good conductivity. A porous and electro-conductive Cu{Cu(pdt)$_2$}[150] (pdt = pyrazine-2,3-dithiolate) MOF structures was reported in 2009, showing an electrical conductivity of 6×10^{-4} S cm^{-1} at 300 K.[151] However, the porosity of such a structure could not be measured because of the collapse of the framework. In 2010, Kobayashi *et al.*[152] reported a MOF structure Cu{Ni(pdt)$_2$}, Fig. 8a–b, displaying an electronic conductivity, doping and redox behaviour. The synthesis of Cu{Ni(pdt)$_2$} was similar to Cu{Cu(pdt)$_2$}.[152] By mixing Na{Ni(pdt)$_2$}·2(H_2O) with CuI in acetonitrile, a dark red precipitation could be obtained, as revealed by the isostructure of Cu{Cu(pdt)$_2$}.[152] The solvated Cu{Ni(pdt)$_2$} could remain porous as assessed by BET (Brunauer–Emmett–Teller) with a surface area of 385 m^2 g^{-1} after desolvation with a conductivity of 1×10^{-8} S cm^{-1} at room temperature.[152] The value of conductivity was enhanced to 1×10^{-4} S cm^{-1} because of oxidative doping through I_2, Fig. 8c–d.[152] The redox couple half-wave potentials ($E_{1/2}$) for $\{Cu(pdt)_2\}^{2-/1-}$ and $\{Ni(pdt)_2\}^{2-/1-}$ were, respectively, −0.648 V and 0.391 V *vs.* Ag/Ag$^+$, as shown in Fig. 8e–f. It can be seen from the figures that a better electrochemical response can be observed in the smaller anion, *viz.*, tetra-*n*-butylammonium bromide (TBABr), with respect to tetra-*n*-butylammonium hexafluorophosphate (TBAPF$_6$). A possible explanation for such a phenomenon is the larger particle size of Cu{Cu(pdt)$_2$}, in comparison to Cu{Ni(pdt)$_2$}, leading to more important bulk diffusion throughout the pores. Moreover, diamond-like 3D networks of $\{Rh_2(acam)_4\}2I\cdot 6nH_2O$ (acam = CH_3CONH_2),[153] $Cu((CH_3)_2DCNQI)_2$ (DCNQI = *N*,*N'*-dicyano-quinonediimine),[154] $Cu_6(\mu 4\text{-SpyH})_4I_6$,[155] $Ag(H_2btc)_2Ag_2(Hbtc)$ (btc = benzene-1,3,5-tricarbonxylate),[156] and $Ag_4(O_3PCH_2CH_2PO_3)$ were reported as electron conducting MOFs.[157] The reported conductivity for 1D, 2D and 3D MOF structures in the literature is summarized in Table 1.

Another approach to enhance the conductivity of highly porous MOFs is *via* the adsorption of conducting organic molecules into the pores of a framework. Talin *et al.*,[158] reported an improved conductivity of a MOF thin film: $Cu_3(btc)_2\cdot nH_2O$ (HKUST-1), Fig. 9a–b, on Si wafer. HKUST-1 is a highly porous MOF material. The nitrogen adsorption isotherm showed BET specific surface area >1050 m^2 g^{-1}.[159] The 7,7,8,8-tetracyanoquinododimethane (TCNQ) is known as an electron acceptor, and is widely used to prepare charge transfer salts, Fig. 9a. The XRD pattern was barely changed between as-prepared, dried HKUST-1 powder, and TCNQ infiltrated HKUST-1 thin film, indicating that the MOF crystalline structure remained unchanged, and the TCNQs penetrated the framework, acting as guest molecules, Fig. 9c. As shown in the Fig. 9d, the conductivity of TCNQ@HKUST-1 thin film was enhanced to 7 S m^{-1}, higher than HKUST-1 thin film. Thus, the conductivity was tuneable, *via*

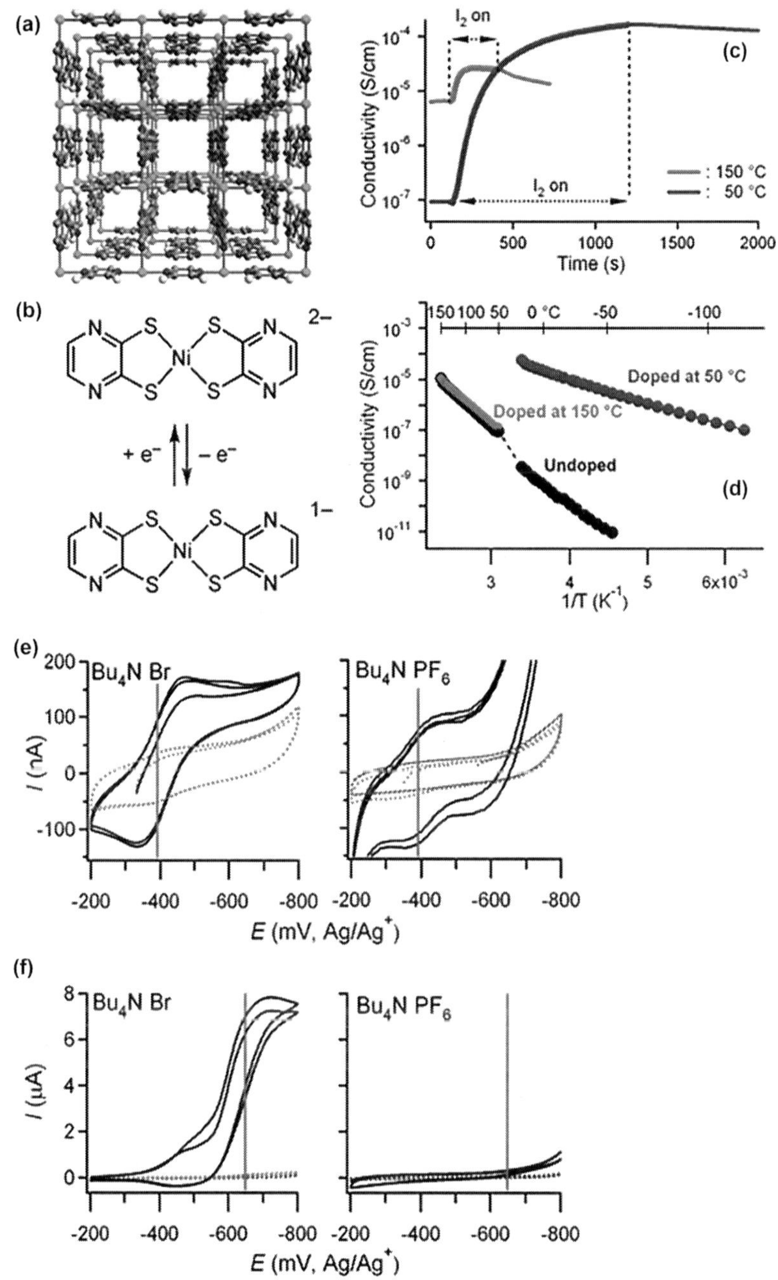

Fig. 8 (a–b) Crystallographic structure and (c–d) conductivity of Cu{Ni(pdt)$_2$} (pdt = pyrazine-2,3-dithiolate). Cyclic voltammograms recorded at 10 mV s^{-1} in TBABr and TBAPF$_6$ + 0.1 M MeCN, for (e) Cu{Ni(pdt)$_2$}, and (f) Cu{Cu(pdt)$_2$} on a Pt disk electrode. The dotted blue line is the bare Pt disk electrode, the red dotted line is the Pt electrode with MOF. The red vertical line indicates $E_{1/2}$ for the dithiolene complex in solution. Reprinted with permission from Y. Kobayashi, B. Jacobs, M. D. Allendorf and J. R. Long, Conductivity, Doping, and Redox chemistry of a Microporus Dithiolene-Based Metal-Organic Framework, *Chem. Mater.*, 2010, **22**, 4120–4122. Copyright (2010) American Chemical Society.

Table 1 Conductivity of 1D, 2D and 3D MOF.[a]

MOF	Dimension	Conductivity/S cm^{-1}	Ref.
$\{Pt(en)_2\}\{Pt(en)_2I_2\}(BF)_4$	1D	1.6×10^{-9}	136
$\{Pt(en)_2\}\{Pt(en)_2Cl_2\}(ClO_4)_4$	1D	2×10^{-15}	136
$\{Pt(en)_2\}\{Pt(en)_2Br_2\}(ClO_4)_4$	1D	3×10^{-11}	136
$\{Pt(en)_2\}\{Pt(en)_2I_2\}(ClO_4)_4$	1D	1.8×10^{-8}	136
$\{Pd(en)_2\}\{Pd(en)_2Cl_2\}(ClO_4)_4$	1D	2×10^{-12}	137
$\{Pd(en)_2\}\{Pd(en)_2I_2\}(ClO_4)_4$	1D	1.1×10^{-8}	137
$\{Pd(en)_2\}\{Pt(en)_2Cl_2\}$	1D	4×10^{-18}	138
$\{Pd(en)_2\}\{Pt(en)_2Br_2\}$	1D	1.4×10^{-15}	138
$\{Pd(en)_2\}\{Pt(en)_2I_2\}$	1D	4.5×10^{-11}	138
$\{Ni(en)_2\}\{Ni(en)_2Cl_2\}(ClO_4)_4$	1D	1.3×10^{-8}	139
$Ni(chxn)_2Br_3$	1D	6.9×10^{-4}	140
$Ni(chxn)_2Cl_{1.28}Br_{1.72}$	1D	1.9×10^{-5}	140
$Ni(chxn)_2Br(NO_3)_2$	1D	2.4×10^{-6}	140
$\{Ni(en)_2\}\{Pt(en)_2Cl_2\}(ClO_4)_4$	1D	3.2×10^{-17}	141
Ag_2CA	2D	5×10^{-3}	142
$Ni_2(C_4N_2H_3S)_4$	2D	5×10^{-3}	143
$Ag(C_5H_4NS)$	2D	2×10^{-5}	144
$Cu_2Br(IN)_2$	2D	1.2×10^{-5}	146
$Ni_3(HITP)_2$	2D	2 (Powder) and 40 (thin film)	147
CuHT	2D	120	149
$Cu\{Cu(pdt)_2\}$	3D	6×10^{-4}	151
$Cu\{Ni(pdt)_2\}$	3D	1×10^{-8}	152
$Rh_2(acam)_4(\mu_3\text{-}Cl)_2\cdot4nH_2O$	3D	2×10^{-7}	153
$Cu((CH_3)_2DCNQI)_2$	3D	1.0×10^{-3}	154
$\{Ag(H_2btc)_2\}\{Ag_2(Hbtc)\}$	3D	1.0×10^{-6}	156
$Ag_4(O_3PCH_2CH_2PO_3)$	3D	7.0×10^{-7}	157
$Cu_3(btc)_2\cdot(TCNQ)$	3D	7.0×10^{-2}	158

[a] en = ethylenediamine; chxn = 1*R*, 2*R*-diaminocyclohexane; CA = cyanuric acid; $C_4N_2H_3S$ = pyrimidine-2-thiol; C_5H_4NS = pyridine-2-thiolate; btc = benzene-1,3,5-tricarboxylate; acam = acetamidato; IN = isonicotinate; HITP = 2,3,6,7,10,11-hexaiminotriphenylene; HT = 4-hydroxythiophenol; pdt = pyrazine-2,3-dithiolate; DCNQI = *N*,*N*′-dicyano-quinonediimine; TCNQ = 7,7,8,8-tetracyanoquinododimethane.

the control of the time of reaction with TCNQ (Fig. 9e). The BET surface area of dried HKUST-1 powder obtained in this work was 1844 ± 4 m^2 g^{-1}, demonstrating almost no collapse of the porous structure and no residual reactant in the porosities.[160] After infiltration, the BET surface area was 214 ± 0.5 m^2 g^{-1}, suggesting 8 TCNQ molecules per unit cell or 1 per MOF pore. In fact, the colour of HKUST-1 thin film turns to green from the original gold upon exposure under TCNQ (Fig. 9b). Such a phenomenon implies an electronic modification of the MOF structure. In order to understand the nature of the interaction between MOF and TCNQ, the ultraviolet-visible (UV-Vis) spectroscopy of HKUST-1, TCNQ@HKUST-1 and TCNT were collected, see Fig. 9f. New absorption bands, at around 700 and 800 nm in TCNQ@HKUST-1, support the charge transfer phenomenon.[161,162] From Raman spectroscopy, one can identify new peaks at 1325 and 1296 cm^{-1} for TCNQ@HKUST-1, suggesting that TCNQ interacts with the Cu^{2+} ion in the HKUST-1 structure. The importance of interaction was probed *via* comparison of

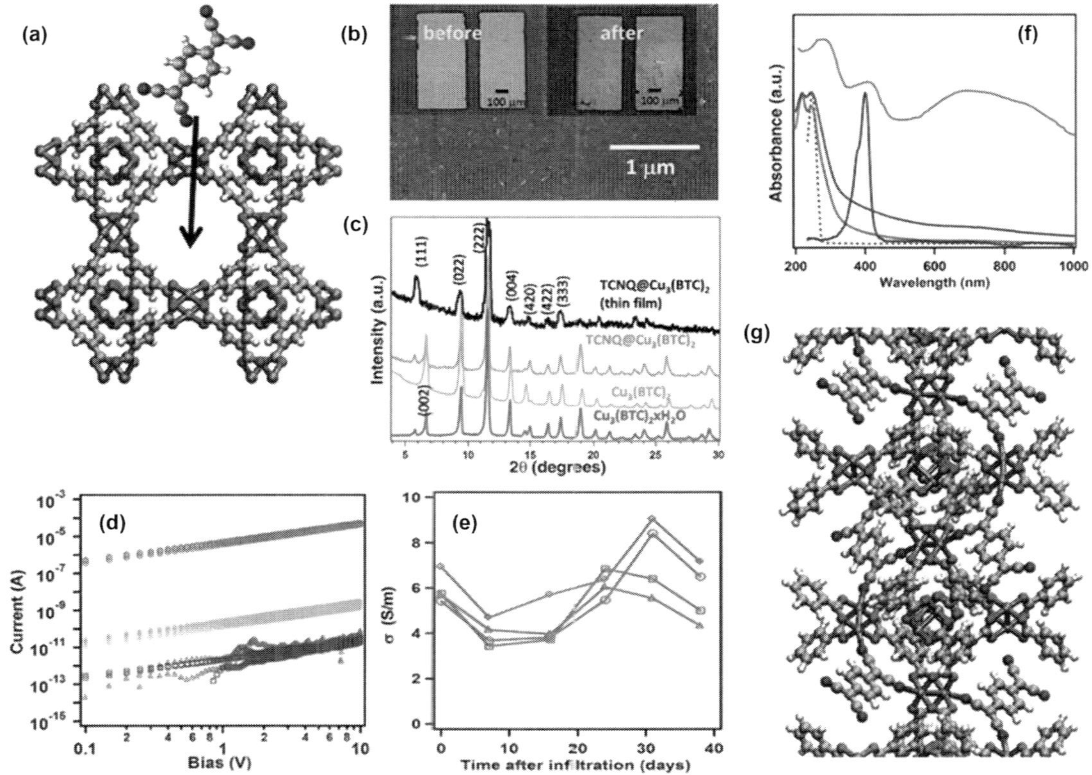

Fig. 9 (a) Structure of HKUST-1 MOF and TCNQ (7,7,8,8-tetracyanoquinododimethane) molecule. White, hydrogen; blue, nitrogen; cyan, carbon; red, oxygen; light brown, copper. (b) SEM image of TCNQ-treated MOF thin film. Insets: before and after TCNQ-treatment HKUST-1 thin film. (c) XRD patterns for as-prepared, dried, TCNQ-treated HKUST-1 powder and TCNQ-treated HKUST-1 film. (d) Current–potential curves for thin film before (red) and after TCNQ (green), F4-TCNQ (gold) and H4-TCNQ (purple) treatment. (e) Stability of conductivity over time for several TCNQ-treatment HKUST-1 thin films. (f) Ultra-violet spectra for thin film before (red) and after TCNQ (green), H4-TCNQ (purple) treatment and TCNQ (blue). (g) Illustration of possible mechanism for conducting TCNQ treated HKUST-1 thin film. Reprinted with permission from A. A. Talin, A. Centrone, A. C. Ford, M. E. Foster, V. Stavila, P. Haney, R. A. Kinney, V. Szalai, F. El Gabaly, H. P. Yoon, F. Léonard and M. D. Allendorf, Tunable Electrical Conductivity in Metal-Organic Framework Thin-Film Devices, *Science*, **343**, 66–69. Copyright (2014) The American Association for the Advancement of Science.

cyclohexane-(1,4-diylidene)dimalononitrile (H4-TCNT) and 2,3,5,6-tetrafluoro-7,7,8,8-tetracyanoquinodimethane (F4-TCNQ) treated HKUST-1 thin film. H4-TCNT is a molecule that lacks a conjugated π electron network, while F4-TCNQ possesses a higher electron affinity with respect to TCNQ. The conductivity of H4-TCNQ@HKUST-1 and F4-TCNQ@HKUST-1 (1 H4-TCNQ or 1 F4-TCNQ molecule per pore) is less than TCNQ@HKUST-1, similar to that of HKUST-1. These results indicated that the guest molecule of TCNQ accepts the charge, instead of H4-TCNQ and F4-TCNQ. Finally, the ab initio calculations suggested that TCNQ molecule strongly binds to MOF with a binding energy of 83.9 $kJ\,mol^{-1}$. The result of the calculation showed that the charge transfer should be induced by bonded TCNQ, illustrated in Fig. 9g, where the MOF HOMO–LUMO gap was modified.

2.1.2 MOF/carbon composite. The composites of MOF/carbon has been largely developed, aiming at applying these materials as electrocatalysts. The electrical conductivity can be improved if carbon materials are introduced.[163] The chemical bonds between MOF and carbon materials in a composite can favor the charge transfer, leading to conductive MOF/carbon composites. The composite can be prepared through various chemical routes. The most commonly used methods are the:[164] (1) direct mixing and self-assembly method; (2) one-pot synthesis; (3) stepwise synthesis; (4) *in situ* polymerization method.

The electrical performance of MOF can be promoted by direct mixing with carbon material. For example, supercapacitors based on MOF/carbon composites could be prepared *via* a direct dispersion of MOFs on reduced graphenes (rGOs) in organic solvents.[165,166] MOF@rGO composite, where MOF = MIL-101(Cr), prepared by mixing MOF, rGO, and CS_2 in a two-step route, was applied as cathode for Lithium–Sulfur battery.[167] During the direct mixing process, it is possible that MOF and carbon material integrate *via* hydrogen bonding and π–π stacking.[168–171] For example, Lin *et al.*[168] reported that magnetic Co/GO nanocomposite could be derived from a self-assembly of MOF/GO composite (MOF = ZIF-67). The nanocomposite of Cu(tpa)/GO (tpa = terephthalate) was synthesized from self-assembly of MOF on GO through simple ultrasonication.[169] The physical characterizations revealed that the formation of the composite was based on hydrogen bonding, π–π stacking and Cu–O coordination in the interface of Cu(tpa) and GO. Moreover, MOF/carbon composites can be prepared *via* one-pot synthesis. It is a facile, energy and time saving approach, and the synthetic process is a classical hydrothermal method. Briefly, carbon materials were added into an autoclave containing solution of dissolved MOF precursors, followed by hydrothermal procedures, Fig. 10a. It has been reported that frameworks of MOF-5, MIL-100(Fe), HKUST-1, NH_2-MIL-101(Al) and ZIF-8 could be deposited onto carbon materials of graphite, graphite oxide, carbon nanotube (CNT), vertically aligned CNT (VACNT), macroporous carbon (MPC), and functionalized carbon nanofiber (FCNF).[172–186] The functional groups and defects of carbon materials served as nucleation sites

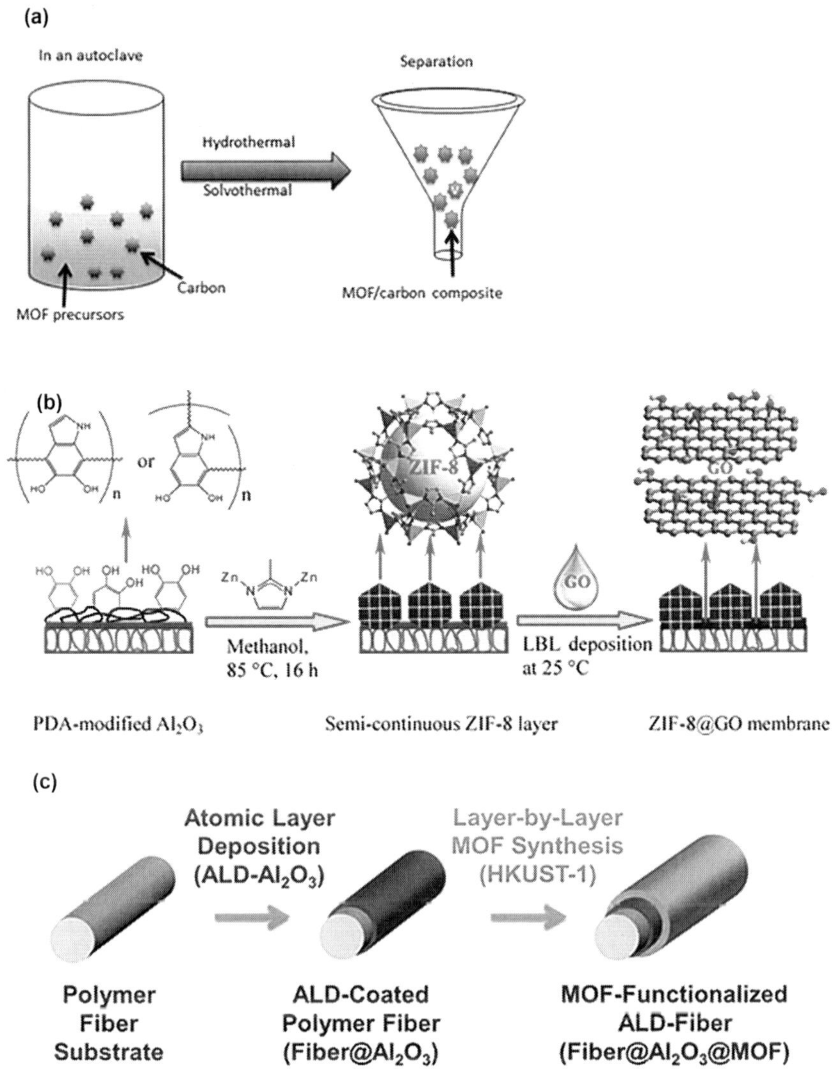

Fig. 10 (a) One-pot synthesis for MOF/carbon composites. (b) Layer-by-layer route to synthesize bi-continuous ZIF-8/GO composite, ZIF = zeolitic imidazolate framework, ZIF-8 = Zn(2-methyl imidazolate)$_2$, GO = graphene oxide. Reprinted with permission from A. Huang, Q. Liu, N. Wang, Y. Zhu and J. Caro, Bicontinuous Zeolitic Imidazolate Framework ZIF-8@GO Membrane with Enhanced Hydrogen Selectivity, *J. Am. Chem. Soc.*, 2014, **136**(42), 14686–14689. Copyright (2014) American Chemical Society. (c) Layer-by-layer route to deposit HKUST-1 onto carbon fiber surface modified by Al_2O_3. Reprinted from ref. 190 with permission from the Royal Society of Chemistry. (d) Seed growth method of ZIF-8 membrane on the surface of macroporous carbon tubes Reprinted from ref. 192 with permission from the Royal Society of Chemistry. (e) *In situ* growth of ZIF-8 onto on ZnO modified CNTs surface. Reprinted from Y. Yue, B. Guo, Z.-A. Qiao, P. F. Fulvio, J. Chen, A. J. Binder, C. Tian and S. Dai, Multi-wall carbon nanotube@zeolite imidazolate framework composite from a nanoscale zinc oxide precursor, *Microporous Mesoporous Mater.*, **198**, 139–143. Copyright 2014 with permission from Elsevier B.V. (f) *In situ* growth of Co-phorphyrin MOF onto on MWCNTs surface. Reprinted with permission from H. Jia, Z. Sun, D. Jiang and P. Du, Covalent Cobalt Porphyrin Framework on Multiwalled Carbon Nanotubes for Efficient Water Oxidation at Low Overpotential, *Chem. Mater.*, 2015, **27**, 4586–4593. Copyright (2015) American Chemical Society.

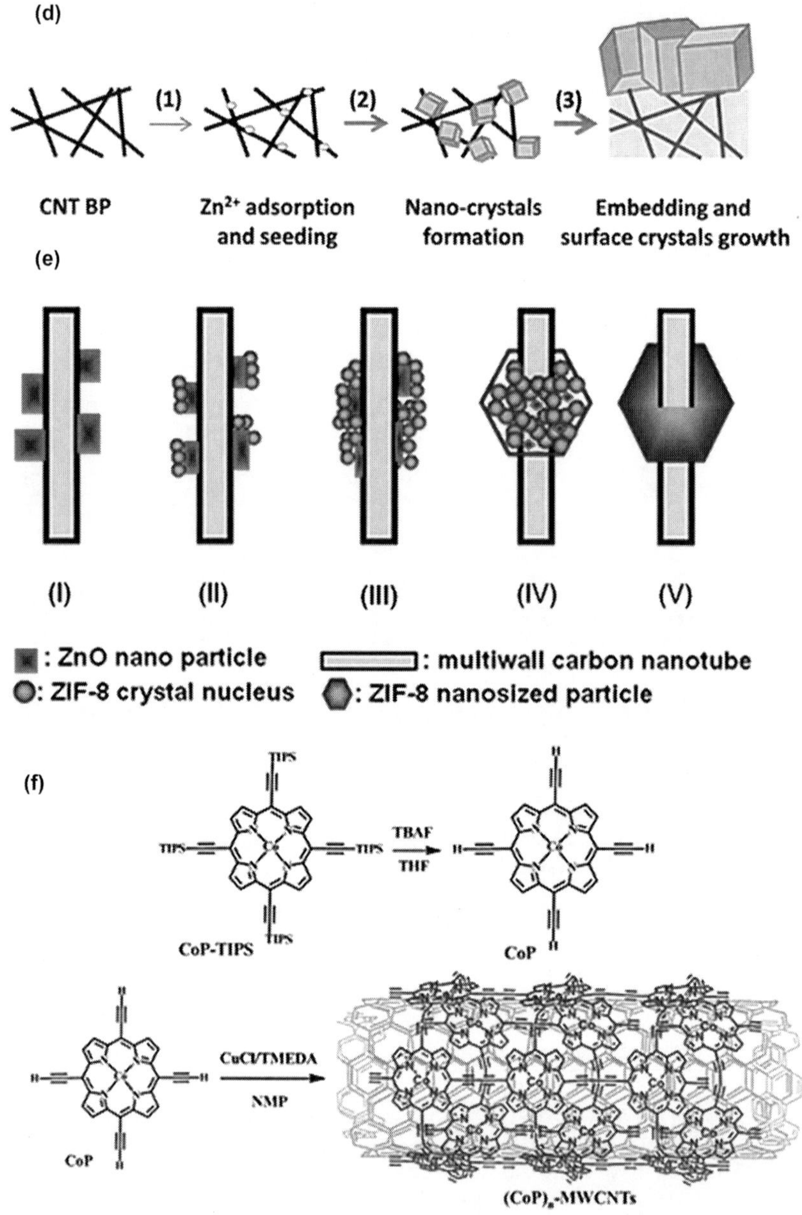

Fig. 10 Continued.

for the growth of MOF crystals.[172–180] Thus, the formation of the composite *via* one-pot synthesis is highly dependent on the surface functionalization and defects of carbon materials. It becomes difficult to employ such a method if the surface of carbon materials lacks anchoring sites. Therefore, stepwise synthesis, based on layer-by-layer assembly and secondary growth, should be an ideal approach for the growth of MOF crystals on carbon materials. Layer-by-layer assembly is a strategy generally adopted by researchers to fabricate MOF thin films, in controlling homogeneity and thickness of MOF layer on substrates.[187,188] Generally, the supporting materials should be firstly grafted by anchoring molecules. Then they should be successively immersed into a solution of metal and ligand precursor. For example, ZIF-8/GO composite could be prepared onto a polydopamine modified Al_2O_3 support, Fig. 10b, where ZIF = zeolitic imidazolate framework, ZIF-8 = Zn(2-methylimidazolate)$_2$, GO = graphene oxide.[189] HKUST-1 could be deposited *via* layer-by-layer route onto carbon fiber surface modified by Al_2O_3 *via* atomic layer deposition, Fig. 10c.[190] During the procedure of secondary growth of MOF on carbon materials, seeds are firstly deposited onto support, facilitating the further growth of MOF crystals.[191] In the seeding process, the key step for secondary growth, could be achieved *via* several techniques, such as heating, wiping, spin-coating and so on.[191] For example, ZIF-8 membrane could be prepared on the surface of macroporous carbon tubes that were previously dip-coated by graphite composites and ZIF-8 nanoparticles as seeds.[192] Then a dense layer of ZIF-8 could be formed under solvothermal conditions from the seeds, Fig. 10d.[192] An alternative approach is an *in situ* growth of MOF onto an oxide modified carbon surface. For example, it was reported that ZIF-8 could grow on a ZnO modified CNTs surface.[193] In such a route, ZnO acting as metal precursor, was reacted with protonated 2-methylimidazole ligands.[193] Figure 10e illustrates such a secondary growth of a MOF layer onto carbon support.[193]

Jia *et al.*,[194] recently reported an *in situ* chemical route to prepare Co-porphyrin-based MOF on MWCNT, *via* butadiene bridging, leading to a stable core–shell structure, Fig. 10f.[194] The *in situ* polymerization process of Co ions and porphyrin was catalyzed by CuCl and TMEDA (tetramethylethylenediamine), leading to a templated core–shell structure on MWCNT.[194]

2.1.3 MOF as a catalytic centre for energy conversion. As discussed in the introduction, MOFs are potential polymerized molecular electrocatalysts, since they possess redox metal centres that serve as catalytic centres. When the molecules, such as CO_2 and O_2, were adsorbed on the electrode surface, the electron transfer between active sites and molecules could be established, leading to a catalytic reaction. The electrocatalytic process illustrated in Fig. 11 shows that a charge-transfer (CT) coordination polymer is crucial. As the electron-transfer occurs *via* coordination bonds, between ligand and metal, the design and selection of the ligand is the key. In the domain of energy conversion, MOF-based electrocatalysts are divided into 4 categories, according to the type of coordination bonds of the metal centre: (1) metal–nitrogen, (2) metal–oxygen, and (3) metal–sulphur bonds which favour the electrocatalytic process.

Considering a MOF formed *via* metal–nitrogen coordination, a ligand of porphyrin macrocycle is able to achieve a CT with a metal site. This is the reason why metallo-porphyrin complexes were widely studied for energy conversion.[74,94–96] It was proven that a Co-based metallo-porphyrin MOF thin film on F-doped tin oxide (CoPIZA/FTO) could be

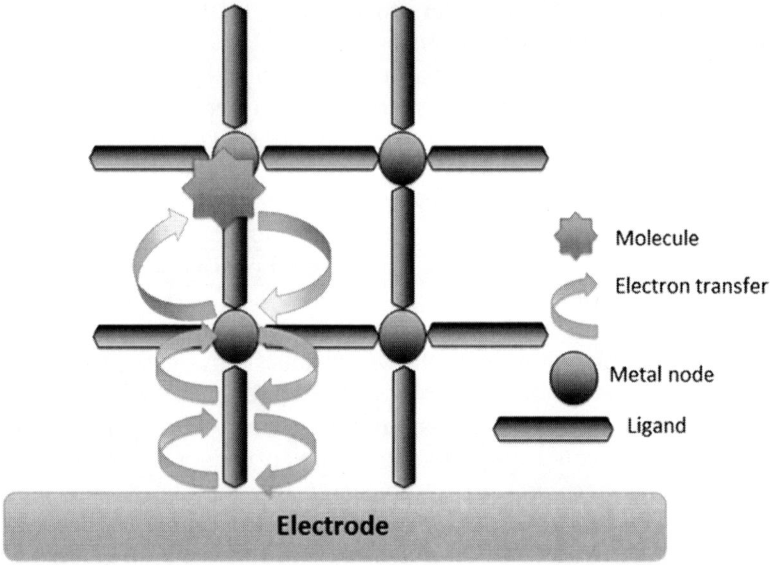

Fig. 11 A possible mechanism for the electrocatalytic process on MOFs.

reduced and remained stable for hours, based on charge transport of redox hopping.[195] Therefore, many MOFs based on porphyrin derivatives were considered as candidates for electrocatalytical process, as listed in Table 2. For example, Jahan *et al.*,[76] reported a graphene–porphyrin MOF composite, Fig. 12a, which was active for ORR in 0.1 M KOH. Such a composite was prepared by a hybrid Fe-porphyrin MOF with dye-functionalized rGO sheets.[76] It revealed that the functionalizing group of pyridinium on a GO sheet favoured the 4-electron exchanged ORR on Fe-porphyrin MOF, as shown in Fig. 12b–c.[76] Recently, a carbon nanotube templated covalent Co(II)/meso-tetraethynylporphyrin framework was reported for ORR in 0.5 M H_2SO_4.[196] The number of electrons transferred in this work was close to 4.[196] A 3D non-interpenetrating porous MOF based on Pd and porphyrin ligand (Pd-TCPP), namely $Pd_2(HTCPP)\cdot 4DMF\cdot H_2O$ (H_6TCPP = 5,10,15,20-tetra(carboxyphenyl)porphyrin, DMF = dimethylformamide), depicted in Fig. 13 inset, was reported as an electrocatalyst for the oxygen evolution reaction (OER).[197] The potential at 10 mA cm^{-2} on Pd-TCPP for OER was 1.70 V/RHE, and the mass activity at an overpotential of 1.2 V/RHE achieved 18.26 A g^{-1} in 1.0 M KOH.[197] These values were enhanced with respect to those of the mixture of Pd(II) ions and TCCP ligand.[197] This revealed that the improved charge-transfer efficiency in Pd(II) hybrid porphyrin ligand should be responsible for the OER activity.[197] The TOF value for OER on Pd-TCPP was 5×10^{-4} s^{-1} at 1.2 V/RHE.[197] Moreover, covalent organic frameworks and MOF containing Co-porphyrin entities showed activity for the electroreduction of CO_2.[198,199] In these structures, the catalytic sites of Co-porphyrin were connected by a multi-carboxylate ligand *via* covalent bonds or coordination bonds between an organic ligand and a secondary metal centre.[198,199] The catalysts COF-366-Co and COF-367-Co (Fig. 14a) were active for the CO_2 reduction

Table 2 Examples of MOFs as catalytic center for electrocatalytic energy conversion.[a]

MOF	Supporting material	Application	Ref.
Fe(porphyrin)	Dye-functionalized rGO	ORR	76
Co(II)/meso-tetraethynylporphyrin	MWCNT	ORR	196
$Pb_2(H_2TCPP)\cdot 4DMF\cdot H_2O$	—	OER	197
COF-366-Co	—	CO_2 reduction	198
COF-367-Co			
$Al_2(OH)_2TCPP'$-Co	—	CO_2 reduction	197
Fe_MOF-525	—	CO_2 reduction	200
$Ni_3(HITP)_2$	—	ORR	201
Cu(bpy)(benzene-1,3,5-tricarboxylate)	—	ORR	202
M_2\{4-(5-(pyridine-4-yl)-4*H*-1,2,4-triazol-3-yl)benzoate\}$_4\cdot 3H_2O$ (M = Co, Cu, Zn)	—	HER and OER	204
Co_4\{4-(5-(pyridine-4-yl)-4*H*-1,2,4-triazol-3-yl)benzoic\}$_2$(4,4-bipyridine)$(H_2O)_6\cdot 3.5H_2O$ and Co_2\{4-(5-(pyridine-4-yl)-4*H*-1,2,4-triazol-3-yl)benzoic\}$_2$\{(*E*)-1,2-di(pyridine-4-yl)di-azene\}$(H_2O)_3\cdot$ DMF	—	HER and OER	205
$Co(Hmim)_2$	—	OER	206
$Cu_3(Mo_8O_{26})(H_2O)_2(OH)_2(L1)_4$ (L1 = 4*H*-4-amino-1,2,4-triazole)	—	HER	207
$Ag_4(Mo_8O_{26})(L2)_{2.5}(H_2O)$ (L2 = 3,5-dimethyl-4-amino-4*H*-1,2,4-triazole)			
Co-modified $Zr_6(\mu_3\text{-}O)_4(\mu_3\text{-}OH)_4(\text{-}OH)_4(\text{-}OH_2)_4\ (TBAPy)_2$	—	OER	208
$(TBA)_3\{PMo^{V}{}_8Mo^{VI}{}_4O_{36}(OH)_4Zn_4\}\{C_6H_3(COO)_3\}_{4/3}\cdot 36H_2O$	—	HER	210
M(benzene-1,3,5-tricarboxylate) (M = Fe and Co)	Carbon black	ORR and OER	211
N,*N*′-bis(2hydroxyethyl) dithiooxamidatocopper(II)	Carbon black	EOR	212

[a] rGO = reduced graphene sheet; GO = graphene sheet; H_6TCPP = 5,10,15,20-tetra(carboxyphenyl)porphyrin; DMF = dimethylformamide; MWCNT = multi-walled carbon nanotube; COF = covalent organic framework; TCPP′ = 4,4′,4″,4‴-(porphyrin-5,10,15,20-tetrayl)tetrabenzoate; MOF-525 = $Zr_6O_4(OH)_4(TCPP'\text{-}H_2)_3$; FTO = F-doped tin oxide; HITP = 2,3,6,7,10,11-hexaiminotriphenylene; bpy = 2,2′-bipyridine; Hmim = 2-Methylimidazole; TBAPy = 1,3,6,8-tetrakis(*p*-benzoic acid)pyrene.

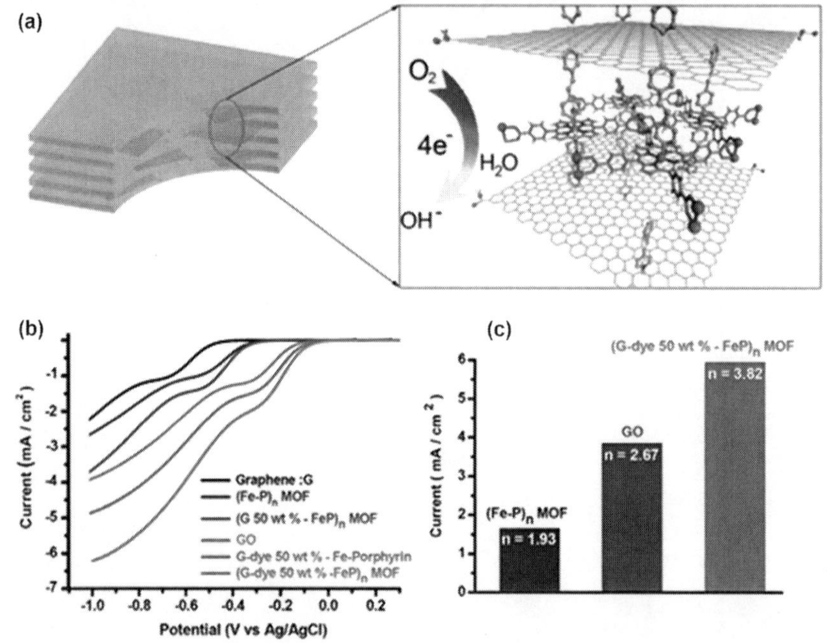

Fig. 12 (a) The composite made of pyridinium functionalized graphene oxide (G-dye GO) and Fe-phorphyrin (Fe-P). (b) ORR polarization curves, recorded at 10 mV s^{-1} in 0.1 M KOH at 2000 rpm on rotating disk electrode, for graphene (G), GO, Fe-P, G/Fe-P composites, mixture of G-dye GO and Fe-P and composite of G-dye GO/Fe-P favoured the 4-electron exchanged ORR on Fe-porphyrin MOF. (c) Number of electron transfer for GO, Fe-P and composite of G-dye GO/Fe-P. Reprinted with permission from M. Jahan, Q. Bao and K. P. Loh, Electrocatalytically Active Graphene-Porphyrin MOF Composite for Oxygen Reduction Reaction, *J. Am. Chem. Soc.*, 2012, **134**, 6707–6713. Copyright (2012) American Chemical Society.

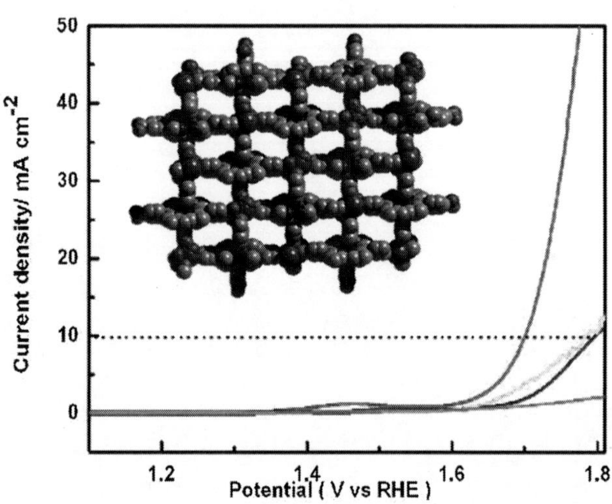

Fig. 13 OER curves at 5 mVs^{-1} for $Pd_2(HTCPP)\cdot 4DMF\cdot H_2O$ (H_6TCPP = 5,10,15,20-tetra(carboxyphenyl)porphyrin, DMF = dimethylformamid), namely, Pd-TCPP, in 1.0 M (red) and 0.1 M KOH (blue), in 0.2 M PBS (phosphate buffer saline, black), mixture of Pd ions and TCPP (cyan), H_6TCPP (pink) in 1.0 M KOH. Inset, the structure of Pd-TCPP. Reproduced from ref. 197 with permission from the Royal Society of Chemistry.

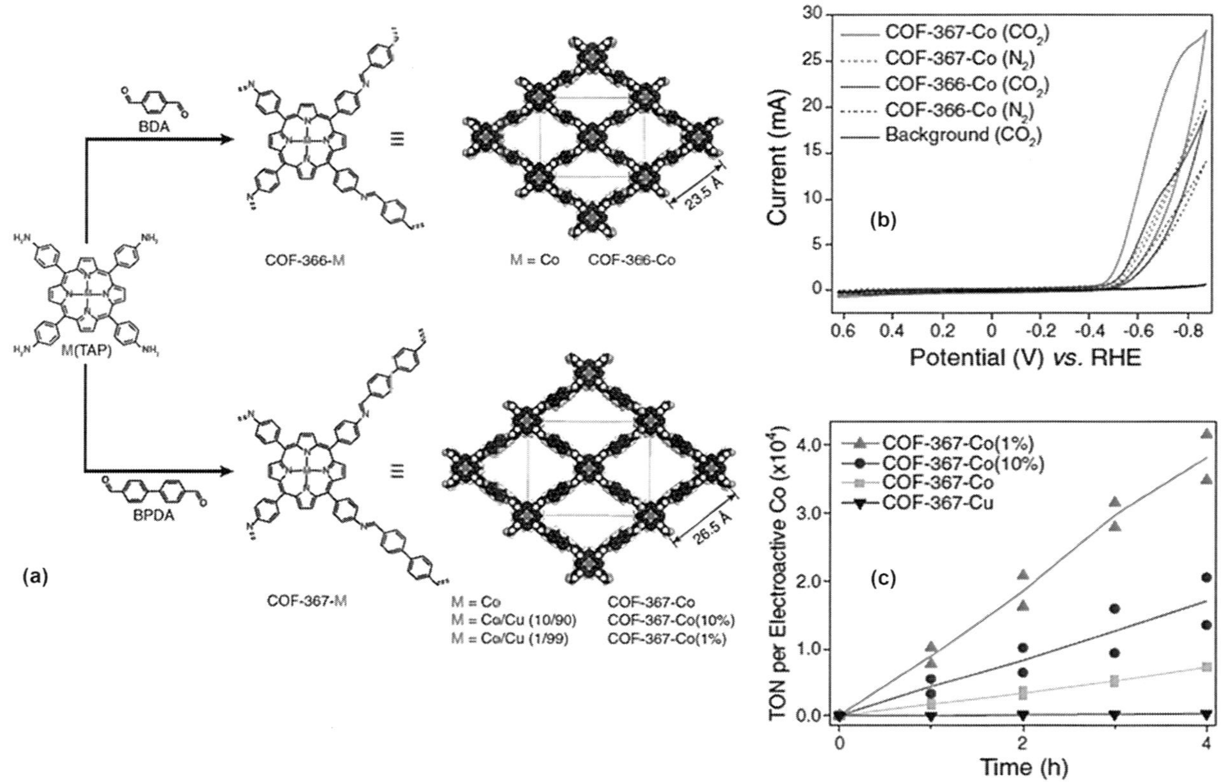

Fig. 14 (a) The synthesis and structure of COF-366-Co and COF-367-Co(Cu). (b) Cyclic voltammogramms recorded in 0.2 M potassium phosphate buffer (pH = 7.2) with 0.5 M $KHCO_3$ under CO_2 or 0.5 M $NaClO_4$ under N_2. (c) Turnover frequency (TOF) for converting CO_2 to CO on different MOF catalysts. Reprinted with permission from S. Lin, C. S. Diercks, Y.-B. Zhang, N. Kornienko, E. M. Nichols, Y. Zhao, A. R. Paris, D. Kim, P. Yang, O. M. Yaghi and C. J. Chang, Covalent organic frameworks comprising cobalt porphyrins for catalytic CO_2 reduction in water, *Science*, **349**, 1208–1213. Copyright (2015) The American Association for the Advancement of Science.

reaction (Fig. 14b), converting CO_2 to CO with 90% of Faraday efficiency and 9400 h^{-1} of turnover frequency (TOF), Fig. 14c.[198] In addition, electro-reduction of CO_2 could also be efficient on a Fe-porphyrin based MOF structure.[200] In this work, the Faraday efficiency of electroreduction was 100%, leading to final products of equal amount of CO and H_2 (syngas).[200] The mechanism for catalytic activity on these MOF materials should be attributed to the M–N_4 units, which have been proven to be the active site elsewhere. For example, Miner *et al.*[201] developed a MOF structure of $Ni_3(HITP)_2$ (HITP = 2,3,6,7,10,11-hexaiminotriphenylene), Fig. 15a. Such a 2D conducting framework containing an Ni–N_4 unit electrochemically activating the reduction of oxygen to water, and H_2O_2.[147,201] The onset potential for ORR on this material was 0.82 V/RHE (*vs.* 1.0 V/RHE on Pt) in 0.1 M KOH, Fig. 15b.[201] The ORR current remained 88% of the initial current at 0.77 V for 8 h.[201]

MOFs involving pyridine ligand were also largely studied as electrocatalysts for the energy conversion. In these materials, metal coordinated to nitrogen atom(s) of pyridine ligand served as catalytic centre for electrocatalytic processes, summarized in Table 2. Mao *et al.*,[202] reported a MOF composed by Cu(II) connected to BTC and bpy ligands (Cu-bpy-BTC), Fig. 16a, displays an ORR activity *via* a four-electron transfer pathway. Compared with the MOF structure of Cu-BTC (also named HKUST-1), the electrochemistry of Cu-bpy-BTC was more stable in buffer solution (pH = 6).[202] Although the authors claimed that the crystallographic structure of Cu-bpy-BTC is the same as the structure reported in the literature,[202,203] the powder X-Ray diffraction (pXRD) pattern did not match that reported in the literature (Fig. 16b).[202] The ORR activity on Cu-bpy-BTC, recorded by the rotating ring-disk electrode (RRDE) in phosphate buffer solution (pH = 6), was a four-electron process (water formation), as shown in Fig. 16c–d.[202] Using pyridine ligand, namely, 4-(5-(pyridine-4-yl)-4*H*-1,2,4-triazol-3yl)benzoic acid, two isostructures, namely, Co and Cu containing 2D MOFs, could be obtained, Fig. 17a–b, showing the electrocatalytic activity towards water splitting (HER and OER), Fig. 17c–d.[204] However, the isostructural Zn-based compound showed no catalytic activity.[204] It was also found that the Co-based compound performed better than the Cu one,[204] probably due to CT between the ligand and the unsaturated coordinated metal centre. Gong *et al.*[205] reported two Co-based MOFs, comprising multi-carboxylate of 4,5-di(4′-carbonxylphenyl)-phthalate and bpy of 4,4′-bipyridine or (*E*)-1,2-di(pyridine-4-yl)di-azene ligands. In these structures, the Co centre was coordinated to O atoms from carboxylate ligands, water ligands, and N atoms from bpy ligands.[205] Both structures showed water-splitting ability under applied electrode potential.[205] Better catalytic activity was observed on azene involving MOF with respect to the 4,4′-bpy one.[205] Metal–nitrogen coordination based MOFs can also be formed from azole ligands. They were reported for their electrocatalytic activity for energy conversion. For example, highly porous ZIF-67 composed of Co ions and methylimidazolate ligands was active for electrocatalytic OER in aqueous solution

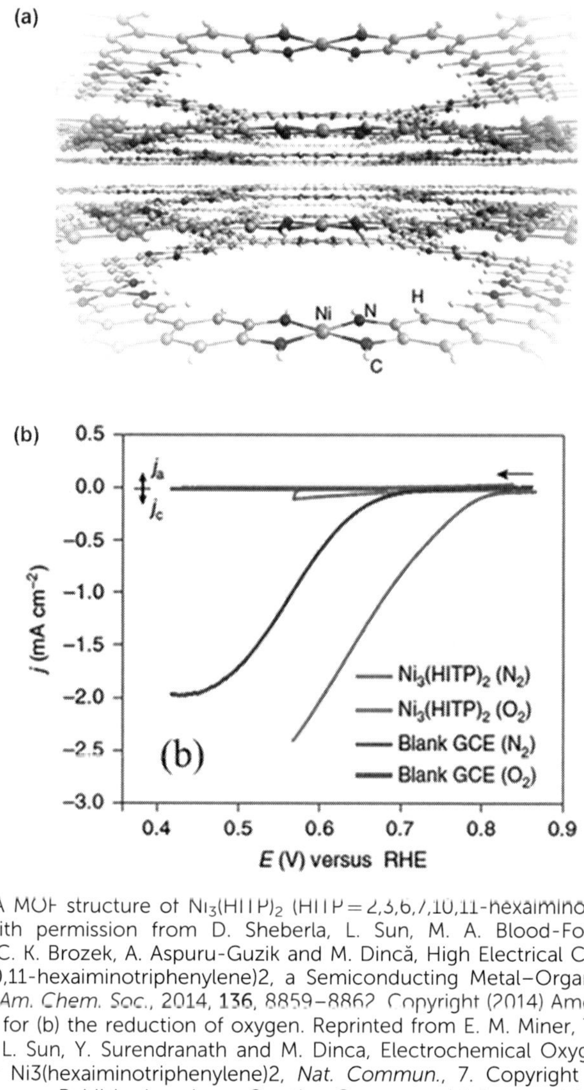

Fig. 15 (a) A MOF structure of $Ni_3(HITP)_2$ (HITP = 2,3,6,7,10,11-hexaiminotriphenylene), (reprinted with permission from D. Sheberla, L. Sun, M. A. Blood-Forsythe, S. Er, C. R. Wade, C. K. Brozek, A. Aspuru-Guzik and M. Dincă, High Electrical Conductivity in Ni3(2,3,6,7,10,11-hexaiminotriphenylene)2, a Semiconducting Metal–Organic Graphene Analogue, *J. Am. Chem. Soc.*, 2014, **136**, 8859–8862. Copyright (2014) American Chemical Society.) for (b) the reduction of oxygen. Reprinted from E. M. Miner, T. Fukushima, D. Sheberla, L. Sun, Y. Surendranath and M. Dinca, Electrochemical Oxygen reduction catalysed by Ni3(hexaiminotriphenylene)2, *Nat. Commun.*, 7. Copyright 2016 Nature Publishing Group. Published under a Creative Commons Attribution 4.0 International License (http://creativecommons.org/licenses/by/4.0/).

with pH varying from 2.1 to 13.1.[206] One could observe a pH-dependent OER activity on ZIF-67 catalysts. Specifically, the OER overpotential decreased with increasing pH.[206] The OER current density remained stable after 5 h and 24 h of operation in acid and neutral solutions, respectively.[206] Two MOF structures, reported by Gong *et al.*,[207] were active for HER from water. Both $Cu_3(Mo_8O_{26})(H_2O)_2(OH)_2(L1)_4$ (L1 = 4*H*-4-amino-1,2,4-triazole) and $Ag_4(Mo_8O_{26})(L2)_{2.5}(H_2O)$ (L2 = 3,5-dimethyl-4-amino-4*H*-1,2,4-triazole) were composed of bi-metal centres.[207] In both cases, the Mo ions were coordinated to O atoms, whereas Cu or Ag ions were connected to N atoms from azole ligands and O from OH or

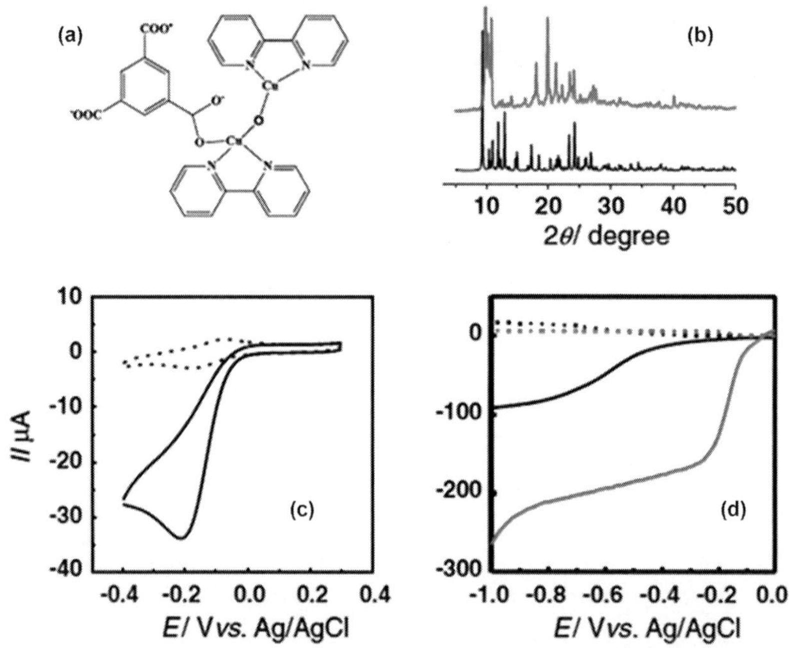

Fig. 16 (a) The proposed coordination mode of Cu-bpy-BTC. (b) XRD patterns for (red) Cu-bpy-BTC in this work and (black) that in the ref. 203. (c) CV curves recorded at 20 mV s^{-1} in (dotted) N_2- and (solid) O_2-saturated phosphate buffer solution (pH = 6); (d) ORR curves recorded on the (solid) disk with (red) and without (black) MOF and on the (dotted) ring of RRDE, 400 rpm. Reprinted from J. Mao, L. Yang, P. Yu, X. Wei and L. Mao, Electrocatalytic four-electron reduction of oxygen with Copper (II)-based metal-organic frameworks, *Electrochem. Commun.*, **19**, 29–31. Copyright 2012 with permission from Elsevier B.V.

H_2O ligands.[207] The HER activity on Cu/Mo-based MOF was better than on Ag/Mo-based MOF.[207]

Beside MOFs constituted by metal–nitrogen coordination, those composed by metal–oxygen bonds could also be used as electrocatalysts for energy conversion. As listed in Table 2, a MOF thin film, *i.e.*, Co-modified NU-1000, showed promising OER activity in 0.1 M Na_2SO_4/NaOH from pH = 8 to pH = 11.[208] The NU-1000 skeleton consisted of Zr, and 1,3,6,8-tetrakis(*p*-benzoic acid)pyrene (TBAPy) ligand,[209] modified by Co(II) ions, Fig. 19a.[208] It revealed that the Co ions were installed into the channel of NU-1000, bonded to TBAPy ligand (Fig. 18b).[208] Further studies showed that the addition of Co ions did not affect the robust MOF structure. The Co-modified NU-1000 showed pH-dependent OER activity, and the onset potential of OER was negatively shifted from ~1.8 V/RHE in pH = 8.2 to 1.4 V/RHE in pH = 11 solution (Fig. 18c–d).[208] The active sites were recognised as Co–O coordination, since the NU-1000 film alone did not display OER activity.[208] A 3D polyoxometalate-based MOF (POMOF), $(TBA)_3\{PMo^{V}_8Mo^{VI}_4O_{36}(OH)_4Zn_4\}\{C_6H_3(COO)_3\}_{4/3}\cdot 36H_2O$, showed higher electrocatalytic HER activity compared to Pt catalyst.[210] It achieved a turnover number of 1.2×10^5 after 5 h.[210] Wang *et al.*[211] observed the bi-functional catalytic properties on MOF composed by hetero-polynuclear

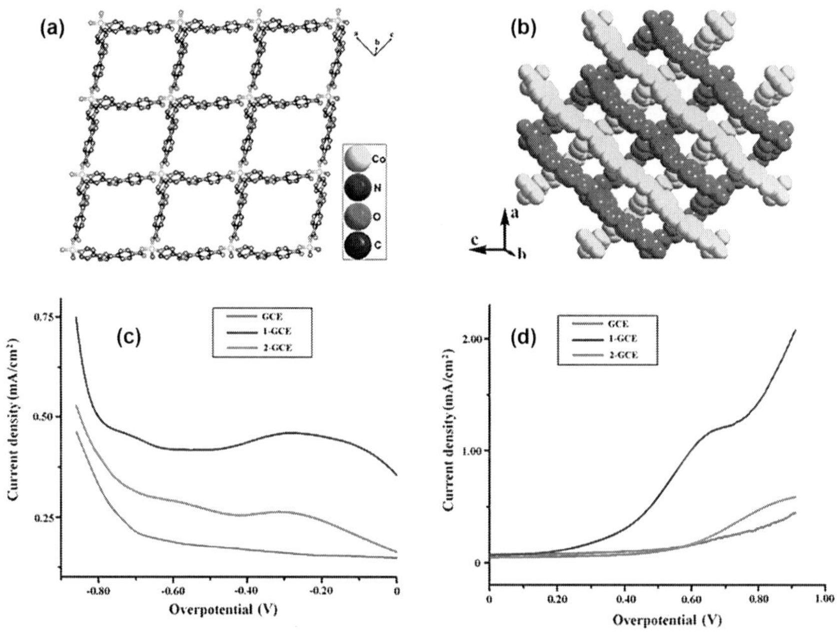

Fig. 17 (a–b) A MOF structure composed by (1) Co or (2) Cu and 4-(5-(pyridine-4-yl)-4*H*-1,2,4-triazol-3yl)benzoic ligand, herein only the Co-based structure is presented, for electrocatalytic (c) HER and (d) OER in 0.05 M phosphate recorded at 10 mV s^{-1}. Reprinted with permission from Y. Gong, H.-F. Shi, P.-G. Jiang, W. Hua and J.-H. Lin, Metal(II)-Induced Coordination Polymer Based on 4-(5-(Pyridin-4-yl)-4H-1,2,4-triazol-3-yl)benzoate as an Electrocatalyst for Water Splitting, *Cryst. Growth Des.*, 2014, **14**, 649–657. Copyright (2014) American Chemical Society.

Fe/Co hybrid benzene-1,3,5-tricarboxylate ligand towards ORR and OER in alkaline medium.[211] Although the crystallographic structure of MOF was not revealed by Wang *et al.*, the catalytic activity of composites of carbon black (super P, SP) and Fe/Co MOF was improved towards ORR, and OER with respect to carbon black alone, Fig. 14.[211]

Concerning MOFs containing metal–sulfur coordination bonds, Yang *et al.*[212] reported, that a MOF of *N,N′*-bis(2hydroxyethyl) dithiooxamidatocopper(II) also called $(HOC_2H_4)_2$dtoaCu, was a catalyst for electrooxidation of ethanol in acid medium. Ethanol is an important fuel for fuel cells, because it is a low-toxic fuel with a large number of electron transfer (12 e^-). In this MOF, Cu ions were coordinated to S, and N atoms as shown in Fig. 19a.[212] Yang *et al.*,[212] found that the catalytic activity of this MOF depended on the concentration of ethanol in the electrolyte, Fig. 19c. The anodic current density could achieve 3 mA cm^{-2} when the ethanol concentration reached 2 M.[212] Although such an activity was lower than that on Pt-based materials, the MOF-based catalyst was less poisoned by the intermediates, *e.g.*, CO, during the oxidation of ethanol.[213] It was assumed, based on a DFT calculation, that the ethanol loses H atoms from the OH group during the oxidation process.[212] Then, the H atoms attached to N atoms in $(HOC_2H_4)_2$d toaCu, *cf.* Fig. 19a and b, and O atom of ethanol interacted with the Cu

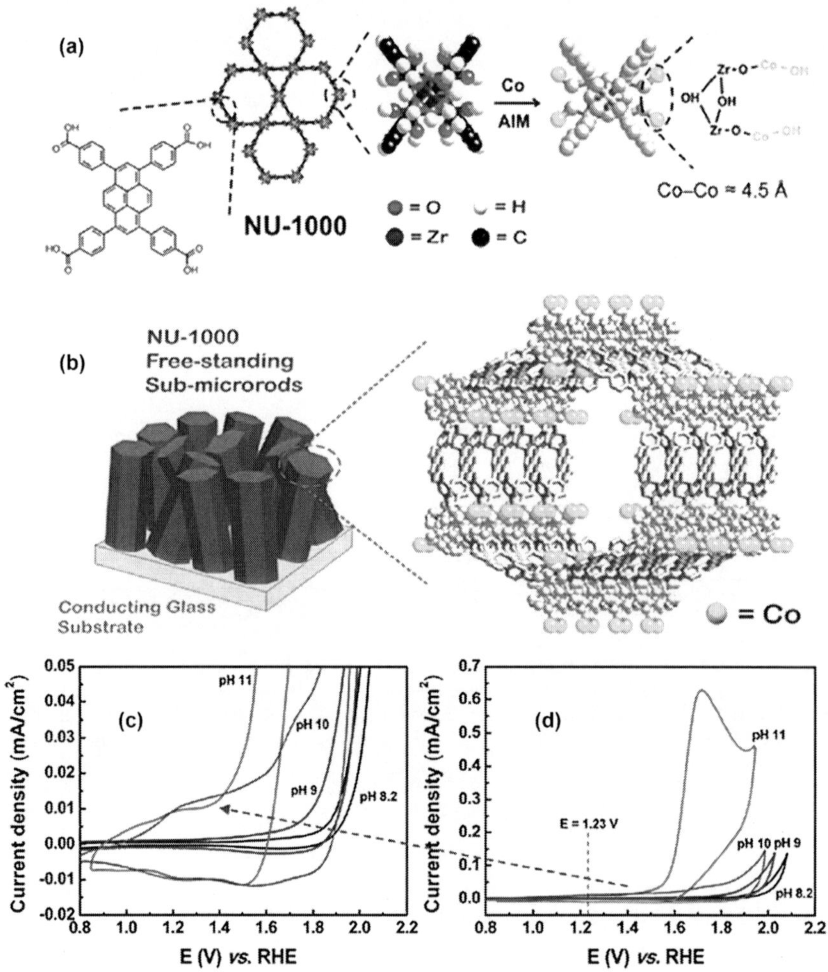

Fig. 18 (a) The NU-1000 skeleton consisted of Zr, and 1,3,6,8-tetrakis(*p*-benzoic acid)-pyrene (TBAPy) ligand, modified by Co(II) ions (Co-NU-1000). (b) Illustration of MOF thin film on conducting glass as electrode. (c) (d) CV recorded on Co-NU-1000 thin film at 25 mV s^{-1} in 0.1 M Na_2SO_4/NaOH with different pH values. Reprinted with permission from C.-W. Kung, J. E. Mondloch, T. C. Wang, W. Bury, W. Hoffeditz, B. M. Klahr, R. C. Klet, M. J. Pellin, O. K. Farha and J. T. Hupp, Metal-Organic Framework Thin Films as Platforms for Atomic Layer Deposition of Cobalt Ions To Enable Electrocatalytic Water Oxidation, *ACS Appl. Mater. Interfaces*, 2015, **7**, 28223–28230. Copyright (2015) American Chemical Society.

atom in MOF.[212] Thus the charge transfer was realized from ethanol to one Cu atom in $(HOC_2H_4)_2$dtoaCu. Besides, acetaldehyde was found as a byproduct of the oxidation reaction.[212]

2.2 MOF as catalyst supports for energy conversion

The microporous and macroporous MOF materials are an ideal support for catalytic nanoparticles since the large surface area of MOF favours the deposition of nanoparticles.[214] Such a kind of composite has been widely studied in the domain of chemical catalysis, and photo-catalysis, where

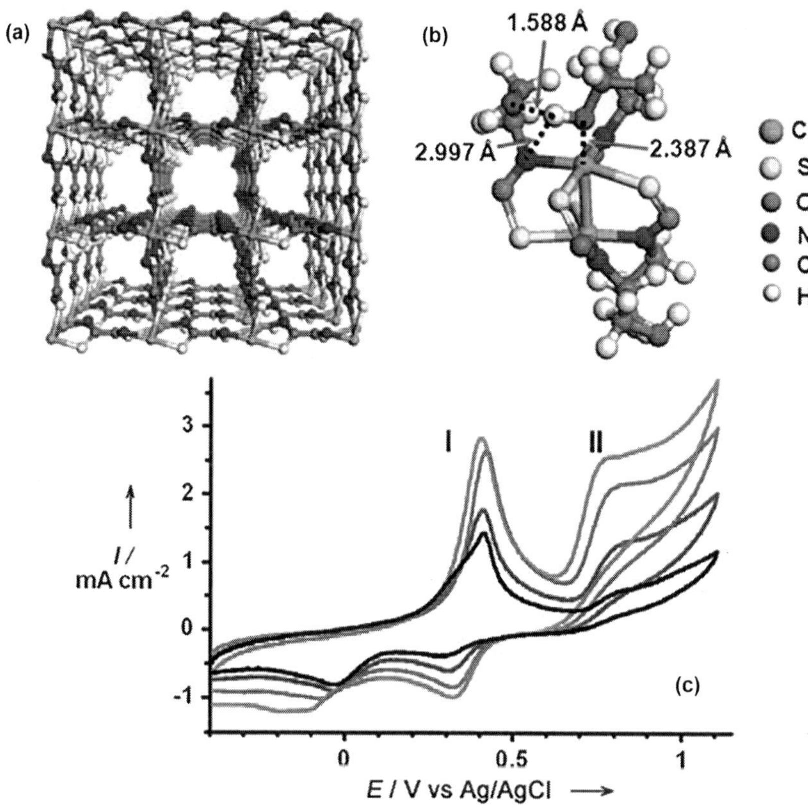

Fig. 19 (a) MOF structure of *N,N′*-bis(2hydroxyethyl) dithiooxamidato-copper(II), namely, $(HOC_2H_4)_2$dtoaCu. (b) Coordination environment of $(HOC_2H_4)_2$dtoaCu. (c) CVs at 0.1 V s^{-1} in 0.5 M H_2SO_4 with 0.0 M (black), 0.5 M (blue), 1.0 M (green) and 2.0 M (red) ethanol. Reprinted with permission from L. Yang, S. Kinoshita, T. Yamada, S. Kanda, H. Kitagawa, M. Tokunaga, T. Ishimoto, T. Ogura, R. Nagumo, A. Miyamoto and M. Koyama, A Metal-Organic Framework as an Electrocatalyst for Ethanol Oxidation, *Angew. Chem.*, 2010, **122**, 5476–5479. Copyright (2010) WILEY-VCH Verlag GmbH & Co. KGaA, Weinheim.

the nanoparticles and/or MOF support serve as a catalytic centre towards the reaction.[215–217] However, this kind of material was less applied in electrocatalysis. One example was Co(II)–salen complex encapsulated into the porosities of MIL-100(Cr), as shown in Fig. 20a.[218] Such a material was active in aqueous solution towards ORR with four-electrons exchanged at pH 6.8, Fig. 20b.[218] In fact, the MOF powder was synthesized first, it was then added into a solution of Co^{2+} and then into a solution of H_2salen.[218] With this method, the formation of a Co–salen complex in the cavities of MIL-100(Cr) was possible.[218] The ORR activity on Co(II)–salen complex was promoted by the MOF structure, since less ORR activity was observed on Co–salen or MOF.[218] Wang *et al.*[219] reported a composite of ε-MnO_2/MOF for ORR in alkaline solution. The MOF(Fe) structure of Fe ions and ligands of 1,3,5-benzene-tricarboxylic acid was synthesized to load ε-MnO_2 nanoparticles (Fig. 21a).[219] It showed an enhanced ORR activity with respect to MOF(Fe), ε-MnO_2

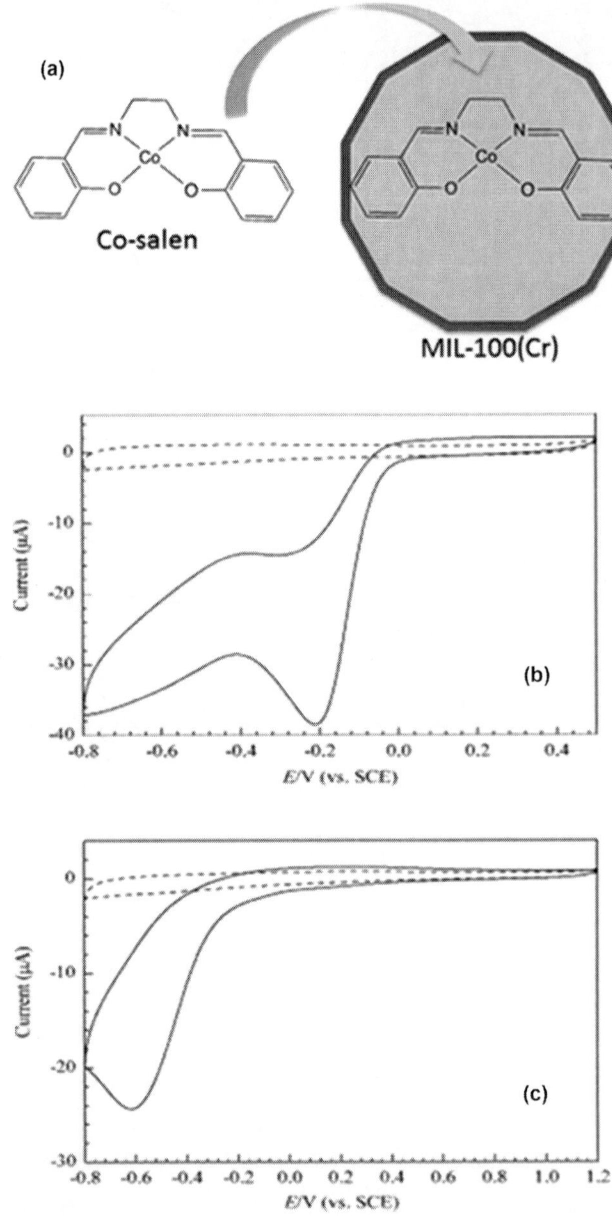

Fig. 20 (a) Illustration of Co(II)–salen complex and MIL-100(Cr). (b) Cyclic voltammograms for (a) composite of Co(II)–salen complex and MIL-100(Cr) supported by glassy carbon electrode, (b) bare glassy carbon electrode in N_2-(dash) and O_2-(solid) saturated pH = 6.8 buffer solution at scan rate of 50 mV s^{-1}. Reprinted from P. Miao, G. Li, G. Zhang and H. Lu, Co(II)-salen complex encapsulated into MIL – 100(CR) for electrocatalytic reduction of oxygen, *J. Energy Chem.*, **23**, 507–512. Copyright (2014) with permission from Elsevier.

and mixed MOF(Fe) + ε-MnO_2 (Fig. 21b).[219] The catalytic reaction was more stable on ε-MnO_2/MOF composite in comparison to ε-MnO_2, and Pt/C catalyst (Fig. 21c).[219]

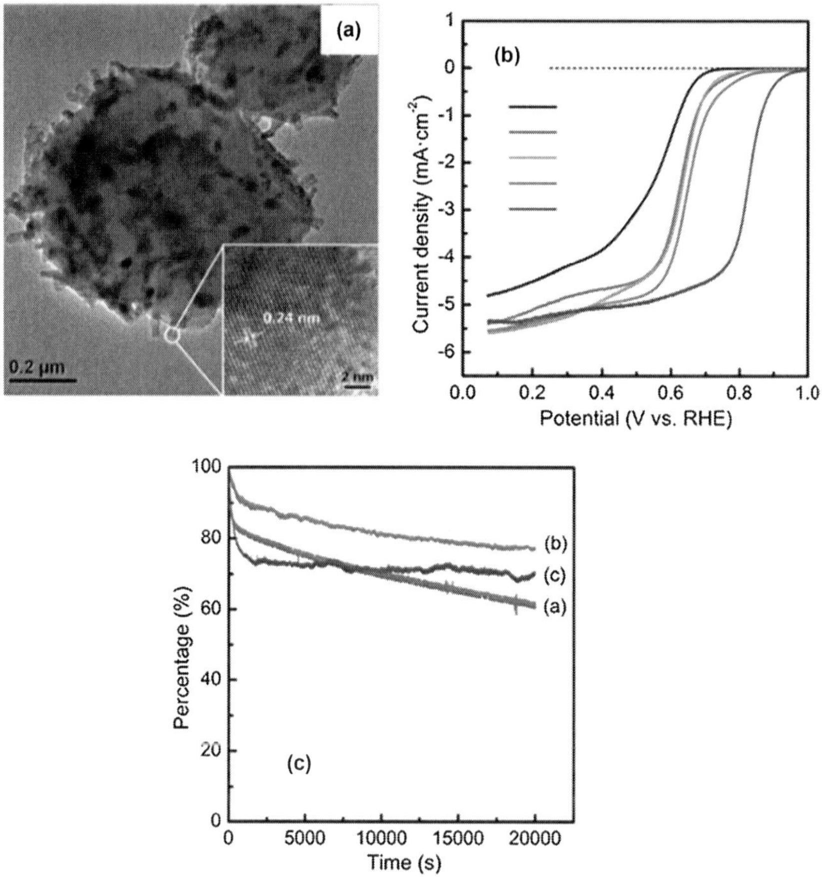

Fig. 21 (a) TEM images for ε-MnO_2/MOF(Fe). (b) ORR polarization curves at 1600 rpm in 0.1 M KOH for MOF(Fe) (black), ε-MnO_2 (green), mixed ε-MnO_2 + MOF(Fe) (orange), ε-MnO_2/MOF(Fe) (red) and 20 wt% Pt/C (blue). (c) Chronoamperometric curves for ε-MnO_2 (green), ε-MnO_2/MOF(Fe) (red), and 20 wt% Pt/C (blue) in at 0.55 V/RHE 0.1 M KOH. Reproduced from ref. 219 with permission from the Royal Society of Chemistry.

2.3 MOF-driving materials for energy conversion

MOF can be used as a precursor to synthesize advanced carbon materials[129,130,220–227] or composites, *e.g.*, metal or oxide NPs supported by carbon.[132,228–232] There are three main methods to prepare MOF-derived materials, Fig. 22: (1) by direct pyrolysis of a MOF compound, (2) by heat-treatment of a composite based on metal nanoparticles, and MOF or (3) by reduction of MOF mixed with salt (metal precursor). In the first synthetic route, MOFs are heat-treated under inert or reducing gas at high temperature, usually higher than 600 °C, finally leading to diverse metal/carbon or oxide/carbon composites.[228,233] The calcination temperature should be higher than the decomposition temperature of a MOF structure, which can be measured by thermogravimetric analysis (TGA). Taking a MOF structure as example: HKUST-1, as shown in Fig. 23,[159] the mass loss at around 100 °C corresponds to the removal of guest molecules, *e.g.*, water or solvent molecules, in the cavities; while the loss

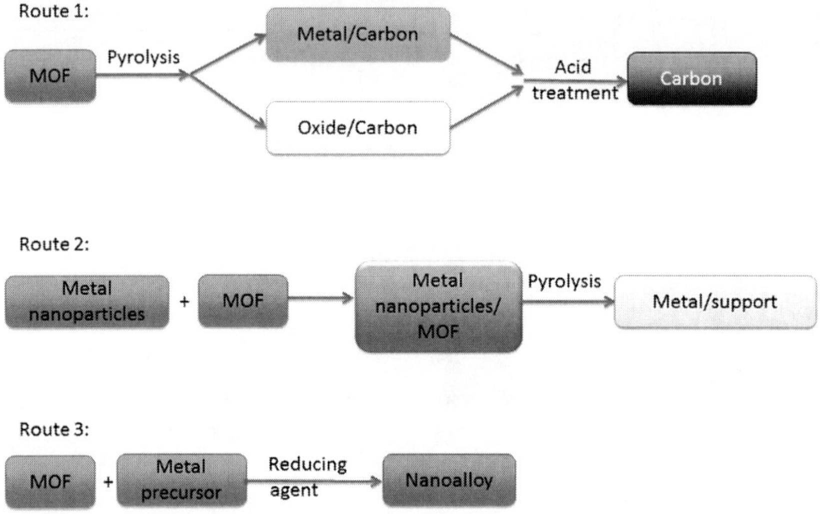

Fig. 22 The synthetic routes for metal/carbon, oxide/carbon composite, carbon materials, carbon supported metal nanoalloy and metal/oxide/carbon composite, using MOF as precursor.

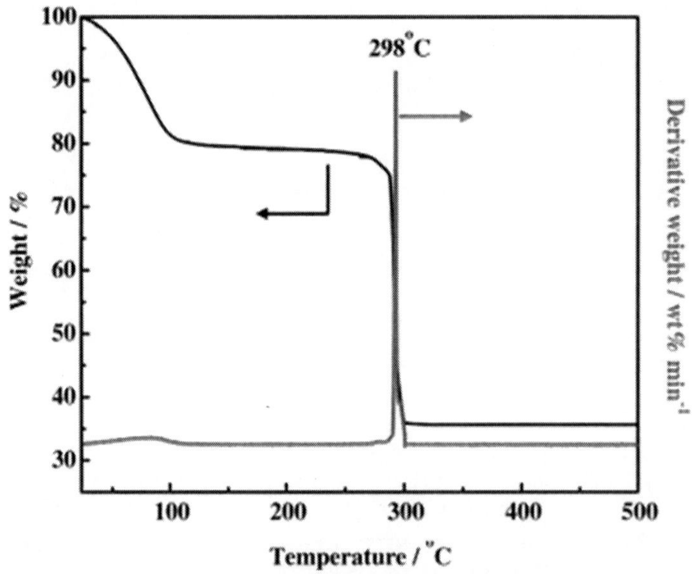

Fig. 23 Thermal gravimetric analysis of $Cu_3(bdc)_2(H_2O)_3$ (HKUST-1). Reprinted from K.-S. Lin, A. K. Adhikari, C.-N. Ku, C.-L. Chiang and H. Kuo, Synthesis and Characterization of porpus HKUST-1 metal organic frameworks for hydrogen storage, *Int. J. Hydrogen Energy*, **37**, 13865–13871. Copyright 2012 Hydrogen Energy Publications, LLC. With permission from Elsevier Ltd.

between 270–300 °C is related to the decomposition of the MOF structure. In such a process, the departure of volatile products, such as CO_2 formed from C and O atoms from the ligand, should be responsible for the loss of the mass. When the temperature is higher than 300 °C, metal oxides are formed, with no further mass change. After acid-treatment, metal or oxide can be removed from the carbon material.[129,130] In the second route, MOFs are used as precursors, mixed with other metal nanoparticles.[132,234] It could be applied for the synthesis of metal/oxide/carbon composites *via* pyrolysis under inert gas at temperatures (*e.g.*, >600 °C).[132,234] The obtained materials have been widely applied in electrocatalysis as well as energy converting devices. Finally, MOF could be mixed with a metal precursor, dissolved or suspended in a solvent.[235] The system is sealed under reducing gas atmosphere, *e.g.*, CO, then nanoalloys supported by carbon materials can be obtained, which could be applied as electrocatalysts.[235]

For the metal/carbon composites, it was reported that both precious, and non-precious metal NPs supported by carbon could be synthesized from calcination of MOFs. As listed in Table 3, *e.g.*, Afsahi *et al.*[232] reported Pt NPs, supported on carbon, were synthesized *via* pyrolysis of a Pt-containing MOF. In fact, the framework, known as MOF-253, composed by Al ions and 2,2′-bipyridine-5,5′-dicarboxylate ligands was firstly synthesized, followed by a post-deposition of $PtCl_2$ in the network, Fig. 24a.[232] Finally, the Pt-MOF was heat-treated between 700 and 1050 °C.[232] The Pt NPs embedded into carbon materials, obtained *via* pyrolysis at 950 °C (C3), and 1050 °C (C4) were applied as electrodes in H_2/air PEMFCs.[232] One can observe that, in Fig. 24b, the assembly with C3 deposited cathode performed better than that with C4, but was still less active with respect to commercial Pt/C catalyst as cathode.[232] From Fig. 24c, the cell with C4 coated anode produced less power with respect to the assembly with commercial Pt/C anode, whereas that with C3 anode delivered a power density close to the commercial one.[232] Recently, Ania *et al.*[228] synthesized a composite (CMGr) of Cu supported by reduced graphene oxide (rGO) from pyrolysis of HKUST-1 with a GO sheet at 800 °C for ORR in alkaline medium. As shown in Fig. 25a–b, it formed Cu NPs dispersed onto the rGO sheet, compared to the agglomerated Cu NPs on carbon prepared *via* a direct heat-treatment of HKUST-1 (CM).[228] The ORR activity on CMGr was high on CM, and rGO, indicating the morphological and support effect towards the catalytic activity (Fig. 26c).[228] However, CMGr was still less active with respect to commercial Pt/C for ORR (Fig. 25c).[228] In addition, the number of electron transfer for ORR on CMGr is *ca.* 3.6, close to *ca.* 3.7 on Pt/C in air saturated solution, Fig. 25d.[228] Another example is Fe/C composite derived from a MOF of $Fe_3O(H_2N\text{-}BDC)$ (H_2N-BDC = 2-aminoterephthalic acid), is known as MIL-88B-NH_3, at 800 °C, Fig. 26a.[236] It showed an enhanced ORR activity and stability with respect to commercial Pt/C, and MIL-88B-NH_3 in 0.1 M KOH, Fig. 26b.[236] Liu's group reported a 3D interconnected porous nano-network (Fe/N/CF) catalyst prepared *via* electro-spin of a polymer solution of tri-1,10-phenanthroline iron(II) perchlorate and MOF, followed by heat-, at 700 °C, and acid-treatment, Fig. 27a.[237] The ORR activity was as high as 3.3 mA cm^{-3} in 0.5 M H_2SO_4, Fig. 27b, higher than the amorphous carbon supported Fe/N catalytic centers (Fe/N/KB) synthesized *via* pyrolysis of the MOF and carbon black.[237] The cathodic

Table 3 Examples of MOF as precursors to generate catalysts for electrocatalytic energy conversion.[a]

Precursor				
MOF	Other	Obtained catalyst	Application	Ref.
Pt-MOF-253: Pt-modified Al(OH)(bpydc)	—	Pt/C	Cathode and anode in a H_2/air PEMFC	232
HKUST-1: $Cu_3(btc)_2$	GO	Cu/rGO	ORR	228
MIL-88B-NH_2: $Fe_3O(H_2N\text{-}bdc)$	—	Fe/N/C	ORR	236
Fe/N/CF	Tri-1,10-phenanthroline iron(II) perchlorate	Fe/N/CF	ORR in H_2/O_2 PEMFC	237
$Co(imidazolate)_2$	—	Co–N	ORR	238
Co{4′-(4pyridyl)-4,2′:6′,4″-terpyridine}	Dicyandiamide and Fe(II) acetate	Fe/N/C	ORR in Li–O_2 battery	130
$Ni(bdc)_2(ted)$	—	Ni/NiN/C	HER	239
Co(naphthalenedicarbonylate)	—	Co_3O_4/C	OER	241
Co-MOF-5: Co-doped $Zn_4O(bdc)_3$	—	Co–ZnO@C	Anode in Li-ion battery	244
MIL-100: $Fe_3X(H_2O)_2O\{(C_6H_3)(CO_2)_3\}_2 \cdot nH_2O$, X = F, OH; $n \sim 14.5$	Carbon black	γ-Fe_2O_3/C	Electrodes in a Li-O_2 battery	245
Fe-modified MOF-5: Fe-$Zn_4O(bdc)_3$	—	ZnO/$ZnFe_2O_4$/C	Electrodes in Li-ion battery	240
MOF-199 (HKUST-1): $Cu_3(btc)_2$	—	CuO	Anode in Li-ion battery	233
$M(btc)_x$	—	MnO/C	Anode in Li-ion battery	246
$Co_3L_2(TPT)_2 \cdot xG$ (G = guest molecules)	—	Co_3O_4	Anode in Li-ion battery	247
$Zn_xCo_{1-x}(MLM)_2$	—	Co/N/C	ORR	248
ZIF-67: $Co(MLM)_2$	—	NCNTF	ORR and OER	131
ZIF-8: $Zn(MLM)_2$	GO	GNPCS	ORR	249
ZIF-8: $Zn(MLM)_2$	Sulfur	MPCP/S	Lithium–sulfur batteries	227
MOF-5: $Zn_4O(bdc)_3$	—	C_3N_4	ORR	224
ZIF-8: $Zn(MLM)_2$	—	Mesoporous carbon	Electrode in Na-ion battery	250
Ce(ATPT)(HATPT)	Na_2PtCl_6 and MWCNT	Pt/CeO_x/C	ORR in H_2/O_2 μLFFC	132
$M(bdc)_x$ M = Fe and Zn	Au NPs	Au@Zn-Fe-C	ORR	249
HKUST-1: $Cu_3(btc)_2$	K_2PdCl_6 and carbon black	PdCu/C	ORR	235

[a] bpydc = 2,2′-bipyridine-5,5′-dicarboxylate; PEMFC = proton exchange membrane fuel cell; btc = benzene-1,3,5-tricarboxylate; (r)GO = (reduced) graphene oxide; H_2N-bdc = 2-aminoterephthalic acid; bdc = 1,4-benzenedicarboxylate; MLM = 2-methylimidazole; NCNTF = N-doped carbon nanotube framework; GNPCS = graphene based N-doped carbon sheet; MWCNT = multi-walled carbon nanotube; MPCP = microporous carbon polyhedron; μLFFC = micro laminar flow fuel cell; NPs = nanoparticles; btc = benzene-1,3,5-tricarboxylate; TPT = 2,4,6-tris(4-pyridyl)-1,3,5-triazine; H3L = 2,4,6-tris{1-(3-carboxylphenoxy)ylmethyl}mesitylene.

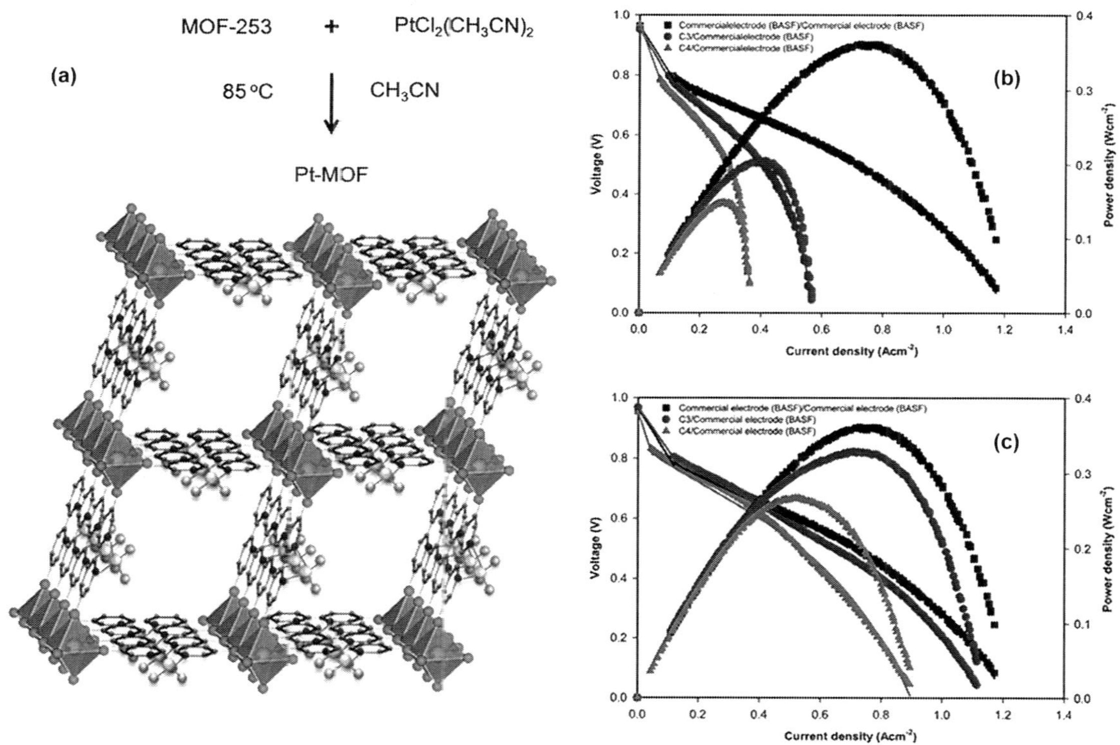

Fig. 24 (a) Sketch for the synthesis of Pt-MOF-253. Test in H_2/air PEMFCs, using C3 (Pt/C catalysts obtained *via* heat-treatment of Pt-MOF-253 at 950 °C), C4 (Pt/C catalysts obtained *via* heat-treatment of Pt-MOF-253 at 1050 °C), and commercial Pt/C catalysts as (b) cathodes (c) anodes. In both cases, the other electrode was deposited by commercial Pt/C, measured at 80 °C with low rate of 120 and 250 mL min^{-1}, respectively. Reprinted from F. Afsahi, H. Vinh-Thang, S. Mikhailenko and S. Kaliaguine, Electrocatalyst synthesized from metal organic frameworks, *J. Power Sources*, **239**, 415–423. Copyright 2013 with permission from Elsevier B.V.

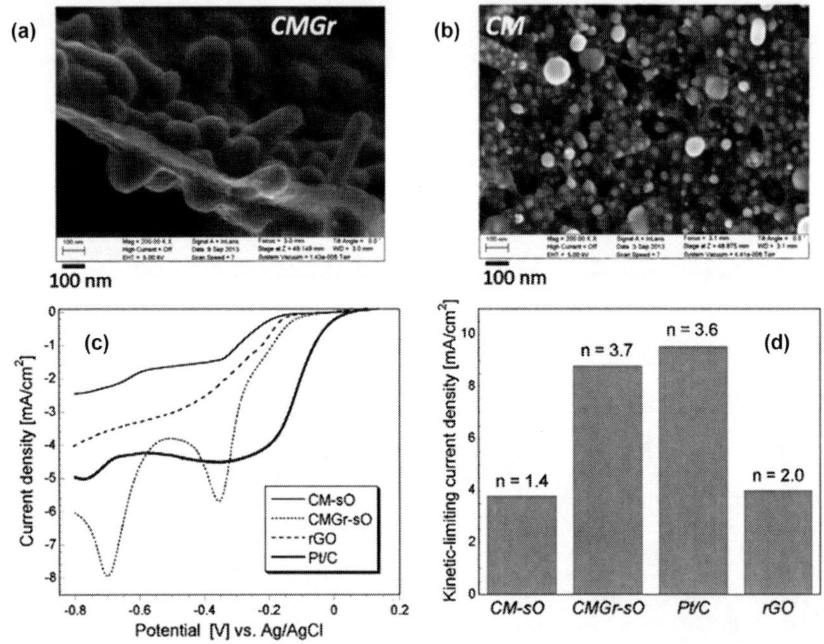

Fig. 25 SEM images for (a) CMGr (Cu/rGO synthesized from pyrolysis of MOF with GO sheet at 800 °C), and (b) CM (Cu/C from heat-treatment of MOF at 800 °C). (c) ORR polarization curves for CMGr, CM, rGO and commercial Pt/C in air-saturated 0.1 M KOH. (d) Number of electron transfer derived from ORR curves. Reprinted from C. O. Ania, M. Seredych, E. Rodriguez-Castellon and T. J. Bandosz, New copper/GO based material as an efficient oxygen reduction catalyst in an alkaline medium: The role of unique Cu/rGO architecture, *Appl. Catal., B*, **163**, 424–435. Copyright 2015 Elsevier B.V.

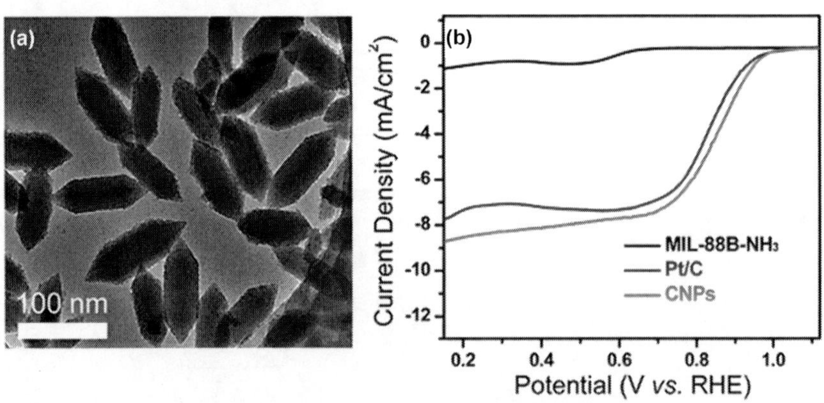

Fig. 26 (a) TEM images for Fe/C composite (CNPs) derived from a MOF of $Fe_3O(H_2N\text{-}BDC)$ (H_2N-BDC = 2-aminoterephthalic acid), known as MIL-88B-NH_3, at 800 °C. (b) ORR curves for CNPs, MIL-88B-NH_3 and commercial Pt/C in O_2-saturated 0.1 M KOH, at 5 mV s^{-1}, 1600 rpm. Reprinted with permission from S. Zhao, H. Yin, L. Du, L. He, K. Zhao, L. Chang, G. Yin, H. Zhao, S. Liu and Z. Tang, Carbonized Nanoscale Metal-Organic Frameworks as High Performance Electrocatalyst for Oxygen Reduction Reaction, *ACS Nano*, 2014, **8**, 12660–12668. Copyright (2014) American Chemical Society.

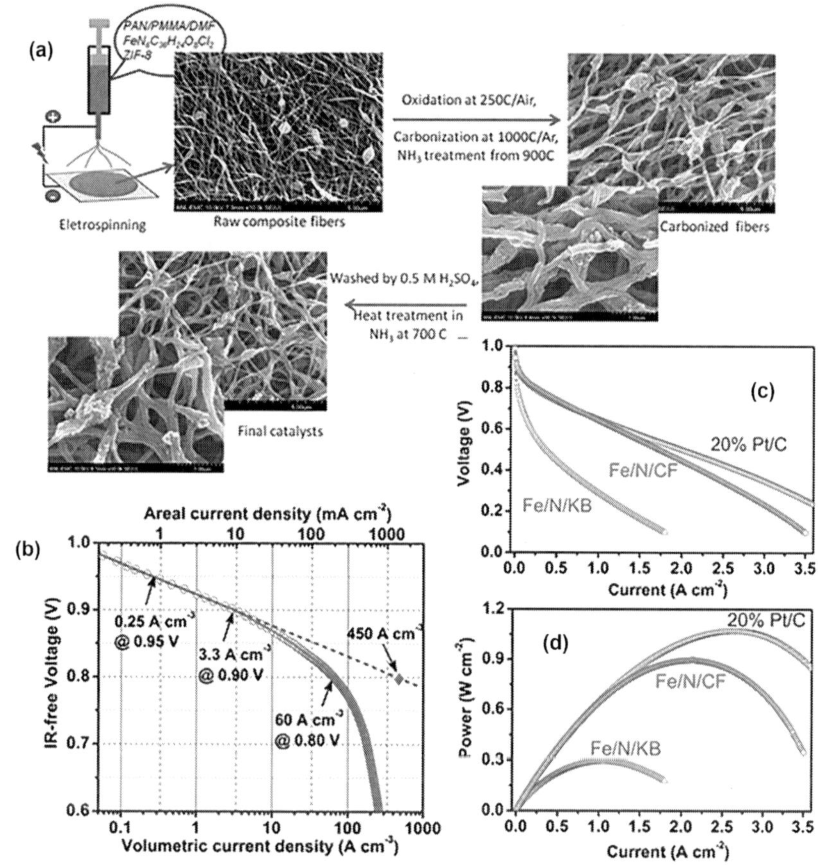

Fig. 27 (a) Synthesis of 3D interconnected porous nano-network (Fe/N/CF) catalyst prepared *via* electro-spin of a polymer solution of tri-1,10-phenanthroline iron(II) perchlorate and MOF, followed by heat-treatment and acid-treatment. (b) The Tafel plot for ORR on Fe/N/CF in 0.5 M H_2SO_4. H_2/O_2 fuel cell test for Pt/C (20 wt%), Fe/N/CF and Fe/N/KB (prepared *via* the same method as Fe/N/CF, but carbon black as support): (c) cathodic and (d) cell performance. In all cases, the anode was 0.3 mg cm^{-2} of Pt, gas pressure and temperature was 2 bars and 80 °C, respectively. Reproduced from J. Shui, C. Chen, L. Grabstanowicz, D. Zhao and D.-J. Liu, Highly efficient nonprecious metal catalyst prepared with metal-organic framework in a continuous carbon nanofibrous network, *Proc. Natl. Acad. Sci.*, 2015, **112**, 10629–10634. With permission from PNAS.

performance of Fe/N/CF was very close to a Pt/C (20 wt%), higher than Fe/N/KB in H_2/O_2 fuel cell test, Fig. 27c.[237] This group also reported that Co-N active sites could be synthesized from a Co-imidazolate MOF structure *via* pyrolysis at a temperature >750 °C, showing ORR activity in 0.1 M $HClO_4$ solution.[238] Li *et al.*,[130] reported that a composite of Fe nanoparticles, supported by N-doped graphene/graphene tube from CoL (L = 4′-(4pyridyl)-4,2′:6′,4″-terpyridine), showed less ORR activity in acid but higher ORR kinetic in alkaline medium with respect to Pt/C. Other metal nanoparticles supported by carbon materials reported as electrocatalysts for energy conversion are summarized in Table 3.[239,240]

Further, oxide/carbon materials could be derived from MOFs. For example, Ma *et al.*[241] reported a Co_3O_4/C array prepared from calcination of a MOF array of Co(naphthalenedicarbonylate) grown on Cu foil, Fig. 28a–b). The porous composite (251 $m^2\,g^{-1}$) with carbon content of 52 wt% can be directly used as electrode for OER in 0.1 M KOH.[241] It showed an onset potential of 1.47 V/RHE, and stable current density of 10 $mA\,cm^{-2}$ at 1.52 V/RHE for 30 h for OER, Fig. 28c.[241] The OER activity was higher than Co_3O_4 nanoarrays (Co_3O_4-NA), Co-based MOF and IrO_2/C.[241] Moreover, such a material was also active for ORR, indicating it could serve as a reversible oxygen electrode, Fig. 28d.[241] The $\Delta E = E_{j=10}$ (potential at current density = 10 $mA\,cm^{-2}$ for OER) − $E_{1/2}$ (half-wave potential for ORR) = 0.74 V/RHE, was lower than ΔE in the literature.[241–243] Yue *et al.*,[244] developed a composite of Co doped ZnO@C (CZO@C), derived from a Co-doped MOF structure (Co-MOF-5). Compared with ZnO@C prepared from non-doped MOF structure, the conductivity and the performance as anode in a Li-ion battery was improved.[244] Hierarchical mesoporous composite of γ-Fe_2O_3/carbon could be prepared from a Fe-containing MOF (MIL-100) under Ar at 600 °C.[245] After pyrolysis, one could observe that the MOF particles were carbonized but the morphology was retained, Fig. 29a.[245] But γ-Fe_2O_3 formed on the particles surface, Fig. 29b.[245] Using carbon black supported γ-Fe_2O_3/carbon composite as electrodes in a Li–O_2 battery, the charge-discharge capacity was 72%, whereas the capacity was 57.5% in a battery using carbon black as electrode, Fig. 30c.[245] From the cyclic voltammograms, Fig. 29d–e, the onset potential for ORR on γ-Fe_2O_3/carbon composite was positively shifted with respect to carbon black.[245] The onset potential of OER on the composite was negatively shifted in comparison to carbon black, Fig. 29f.[245] Other oxide/carbon composites derived from MOF are summarized in Table 3.[233,246–248]

Advanced carbon materials, such as nitrogen doped carbon nanotube framework (NCNTFs) could be synthesized from a MOF precursor (ZIF-67) at 700 °C under reducing gas (Ar/H_2), followed by dissolution of Co in 0.5 M H_2SO_4 for 6 h.[131] The morphology of MOF could be retained after heat-treatment, where the surface was covered by tiny CNT, Fig. 30a.[131] Some particles showed open voids, supporting the hollow structure formed during heat-treatment, Fig. 30b.[131] It is worth to note that after heating in absence of H_2 the material is a porous composite of Co and carbon.[131] This indicates that the *in situ* formation of NCNTFs should be associated with hydrogen.[131] The ORR activity and stability on NCNTFs was enhanced with respect to Pt/C in 0.1 M KOH (Fig. 30c–d).[131] Moreover, the OER activity as well as stability on such an advanced carbon material was higher than Pt/C, Fig. 30e–f.[131] Moreover, Zhong *et al.*,[249] reported a graphene-based N-doped porous carbon sheet from a MOF (ZIF-8)/GO *via* pyrolysis at 800 °C. They firstly obtained the growth of MOF on GO sheet, followed by heat- and acid-treatment, and finally the graphene-based N-doped porous carbon sheet (GNPCSs-800) Fig. 31a–b.[249] The ORR activity and stability was improved with respect to Pt/C (Fig. 31c–d).[249] As shown in Fig. 31c, the ORR activity on GNPCSs obtained by pyrolysis at 800 °C (GNPCS-800) was superior to

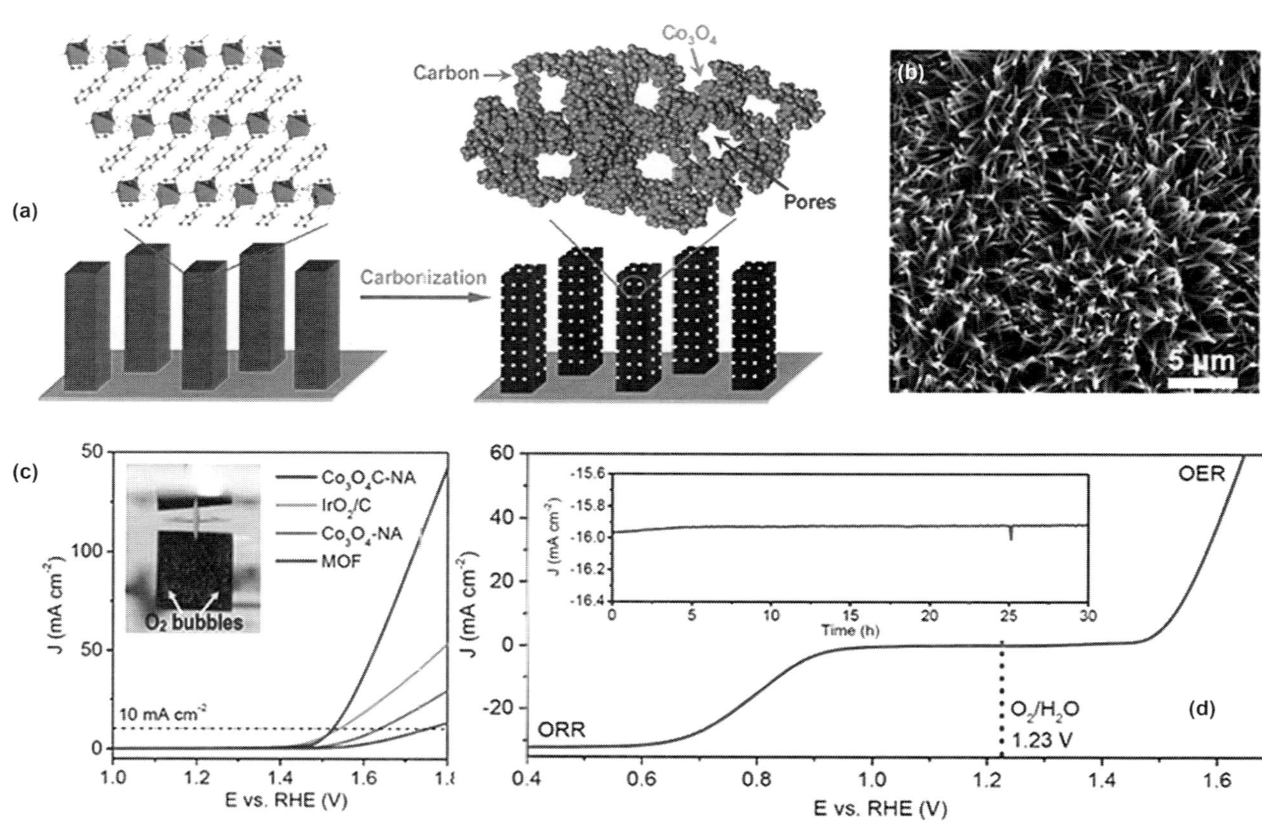

Fig. 28 (a) Illustration for the synthesis of porous Co_3O_4/C arrays (Co_3O_4/C-NA) from pyrolysis of MOF arrays on Cu foil. (b) SEM image for the Co_3O_4/C arrays (Co_3O_4/C-NA). (c) OER curves for Co_3O_4/C-NA, Co_3O_4 arrays (Co_3O_4-NA), IrO_2 and MOF, recorded at 0.5 mV s^{-1} in O_2-saturated 0.1 M KOH. (d) OER and ORR curve for Co_3O_4/C-NA, recorded at 0.5 mV s^{-1} in O_2-saturated 0.1 M KOH. Reprinted with permission from T. Y. Ma, S. Dai, M. Jaroniec and S. Z. Qiao, Metal-Organic Framework Derived Hybrid Co_3O_4-Carbon Porous Nanowire Arrats as Reversible Oxygen Evolution Electrodes, *J. Am. Chem. Soc.*, 2014, **136**, 13925–13931. Copyright (2014) American Chemical Society.

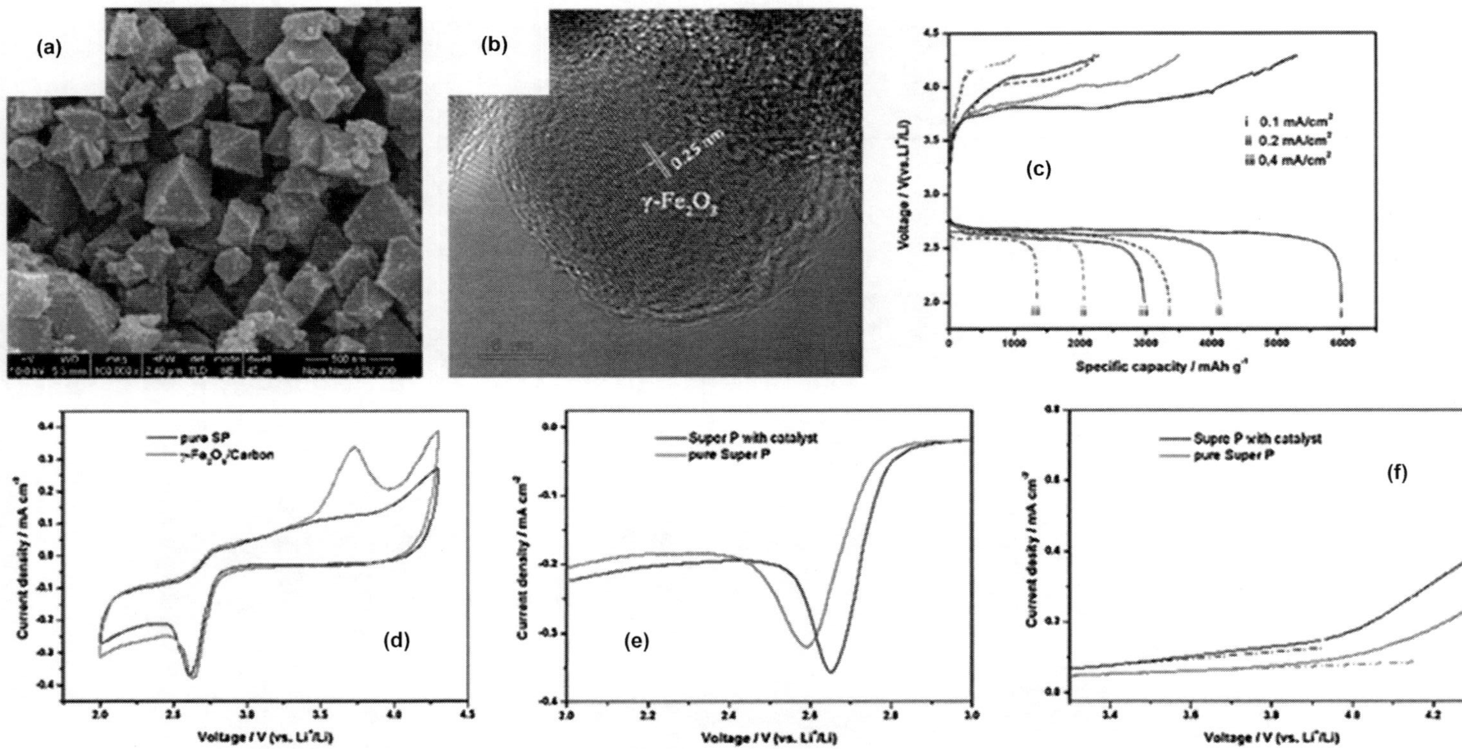

Fig. 29 (a) SEM image for composite of γ-Fe_2O_3/carbon prepared from a Fe-containing MOF (MIL-100) under Ar at 600 °C. (b) TEM image for Fe_2O_3 nanoparticles. (c) Charge–discharge voltage curves for carbon black supported γ-Fe_2O_3/carbon composite (solid), and carbon black (dash) electrode. (d–e) Cyclic voltammograms and (f) OER curves for Fe_2O_3/carbon composite and carbon black in O_2 saturated solution (1 M LiTFSI in tetraethyleneglycol dimethyl ether). Reprinted from W. Chen, Z. Zhang, W. Bao, Y. Lai, J. Li, Y. Gan and J. Wang, Hierarchical mesoporous γ-Fe_2O_3/carbon nanocomposites derived from metal organic frameworks as a cathode electrocatalyst for rechargage Li-O_2 batteries, *Electrochim. Acta*, **134**, 293–301. Copyright 2014 with permission from Elsevier Ltd.

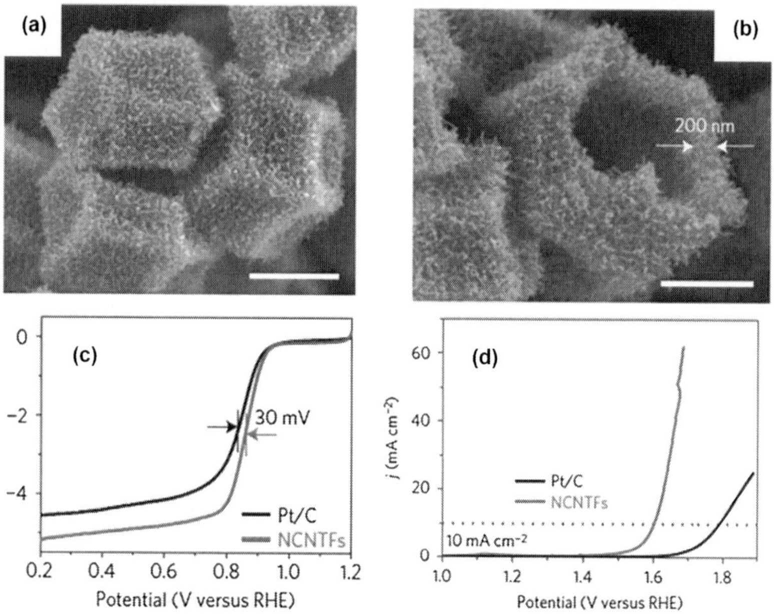

Fig. 30 (a–b) SEM images for nitrogen-doped carbon nanotube (NCNTFs) synthesized from a MOF precursor (ZIF-67) at 700 °C under reducing gas (Ar/H_2), followed by dissolution of Co in 0.5 M H_2SO_4 for 6 h. ORR (c) polarization curves, recorded at 1600 rpm and (d) OER polarization curves in 0.1 M KOH. Reprinted from Macmillian Publishers Ltd: B. Y. Xia, Y. Yan, N. Li, H. B. Wu, X. W. D. Lou and X. Wang, A metal-organic framework derived bifunctional oxygen electrocatalyst, *Nat. Energy*, 2016, **1**, 15006. Copyright 2016 Nature Publishing Group.

the catalyst synthesized at 650 °C (GNPCSs-650) and 950 °C (GNPCSs-950), indicating that the heating temperature had an effect towards catalytic activity.[249] Compared with their counterparts, namely, N-doped carbon (NC), N-doped porous carbon (NPC), and as-prepared GNPCSs, ORR activity on GNPCS-800, was enhanced.[249] Besides, porous carbon derived from MOF was applied as electrodes in Na-ion battery.[250]

Alonso-Vante's group recently reported a nanocomposite Pt/CeO_x/C (C = porous carbon and carbon nanotube), prepared from a Pt NPs supported by Ce-based MOF (Ce(ATPT)(HATPT) · $n$$H_2O$, ATPT = 2-aminoterephthalate) and CNT.[132] Pt NPs were firstly synthesized from direct reduction of Pt precursor, then Ce-MOF and carbon support were added into the system, forming a Pt/MOF/CNT composite. After heat-treatment at 900 °C under Ar, of such a composite, the nanocomposite (SMOF-3) could be obtained, Fig. 32a.[132] Its morphology could be described as a layer of Ce oxide and carbon on the Pt NPs surface.[132] The effect of MOF was studied by comparison of MOF-free samples.[132] It was found that MOF could protect Pt NPs from agglomeration during heat-treatment.[132] The unique morphology of SMOF-3 stabilized both Pt^0 and Ce^{3+} electronic state, inducing a strong interaction between Ce^{3+} and Pt.[132] Therefore, the ORR activity was 10-fold higher than other references

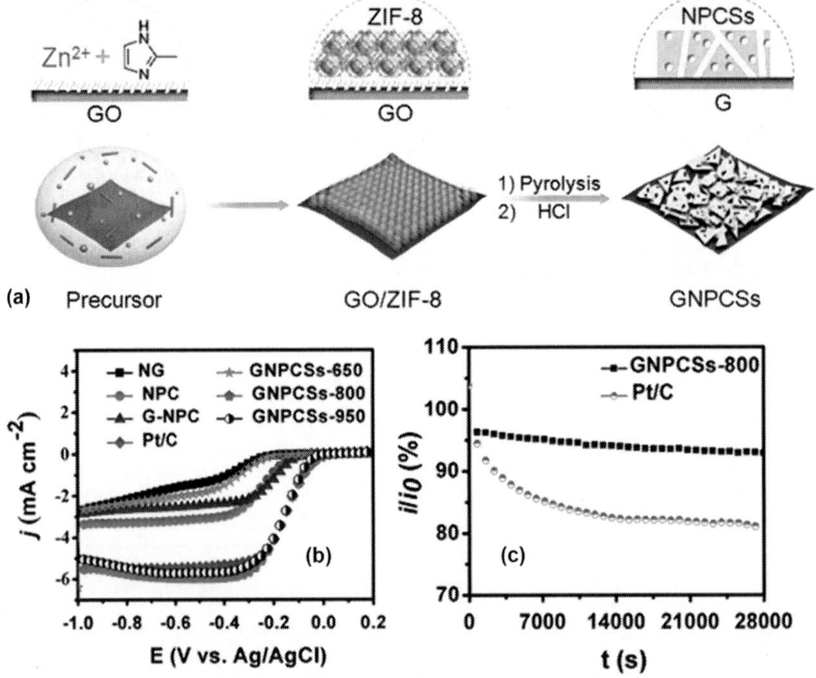

Fig. 31 (a) Sketch of graphene-based N-doped porous carbon sheet from a MOF (ZIF-8)/GO *via* pyrolysis at 800 °C. ORR (b) activity and (c) stability for N-doped porous carbon sheet. Reprinted with permission from H. X. Zhong, J. Wang, Y. W. Zhang, W. L. Xu, W. Xing, D. Xu, Y. F. Zhang and X. B. Zhang, ZIF-8 Derived Graphene-Based Nitrogen-Doped Porous Carbon Sheets as Highly Efficient and Durbale Oxygen Reduction Electrocatalysts, *Angew. Chem., Int. Ed.*, 2014, **53**, 14235–14239. Copyright (2014), John Wiley and Sons.

and commercial Pt/C catalysts, outperformed the DOE 2015/2017 target for ORR on Pt-based catalysts in acid medium, Fig. 32b.[132] The ORR kinetic remained 80% after 16 000 potential cycles on SMOF-3, higher than the remaining value of commercial Pt/C and reference Pt/CNT (SCNT-2) catalysts, Fig. 32c.[132] Chen's group[234] synthesized a composite of Au nanoparticle@Zn-Fe-embedded porous carbons, Fig. 33a. The Zn/Fe-containing MOF were firstly grown on the Au NPs, followed by calcination at 600 °C under Ar.[234] The obtained Au@Zn-Fe-C catalyst was active for both ORR and HER, Fig. 33b–c.[234]

Through direct reduction of a MOF (HKUST-1), and Pd salt, PdCu alloyed nanorods supported by carbon black (PdCu/C-1) could be obtained, Fig. 34a.[235] However, using conventional Cu salt (*e.g.*, $CuCl_2$), the obtained carbon-supported nanorods were rarely alloyed (PdCu/C-2), as shown in Vegards' plot, Fig. 34b.[235,251,252] Compared with commercial Pd/C catalyst, the ORR activity on PdCu/C-1 was increased in both acid and alkaline solutions, Fig. 34c.[235] Although, ORR activity on PdCu/C-1 was lower than commercial Pt/C catalyst, in acid medium, it was higher in alkaline medium, Fig. 34c.[235]

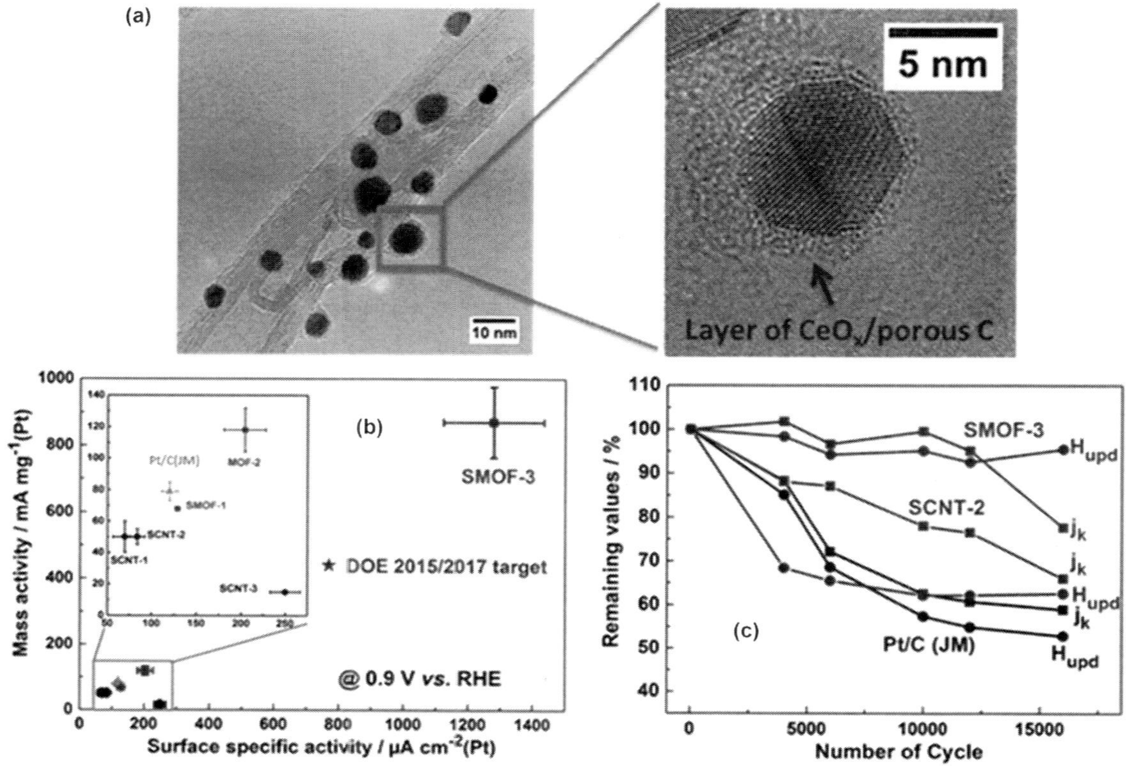

Fig. 32 (a) TEM image for nanocomposite Pt/CeO_x/C (C = porous carbon and carbon nanotube), prepared *via* heat-treatment at 800 °C from a Pt NPs supported by Ce-based MOF (Ce(ATPT)(HATPT) · nH_2O, ATPT = 2-aminoterephthalate) and carbon nanotube (SMOF-1). (b) ORR activity and (c) stability for SMOF-3, reference and commercial Pt/C samples in acid medium. SMOF-2 was 400 °C treated SMOF-1 sample. SCNT-1, SCNT-2 and SCNT-3 are as-prepared, 400 °C, and 900 °C treated Pt NPs supported by CNTs. Reprinted from Y. Luo, L. Calvillo, C. Daiguebonne, M. K. Daletou, G. Granozzi and N. Alonso-Vante, A highly efficient and stable oxygen reduction reaction on Pt/CeOx/C electrocatalyst obtained *via* a sacrifical precursor based on a metal organic framework, *Appl. Catal., B*, **189**, 39–50. Copyright 2016 with permission from Elsevier B.V.

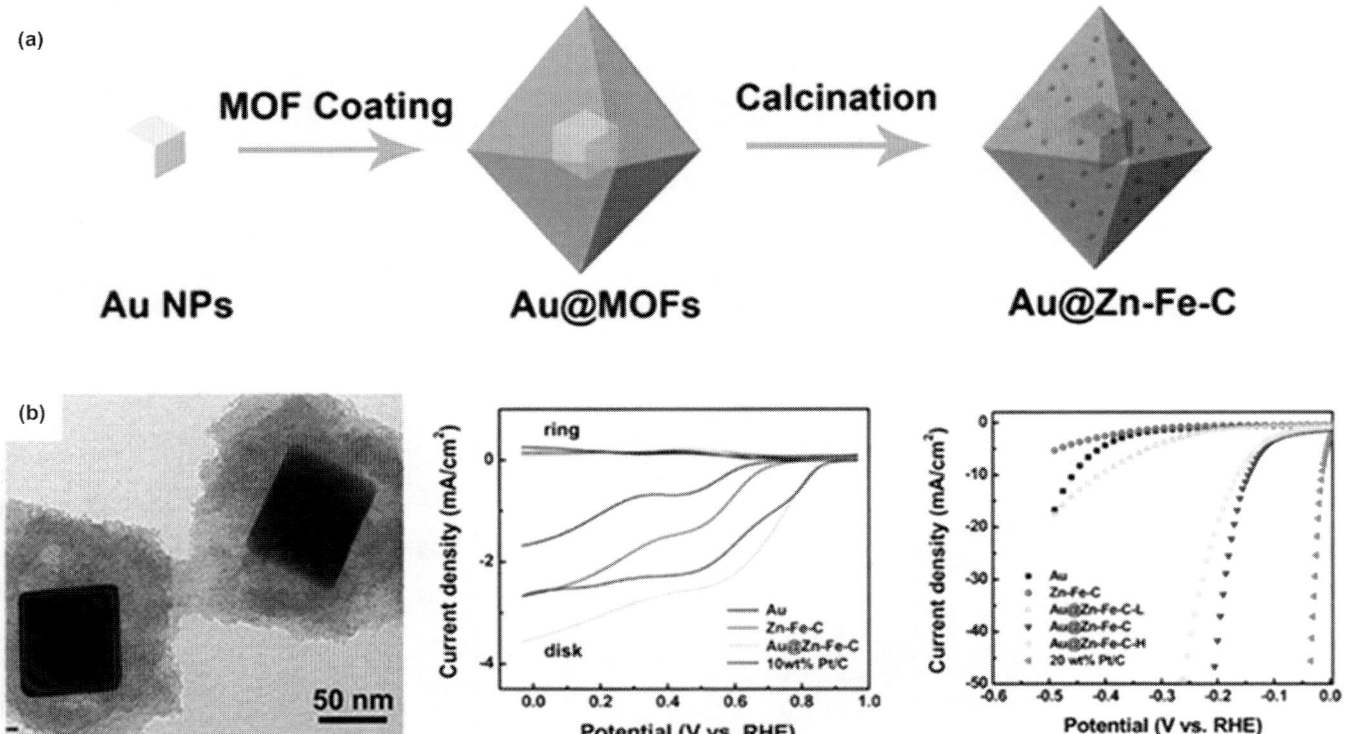

Fig. 33 (a) Illustration of the synthesis of Au nanoparticle@Zn-Fe-embedded in porous carbons (Au@Zn-Fe-C) catalyst, and (b) TEM image for Au@Zn-Fe-C. The catalytic kinetics of Au@Zn-Fe-C towards (c) ORR and (d) HER performance. Reprinted with permission from J. Lu, W. Zhou, L. Wang, J. Jia, Y. Ke, L. Yang, K. Zhou, X. Liu, Z. Tang, L. Li and S. Chen, Core-Shell Nanocomposites Based on Gold Nanoparticle@Zinc-Iron-Embedded Porous Carbons Derived from Metal-Organic Frameworks as Efficient Dual Catalysts for Oxygen Reduction and Hydrogen Evolution Reactions, *ACS Catal.*, 2016, **6**, 1045–1053. Copyright (2014) American Chemical Society.

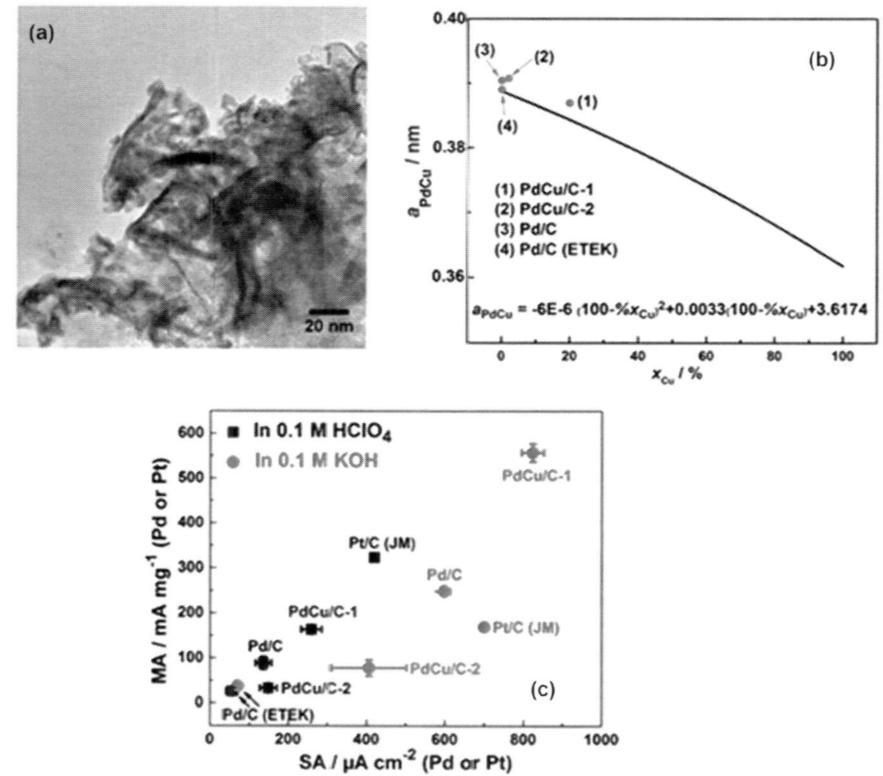

Fig. 34 (a) TEM image for PdCu/C-1 prepared *via* HKUST-1 and K_2PdCl_6 under CO gas. (b) Vegards' plot for PdCu/C-1, PdCu/C-2 (obtained *via* $CuCl_2$ and K_2PdCl_6), and commercial Pd/C. (c) ORR activity in acid and alkaline medium. Reprinted from Y. Luo, L. A. Estudillo-Wong, L. Cavillo, G. Granozzi and N. Alonso-Vante, An easy and cheap chemical route using a MOF precursor to prepare Pd-Cu electrocatalyst for efficient energy conversion cathodes, *J. Catal.*, **338**, 135–142. Copyright 2016 with permission from Elsevier Inc.

3 Outlook and summary

MOF plays an important role in electrocatalytic energy conversion. As shown in this work, this kind of materials have various applications as electrocatalysts *via* (1) direct utilization of MOF as molecular catalyst; (2) as support, and (3) as precursor to generate electrocatalysts. In the first case, the MOF can be recognized as linked coordination complex moieties. Each coordinated metal centre could be an active site, responsible for charge transfer from catalyst to molecule. Therefore, the selection of ligand and metal centre should favour electron transfer. In addition, conducting MOF structure is required. Considering both aspects, the ligand is crucial to realise MOF supramolecular catalysts. In the second case, nanoparticles or complexes can be loaded on the surface or in the porosities of MOFs. The formation of such a composite shows a synergistic effect towards catalytic activity. Obviously, in such a case, MOF is not only a support to accomplish homogenous dispersion of particles or complexes, but also provides active sites. Thus, the tailoring of a MOF as a support should consider both the catalytic activity of moieties and the structure of MOF. Finally, MOF is a significant precursor to generate diverse electrocatalysts. Carbon, metal/carbon, oxide/carbon composite could be derived from MOF precursors. The effect of MOF during the formation of these advanced materials is irreplaceable in controlling the composition and morphology of obtained materials. Hence, more and more efforts from various research groups are devoted to this domain. MOF is becoming an in vogue issue for the electrocatalytic as well as photo-electrocatalytic processes.

Index of ligands

Abbreviation	Name	Image
—	2-Methylimidazole	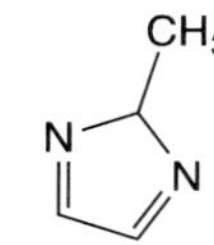
—	2,2′-Bipyridine-5,5′-dicarbonylate	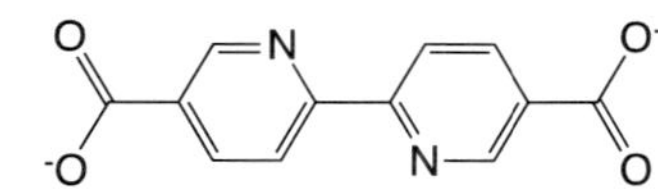
—	2,4,6-Tris{1-(3-carboxylphenoxy)methyl}mesitylene	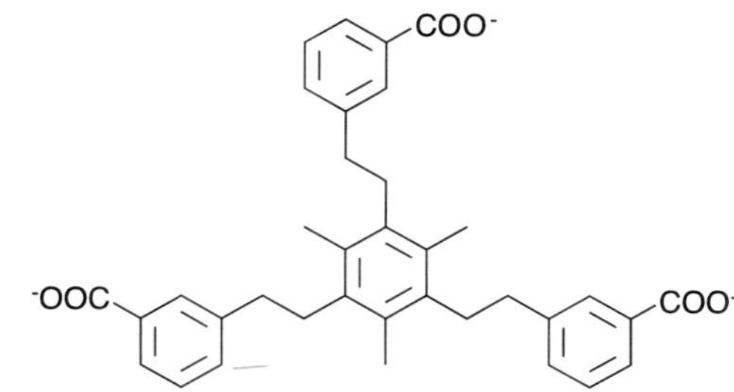
—	4,4′-Pyridine	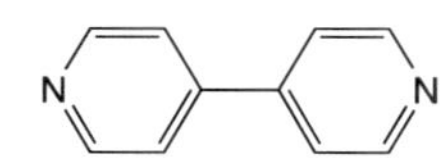
—	4-(5-(Pyridine-4-yl)-4*H*-1,2,4,-triazol-3yl)benzoic acid	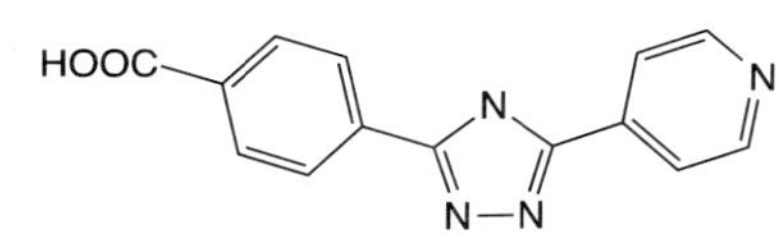

(*Continued*)

Abbreviation	Name	Image
—	4,5-Di(4′-carboxyphenyl)phthalate	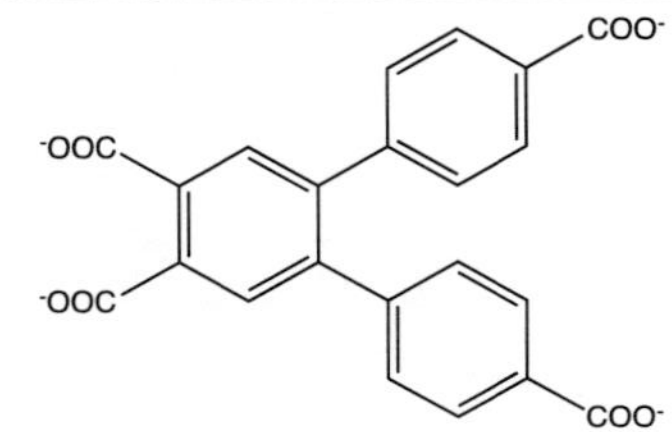
—	*E*-1,2-Di(pyridine-4-yl)diazene	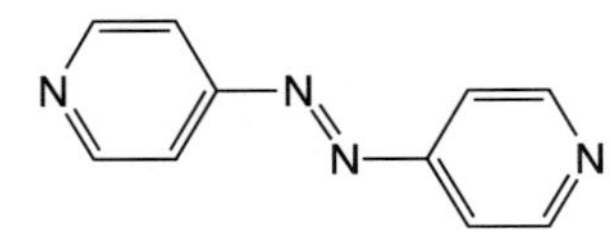
—	Imidazolate	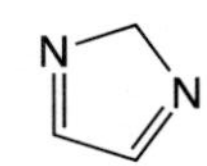
—	Meso-tetrathynyl porphyrin	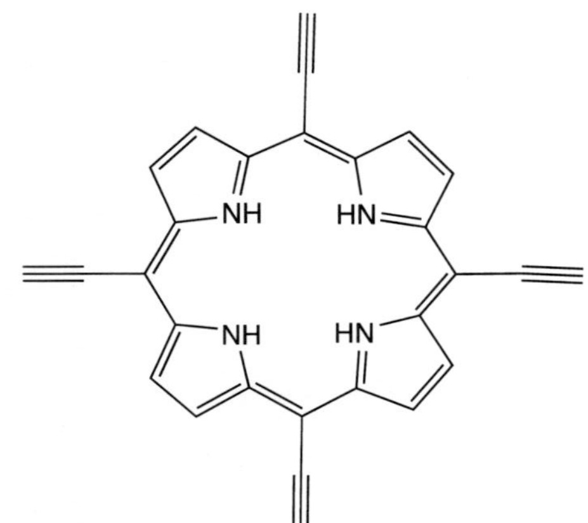

—	Naphthalenedicarboxylate	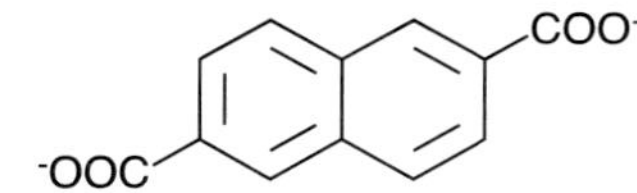
—	Pyridine-2thiolate	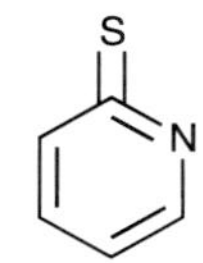
—	Pyrimidine-2-thiol	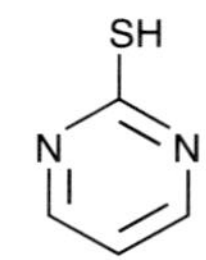
—	Tri-1,10-phenanthroline iron(II)	
acam	Acetamidate	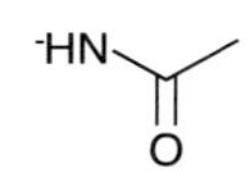

(*Continued*)

Abbreviation	Name	Image
ATPT (NH_2-BDC)	2-Aminoteraphthalate	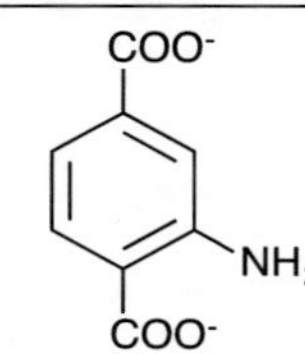
BDC	1,4-Benzenedicarboxylate	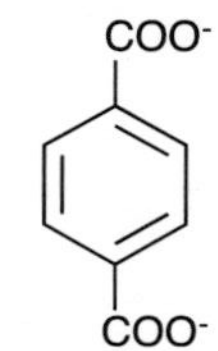
bpy	2,2′-Pyridine	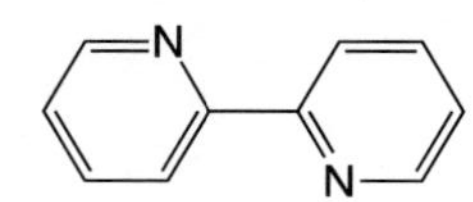
bpydc	2,2′-Bipyridine-5,5′-dicarboxylate	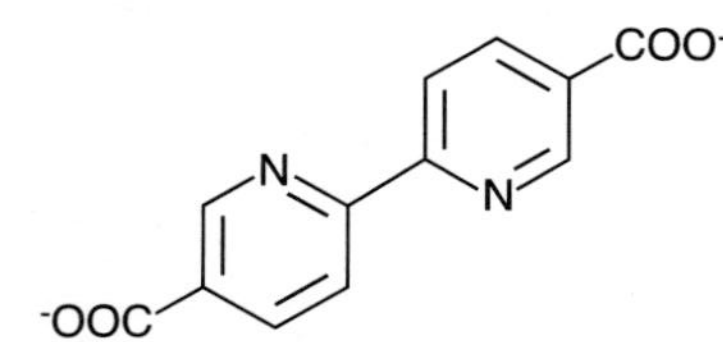
btc	Bezene-1,3,5-tricarbonxylate	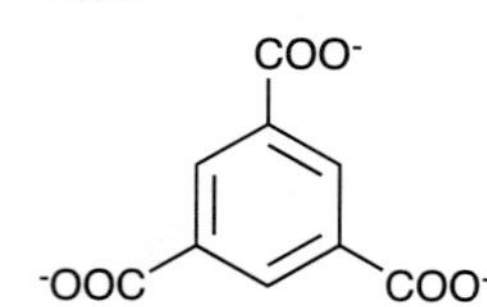

CA	Cyanuric acid	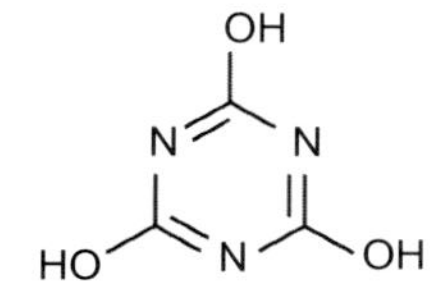
chxn	1*R*,2*R*-Diaminocyclohexane	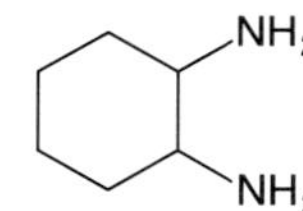
Co(II)–salen	—	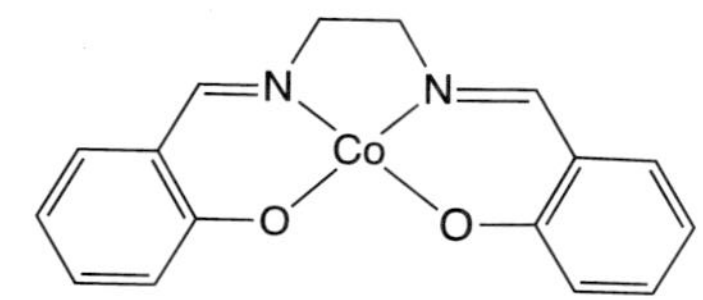
DCNQI	*N*,*N*′-Dicyano-quinonediimine	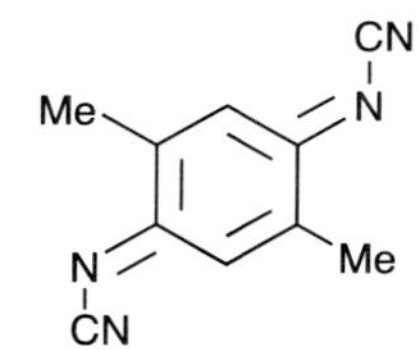
DMF	Dimethylformamide	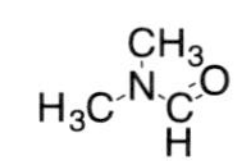
dtoa	Dithiooxamidato	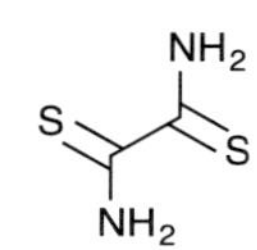
en	Ethylenediamine	

(*Continued*)

Abbreviation	Name	Image
HITP	2,3,6,7,10,11-Hexaiminotriphenylene	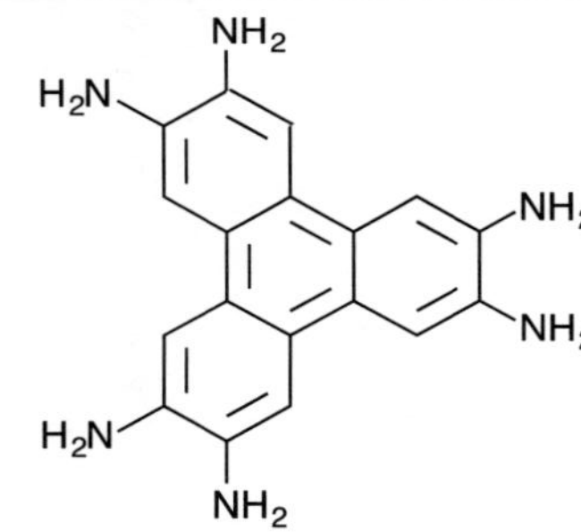
Hmim (MLM)	2-Methylimidazole	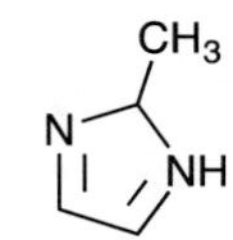
HT	4-Hydroxythiophenol	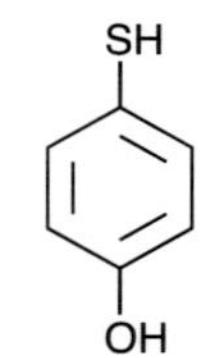
IN	Isonicotinate	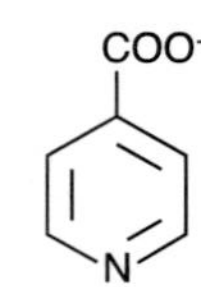
L1	4*H*-4-amino-1,2,4-Triazole	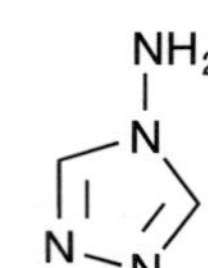

L2	3,5-Dimethyl-4-amino-4*H*-1,2,4-triazole	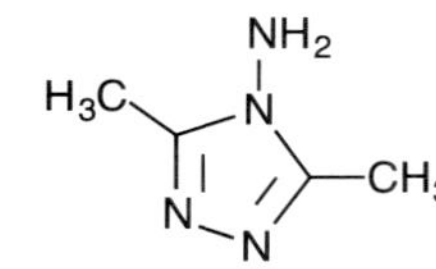
pdt	Pyrazine-2,3-dithiolate	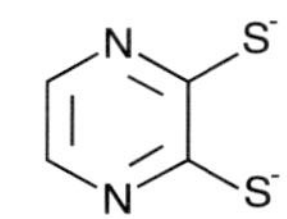
TBAPy	1,3,6,8-Tetrakis(*p*-benzoic acid)pyrene	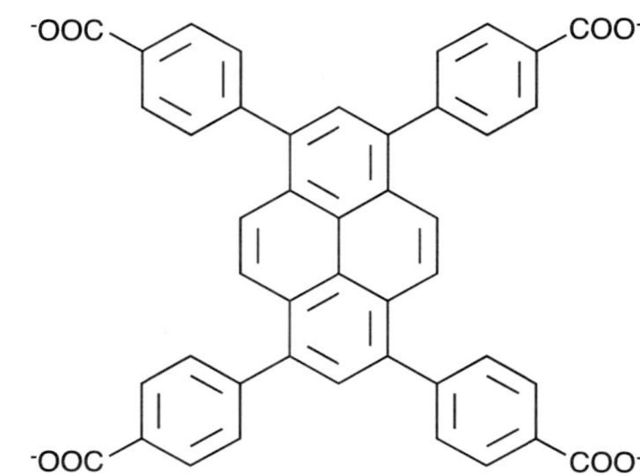
TCNQ	7,7,8,8,-Tetracyanoquinodimethane	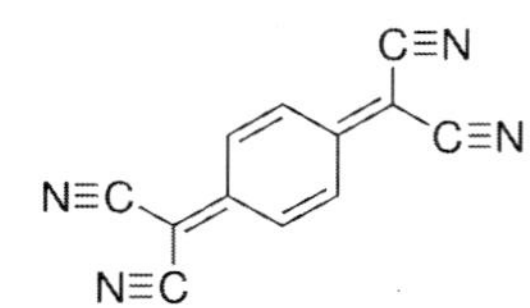

(*Continued*)

Abbreviation	Name	Image
TCPP	5,10,15,20-Tetra(carboxyphenyl)porphyrin	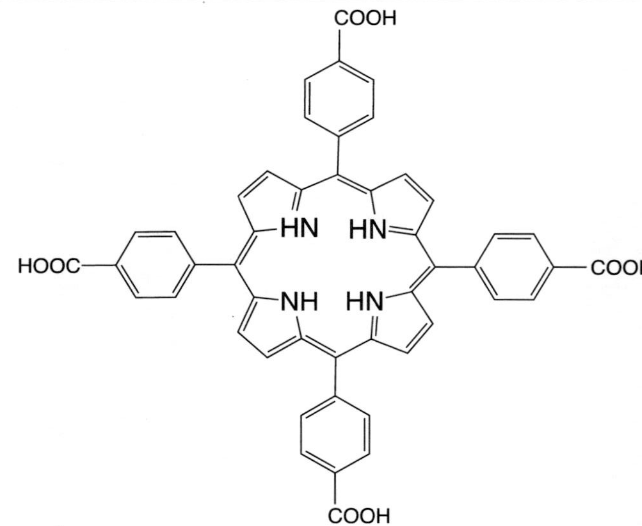
TMEDA	Tetramethylethlenediamine	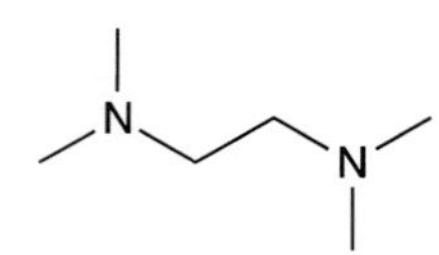
tpa	Terephthalate	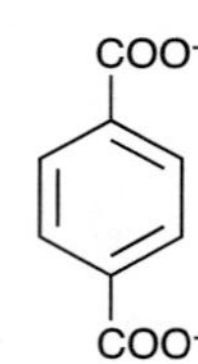

tpp	Tetra-phenyl-porphyrin	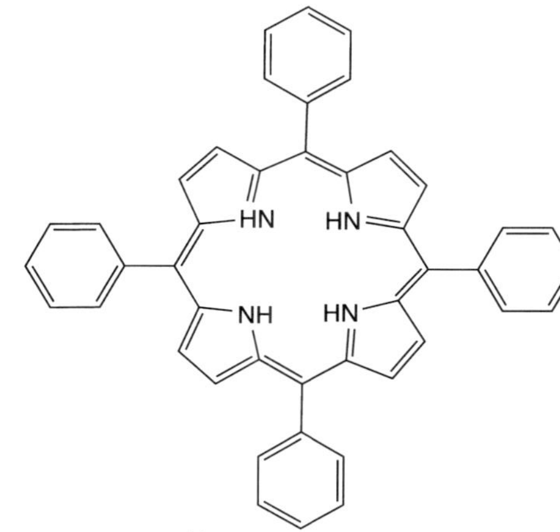
TPT	2,4,6-Tris(4-pyridyl)-1,3,5-triazine	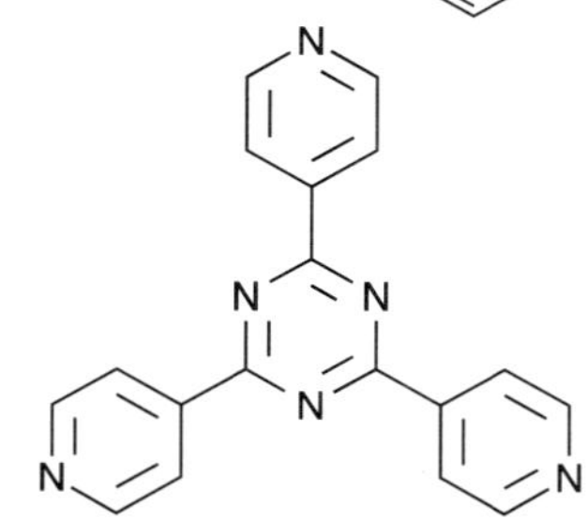
TTF	Tetrathiafulvalene	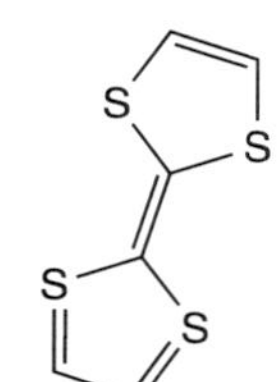

Acknowledgements

The authors acknowledge the support of the University of Poitiers, as well as present and former colleagues present in the list of references. Y. Luo benefited a financial support of a European project: Cathcat.

References

1 J. Zhang, *PEM Fuel Cell Electrocatalysts and Catalyst Layer: Fundamentals and Applications*, Springer-Verlag, London, 2008.
2 J. A. Turner, *Science*, 2004, **305**, 972–974.
3 S. A. Grigoriev, P. Millet, V. I. Porembsky and V. N. Fateev, *Int. J. Hydrogen Energy*, 2011, **36**, 4164–4168.
4 S.-Y. Huang, P. Ganesan, H.-Y. Jung and B. N. Popov, *J. Power Sources*, 2012, **198**, 23–29.
5 G. Chen, S. R. Bare and T. E. Mallouk, *J. Electrochem. Soc.*, 2002, **149**, A1092–A1099.
6 S. Litster and G. McLean, *J. Power Sources*, 2004, **130**, 61–76.
7 V. Mehta and J. S. Cooper, *J. Power Sources*, 2003, **114**, 32–53.
8 A. F. Ghenciu, *Curr. Opin. Solid State Mater. Sci.*, 2002, **6**, 389–399.
9 F. Barbir, in *Fuel Cell Technology: Reaching Towards Commercialization*, ed. N. Sammes, Springer, London, 2006, pp. 27–51.
10 J. Xuan, D. Y. C. Leung, M. K. H. Leung, M. Ni and H. Wang, *Int. J. Hydrogen Energy*, 2011, **36**, 9231–9241.
11 A. Lasia, *Handbook of Fuel Cells*, John Wiley & Sons, Ltd, 2010.
12 J. Greeley, T. F. Jaramillo, J. Bonde, I. B. Chorkendorff and J. K. Nørskov, *Nat. Mater.*, 2006, **5**, 909–913.
13 A. Safavi, S. H. Kazemi and H. Kazemi, *Fuel*, 2014, **118**, 156–162.
14 T. L. Tan, L.-L. Wang, D. D. Johnson and K. Bai, *Nano Lett.*, 2012, **12**, 4875–4880.
15 J. Ma, A. Habrioux and N. Alonso-Vante, *J. Solid State Electrochem.*, 2013, **17**, 1913–1921.
16 D. Hou, W. Zhou, K. Zhou, Y. Zhou, J. Zhong, L. Yang, J. Lu, G. Li and S. Chen, *J. Mater. Chem. A*, 2015, **3**, 15962–15968.
17 D. E. Brown, M. N. Mahmood, M. C. M. Man and A. K. Turner, *Electrochim. Acta*, 1984, **29**, 1551–1556.
18 L. Chen and A. Lasia, *J. Electrochem. Soc.*, 1991, **138**, 3321–3328.
19 J. M. Jakšić, M. V. Vojnović and N. V. Krstajić, *Electrochim. Acta*, 2000, **45**, 4151–4158.
20 W. Sheng, H. A. Gasteiger and Y. Shao-Horn, *J. Electrochem. Soc.*, 2010, **157**, B1529–B1536.
21 L. Birry and A. Lasia, *J. Appl. Electrochem.*, 2004, **34**, 735–749.
22 E. Endoh, H. Otouma, T. Morimoto and Y. Oda, *Int. J. Hydrogen Energy*, 1987, **12**, 473–479.
23 Y. Choquette, L. Brossard, A. Lasia and H. Menard, *J. Electrochem. Soc.*, 1990, **137**, 1723–1730.
24 Y. Choquette, L. Brossard, A. Lasia and H. Menard, *Electrochim. Acta*, 1990, **35**, 1251–1256.
25 J. Divisek, H. Schmitz and B. Steffen, *Electrochim. Acta*, 1994, **39**, 1723–1731.
26 M. Blouin and D. Guay, *J. Electrochem. Soc.*, 1997, **144**, 573–581.
27 Z. Chen, D. Cummins, B. N. Reinecke, E. Clark, M. K. Sunkara and T. F. Jaramillo, *Nano Lett.*, 2011, **11**, 4168–4175.
28 J. C. F. Boodts and S. Trasatti, *J. Appl. Electrochem.*, 1989, **19**, 255–262.

29 Y. Li, H. Wang, L. Xie, Y. Liang, G. Hong and H. Dai, *J. Am. Chem. Soc.*, 2011, **133**, 7296–7299.
30 D. Voiry, H. Yamaguchi, J. Li, R. Silva, D. C. B. Alves, T. Fujita, M. Chen, T. Asefa, V. B. Shenoy and G. Eda, *Nat. Mater.*, 2013, **12**, 850–855.
31 D. Kong, J. J. Cha, H. Wang, H. R. Lee and Y. Cui, *Energy Environ. Sci.*, 2013, **6**, 3553–3558.
32 Q. Gong, L. Cheng, C. Liu, M. Zhang, Q. Feng, H. Ye, M. Zeng, L. Xie, Z. Liu and Y. Li, *ACS Catal.*, 2015, **5**, 2213–2219.
33 T. F. Jaramillo, K. P. Jørgensen, J. Bonde, J. H. Nielsen, S. Horch and I. Chorkendorff, *Science*, 2007, **317**, 100–102.
34 R. M. Q. Mello and E. A. Ticianelli, *Electrochim. Acta*, 1997, **42**, 1031–1039.
35 G. A. Camara, E. A. Ticianelli, S. Mukerjee, S. J. Lee and J. McBreen, *J. Electrochem. Soc.*, 2002, **149**, A748–A753.
36 B. M. Babić, L. M. Vračar, V. Radmilović and N. V. Krstajić, *Electrochim. Acta*, 2006, **51**, 3820–3826.
37 G. Couturier, D. W. Kirk, P. J. Hyde and S. Srinivasan, *Electrochim. Acta*, 1987, **32**, 995–1005.
38 S. Jayaraman and A. C. Hillier, *J. Phys. Chem. B*, 2003, **107**, 5221–5230.
39 M. Shao, *J. Power Sources*, 2011, **196**, 2433–2444.
40 M. T. Giacomini, M. Balasubramanian, S. Khalid, J. McBreen and E. A. Ticianelli, *J. Electrochem. Soc.*, 2003, **150**, A588–A593.
41 W. Sheng, A. P. Bivens, M. Myint, Z. Zhuang, R. V. Forest, Q. Fang, J. G. Chen and Y. Yan, *Energy Environ. Sci.*, 2014, **7**, 1719–1724.
42 I. Nikolov, R. Darkaoui, E. Zhecheva, R. Stoyanova, N. Dimitrov and T. Vitanov, *J. Electroanal. Chem.*, 1997, **429**, 157–168.
43 J. Suntivich, K. J. May, H. A. Gasteiger, J. B. Goodenough and Y. Shao-Horn, *Science*, 2011, **334**, 1383–1385.
44 T. Reier, M. Oezaslan and P. Strasser, *ACS Catal.*, 2012, **2**, 1765–1772.
45 B. Lu, D. Cao, P. Wang, G. Wang and Y. Gao, *Int. J. Hydrogen Energy*, 2011, **36**, 72–78.
46 Y. Li, P. Hasin and Y. Wu, *Adv. Mater.*, 2010, **22**, 1926–1929.
47 C. C. L. McCrory, S. Jung, J. C. Peters and T. F. Jaramillo, *J. Am. Chem. Soc.*, 2013, **135**, 16977–16987.
48 J. Greeley, I. E. L. Stephens, A. S. Bondarenko, T. P. Johansson, H. A. Hansen, T. F. Jaramillo, J. Rossmeisl, I. Chorkendorff and J. K. Nørskov, *Nat. Chem.*, 2009, **1**, 552–556.
49 P. Malacrida, M. Escudero-Escribano, A. Verdaguer-Casadevall, I. E. L. Stephens and I. Chorkendorff, *J. Mater. Chem. A*, 2014, **2**, 4234–4243.
50 P. Hernandez-Fernandez, F. Masini, D. N. McCarthy, C. E. Strebel, D. Friebel, D. Deiana, P. Malacrida, A. Nierhoff, A. Bodin, A. M. Wise, J. H. Nielsen, T. W. Hansen, A. Nilsson, E. L. StephensIfan and I. Chorkendorff, *Nat. Chem.*, 2014, **6**, 732–738.
51 M. Escudero-Escribano, A. Verdaguer-Casadevall, P. Malacrida, U. Grønbjerg, B. P. Knudsen, A. K. Jepsen, J. Rossmeisl, I. E. L. Stephens and I. Chorkendorff, *J. Am. Chem. Soc.*, 2012, **134**, 16476–16479.
52 C. Cui, L. Gan, M. Heggen, S. Rudi and P. Strasser, *Nat. Mater.*, 2013, **12**, 765–771.
53 D. Wang, H. L. Xin, R. Hovden, H. Wang, Y. Yu, D. A. Muller, F. J. DiSalvo and H. D. Abruña, *Nat. Mater.*, 2013, **12**, 81–87.
54 S.-I. Choi, S. Xie, M. Shao, J. H. Odell, N. Lu, H.-C. Peng, L. Protsailo, S. Guerrero, J. Park, X. Xia, J. Wang, M. J. Kim and Y. Xia, *Nano Lett.*, 2013, **13**, 3420–3425.

55 J. K. Nørskov, J. Rossmeisl, A. Logadottir, L. Lindqvist, J. R. Kitchin, T. Bligaard and H. Jónsson, *J. Phys. Chem. B*, 2004, **108**, 17886–17892.
56 M. Shao, J. Odell, M. Humbert, T. Yu and Y. Xia, *J. Phys. Chem. C*, 2013, **117**, 4172–4180.
57 M. Shao, T. Yu, J. H. Odell, M. Jin and Y. Xia, *Chem. Commun.*, 2011, **47**, 6566–6568.
58 N. Alexeyeva, A. Sarapuu, K. Tammeveski, F. J. Vidal-Iglesias, J. Solla-Gullón and J. M. Feliu, *Electrochim. Acta*, 2011, **56**, 6702–6708.
59 K. Jukk, N. Alexeyeva, A. Sarapuu, P. Ritslaid, J. Kozlova, V. Sammelselg and K. Tammeveski, *Int. J. Hydrogen Energy*, 2013, **38**, 3614–3620.
60 M. H. Seo, S. M. Choi, J. K. Seo, S. H. Noh, W. B. Kim and B. Han, *Appl. Catal., B*, 2013, **129**, 163–171.
61 Y. Luo, J. M. Mora-Hernández, L. A. Estudillo-Wong, E. M. Arce-Estrada and N. Alonso-Vante, *Electrochim. Acta*, 2015, **173**, 771–778.
62 T. Jin, S. Guo, J.-L. Zuo and S. Sun, *Nanoscale*, 2013, **5**, 160–163.
63 S. Yin, M. Cai, C. Wang and P. K. Shen, *Energy Environ. Sci.*, 2011, **4**, 558–563.
64 G. Wu, K. L. More, C. M. Johnston and P. Zelenay, *Science*, 2011, **332**, 443–447.
65 X. Yang, S. Li, Y. Liu, X. Wei and Y. Liu, *J. Power Sources*, 2011, **196**, 4992–4995.
66 A. Morozan, B. Jousselme and S. Palacin, *Energy Environ. Sci.*, 2011, **4**, 1238–1254.
67 Z. Cui, R. G. Burns and F. J. DiSalvo, *Chem. Mater.*, 2013, **25**, 3782–3784.
68 H. Zhong, H. Zhang, G. Liu, Y. Liang, J. Hu and B. Yi, *Electrochem. Commun.*, 2006, **8**, 707–712.
69 A. Ishihara, Y. Ohgi, K. Matsuzawa, S. Mitsushima and K.-I. Ota, *Electrochim. Acta*, 2010, **55**, 8005–8012.
70 D. Geng, Y. Chen, Y. Chen, Y. Li, R. Li, X. Sun, S. Ye and S. Knights, *Energy Environ. Sci.*, 2011, **4**, 760–764.
71 K. Gong, F. Du, Z. Xia, M. Durstock and L. Dai, *Science*, 2009, **323**, 760–764.
72 Y. Shao, S. Zhang, M. H. Engelhard, G. Li, G. Shao, Y. Wang, J. Liu, I. A. Aksay and Y. Lin, *J. Mater. Chem.*, 2010, **20**, 7491–7496.
73 N. P. Subramanian, X. Li, V. Nallathambi, S. P. Kumaraguru, H. Colon-Mercado, G. Wu, J.-W. Lee and B. N. Popov, *J. Power Sources*, 2009, **188**, 38–44.
74 J.-M. Savéant, *Chem. Rev.*, 2008, **108**, 2348–2378.
75 M. Lefèvre, E. Proietti, F. Jaouen and J.-P. Dodelet, *Science*, 2009, **324**, 71–74.
76 M. Jahan, Q. Bao and K. P. Loh, *J. Am. Chem. Soc.*, 2012, **134**, 6707–6713.
77 G. Lalande, R. Cote, G. Tamizhmani, D. Guay, J. P. Dodelet, L. Dignard-Bailey, L. T. Weng and P. Bertrand, *Electrochim. Acta*, 1995, **40**, 2635–2646.
78 L. Timperman, A. S. Gago and N. Alonso-Vante, *J. Power Sources*, 2011, **196**, 4290–4297.
79 Y. Feng, A. Gago, L. Timperman and N. Alonso-Vante, *Electrochim. Acta*, 2011, **56**, 1009–1022.
80 Y. Feng, T. He and N. Alonso-Vante, *Chem. Mater.*, 2007, **20**, 26–28.
81 N. Alonso-Vante and H. Tributsch, *Nature*, 1986, **323**, 431–432.
82 C. Zhao, D. Li and Y. Feng, *J. Mater. Chem. A*, 2013, **1**, 5741–5746.
83 Y. Chen, C. W. Li and M. W. Kanan, *J. Am. Chem. Soc.*, 2012, **134**, 19969–19972.
84 D. Gao, H. Zhou, J. Wang, S. Miao, F. Yang, G. Wang, J. Wang and X. Bao, *J. Am. Chem. Soc.*, 2015, **137**, 4288–4291.
85 J. H. Kwak, L. Kovarik and J. n. Szanyi, *ACS Catal.*, 2013, **3**, 2094–2100.

86 Y.-J. Zhang, V. Sethuraman, R. Michalsky and A. A. Peterson, *ACS Catal.*, 2014, **4**, 3742–3748.
87 C. W. Li and M. W. Kanan, *J. Am. Chem. Soc.*, 2012, **134**, 7231–7234.
88 X. Nie, M. R. Esopi, M. J. Janik and A. Asthagiri, *Angew. Chem., Int. Ed.*, 2013, **52**, 2459–2462.
89 W. Zhu, Y.-J. Zhang, H. Zhang, H. Lv, Q. Li, R. Michalsky, A. A. Peterson and S. Sun, *J. Am. Chem. Soc.*, 2014, **136**, 16132–16135.
90 W. Zhu, R. Michalsky, Ö. Metin, H. Lv, S. Guo, C. J. Wright, X. Sun, A. A. Peterson and S. Sun, *J. Am. Chem. Soc.*, 2013, **135**, 16833–16836.
91 Q. Lu, J. Rosen, Y. Zhou, G. S. Hutchings, Y. C. Kimmel, J. G. Chen and F. Jiao, *Nat. Commun.*, 2014, **5**, 3242.
92 K. Chan, C. Tsai, H. A. Hansen and J. K. Nørskov, *ChemCatChem*, 2014, **6**, 1899–1905.
93 K. P. Kuhl, T. Hatsukade, E. R. Cave, D. N. Abram, J. Kibsgaard and T. F. Jaramillo, *J. Am. Chem. Soc.*, 2014, **136**, 14107–14113.
94 C. Costentin, S. Drouet, M. Robert and J.-M. Savéant, *Science*, 2012, **338**, 90–94.
95 I. Bhugun, D. Lexa and J.-M. Savéant, *J. Am. Chem. Soc.*, 1996, **118**, 1769–1776.
96 M. Hammouche, D. Lexa, M. Momenteau and J. M. Saveant, *J. Am. Chem. Soc.*, 1991, **113**, 8455–8466.
97 L. Chen, Z. Guo, X.-G. Wei, C. Gallenkamp, J. Bonin, E. Anxolabéhère-Mallart, K.-C. Lau, T.-C. Lau and M. Robert, *J. Am. Chem. Soc.*, 2015, **137**, 10918–10921.
98 M. Bourrez, F. Molton, S. Chardon-Noblat and A. Deronzier, *Angew. Chem., Int. Ed.*, 2011, **50**, 9903–9906.
99 M. Bourrez, M. Orio, F. Molton, H. Vezin, C. Duboc, A. Deronzier and S. Chardon-Noblat, *Angew. Chem., Int. Ed.*, 2014, **53**, 240–243.
100 J. M. Smieja, M. D. Sampson, K. A. Grice, E. E. Benson, J. D. Froehlich and C. P. Kubiak, *Inorg. Chem.*, 2013, **52**, 2484–2491.
101 H. Ishida, K. Tanaka and T. Tanaka, *Organometallics*, 1987, **6**, 181–186.
102 S. Chardon-Noblat, M. N. Collomb-Dunand-Sauthier, A. Deronzier, R. Ziessel and D. Zsoldos, *Inorg. Chem.*, 1994, **33**, 4410–4412.
103 H. Nagao, T. Mizukawa and K. Tanaka, *Inorg. Chem.*, 1994, **33**, 3415–3420.
104 S. Chardon-Noblat, A. Deronzier, F. Hartl, J. van Slageren and T. Mahabiersing, *Eur. J. Inorg. Chem.*, 2001, **2001**, 613–617.
105 M. R. M. Bruce, E. Megehee, B. P. Sullivan, H. H. Thorp, T. R. O'Toole, A. Downard, J. R. Pugh and T. J. Meyer, *Inorg. Chem.*, 1992, **31**, 4864–4873.
106 C. M. Bolinger, N. Story, B. P. Sullivan and T. J. Meyer, *Inorg. Chem.*, 1988, **27**, 4582–4587.
107 J. M. Smieja and C. P. Kubiak, *Inorg. Chem.*, 2010, **49**, 9283–9289.
108 T. R. O'Toole, B. P. Sullivan, M. R. M. Bruce, L. D. Margerum, R. W. Murray and T. J. Meyer, *J. Electroanal. Chem. Interfacial Electrochem.*, 1989, **259**, 217–239.
109 R. Liu, C. Stephani, J. J. Han, K. L. Tan and D. Wang, *Angew. Chem., Int. Ed.*, 2013, **52**, 4225–4228.
110 E. Torralba-Peñalver, Y. Luo, J.-D. Compain, S. Chardon-Noblat and B. Fabre, *ACS Catal.*, 2015, **5**, 6138–6147.
111 T. Ohno, D. Haga, K. Fujihara, K. Kaizaki and M. Matsumura, *J. Phys. Chem. B*, 1997, **101**, 6415–6419.
112 J. A. Keith and E. A. Carter, *J. Am. Chem. Soc.*, 2012, **134**, 7580–7583.
113 B. Aurian-Blajeni, I. Taniguchi and J. M. Bockris, *J. Electroanal. Chem. Interfacial Electrochem.*, 1983, **149**, 291–293.

114 E. E. Barton, D. M. Rampulla and A. B. Bocarsly, *J. Am. Chem. Soc.*, 2008, **130**, 6342–6344.
115 B. Aurian-Blajeni, M. Halmann and J. Manassen, *Sol. Energy Mater.*, 1983, **8**, 425–440.
116 Y. Qiu, H. Deng, S. Yang, J. Mou, C. Daiguebonne, N. Kerbellec, O. Guillou and S. R. Batten, *Inorg. Chem.*, 2009, **48**, 3976–3981.
117 B. Chen, C. Liang, J. Yang, D. S. Contreras, Y. L. Clancy, E. B. Lobkovsky, O. M. Yaghi and S. Dai, *Angew. Chem., Int. Ed.*, 2006, **45**, 1390–1393.
118 S. S. Kaye, A. Dailly, O. M. Yaghi and J. R. Long, *J. Am. Chem. Soc.*, 2007, **129**, 14176–14177.
119 A. M. Shultz, O. K. Farha, J. T. Hupp and S. T. Nguyen, *J. Am. Chem. Soc.*, 2009, **131**, 4204–4205.
120 F. X. Llabrés i Xamena, A. Abad, A. Corma and H. Garcia, *J. Catal.*, 2007, **250**, 294–298.
121 F. Luo and S. R. Batten, *Dalton Trans.*, 2010, **39**, 4485–4488.
122 D. Ma, B. Li, X. Zhou, Q. Zhou, K. Liu, G. Zeng, G. Li, Z. Shi and S. Feng, *Chem. Commun.*, 2013, **49**, 8964–8966.
123 Y. Tian, J. Cong, S. Shen, Y. Chai, L. Yan, S. Wang and Y. Sun, *Phys. Status Solidi RRL*, 2014, **8**, 91–94.
124 Y. Luo, PhD thesis, 2012, INSA de Rennes.
125 L. Ma, O. R. Evans, B. M. Foxman and W. Lin, *Inorg. Chem.*, 1999, **38**, 5837–5840.
126 T. M. Reineke, M. Eddaoudi, M. Fehr, D. Kelley and O. M. Yaghi, *J. Am. Chem. Soc.*, 1999, **121**, 1651–1657.
127 C. Daiguebonne, O. Guillou, C. Baux, F. Le Dret and K. Boubekeur, *J. Alloys Compd.*, 2001, **323**, 193–198.
128 C. Daiguebonne, Y. Gerault, F. Le Dret, O. Guillou and K. Boubekeur, *J. Alloys Compd.*, 2002, **344**, 179–185.
129 J.-W. Jeon, R. Sharma, P. Meduri, B. W. Arey, H. T. Schaef, J. L. Lutkenhaus, J. P. Lemmon, P. K. Thallapally, M. I. Nandasiri, B. P. McGrail and S. K. Nune, *ACS Appl. Mater. Interfaces*, 2014, **6**, 7214–7222.
130 Q. Li, P. Xu, W. Gao, S. Ma, G. Zhang, R. Cao, J. Cho, H.-L. Wang and G. Wu, *Adv. Mater.*, 2014, **26**, 1378–1386.
131 B. Y. Xia, Y. Yan, N. Li, H. B. Wu, X. W. D. Lou and X. Wang, *Nat. Energy*, 2016, **1**, 15006.
132 Y. Luo, L. Calvillo, C. Daiguebonne, M. K. Daletou, G. Granozzi and N. Alonso-Vante, *Appl. Catal., B*, 2016, **189**, 39–50.
133 A. Morozan and F. Jaouen, *Energy Environ. Sci.*, 2012, **5**, 9269–9290.
134 C. H. Hendon, T. Davide and W. Aron, *Phys. Chem. Chem. Phys.*, 2012, **14**, 13120–13132.
135 G. Gonzalo, A. O. Pilar, C. J. Gómez-García and Z. Félix, *Chem. Soc. Rev.*, 2011, **41**, 115–147.
136 Y. Hamaue, R. Aoki, M. Yamashita and S. Kida, *Inorg. Chim. Acta*, 1981, **54**, L13–L14.
137 M. Haruki, M. Tanaka and S. Kurita, *Synth. Met.*, 1987, **21**, 373–378.
138 M. Yamashita, I. Murase, T. Ito, Y. Wada, T. Mitani and I. Ikemoto, *Bull. Chem. Soc. Jpn.*, 1985, **58**, 2336–2339.
139 M. Yamashita, Y. Nonaka, S. Kida, Y. Hamaue and R. Aoki, *Inorg. Chim. Acta*, 1981, **52**, 43–46.
140 M. Yamashita, T. Manabe, K. Inoue, T. Kawashima, H. Okamoto, H. Kitagawa, T. Mitani, K. Toriumi, H. Miyamae and R. Ikeda, *Inorg. Chem.*, 1999, **38**, 1894–1899.
141 G. C. Papavassiliou and D. Layek, *Z. Naturforsch.*, 1982, **37**, 1406–1409.

142 C. N. R. Rao, A. Ranganathan, V. R. Pedireddi and A. R. Raju, *Chem. Commun.*, 2000, 39–40.
143 Z. Xu, *Coord. Chem. Rev.*, 2006, **250**, 2745–2757.
144 W. Su, M. Hong, J. Weng, R. Cao and S. Lu, *Angew. Chem., Int. Ed.*, 2000, **39**, 2911–2914.
145 Y. Takazaki, Z. Yang, M. Ebihara, K. Inoue and T. Kawamura, *Chem. Lett.*, 2003, **32**, 120–121.
146 P. Amo-Ochoa, L. Welte, R. González-Prieto, P. J. S. Miguel, C. J. Gómez-García, E. Mateo-Martí, S. Delgado, J. Gómez-Herrero and F. Zamora, *Chem. Commun.*, 2010, **46**, 3262–3264.
147 D. Sheberla, L. Sun, M. A. Blood-Forsythe, S. Er, C. R. Wade, C. K. Brozek, A. Aspuru-Guzik and M. Dincă, *J. Am. Chem. Soc.*, 2014, **136**, 8859–8862.
148 R. C. Wheland and J. L. Gillson, *J. Am. Chem. Soc.*, 1976, **98**, 3916–3925.
149 K.-H. Low, V. A. L. Roy, S. S.-Y. Chui, S. L.-F. Chan and C.-M. Che, *Chem. Commun.*, 2010, **46**, 7328–7330.
150 X. Han, F. Cheng, T. Zhang, J. Yang, Y. Hu and J. Chen, *Adv. Mater.*, 2014, **26**, 2047–2051.
151 T. Shinya, H. Miyuki, K. Takashi, M. Hitoshi, Y. Masahiro, N. Yasuyuki, K. Yasutaka, Y. Kizashi, K. Atsushi and K. Hiroshi, *Inorg. Chem.*, 2009, **48**, 9048–9050.
152 Y. Kobayashi, B. Jacobs, M. D. Allendorf and J. R. Long, *Chem. Mater.*, 2010, **22**, 4120–4122.
153 F. Yasuhiro, E. Masahiro, K. Shoichi and K. Takashi, *J. Am. Chem. Soc.*, 2004, **126**, 12238–12239.
154 A. Alexer, E. Peter, K. Gerhard, H. Siegfried, V. S. J. Ulrich and W. Hans-Peter, *Angew. Chem., Int. Ed. Engl.*, 1986, **25**, 740–741.
155 Y. Chen, Z.-O. Wang, Z.-G. Ren, H.-X. Li, D.-X. Li, D. Liu, Y. Zhang and J.-P. Lang, *Cryst. Growth Des.*, 2009, **9**, 4963–4968.
156 D. Sun, R. Cao, J. Weng, M. Hong and Y. Liang, *Dalton Trans.*, 2002, **3**, 291–292.
157 R. Fu, S. Xia, S. Xiang, S. Hu and X. Wu, *J. Solid State Chem.*, 2004, **177**, 4626–4631.
158 A. A. Talin, A. Centrone, A. C. Ford, M. E. Foster, V. Stavila, P. Haney, R. A. Kinney, V. Szalai, F. El Gabaly, H. P. Yoon, F. Léonard and M. D. Allendorf, *Science*, 2014, **343**, 66–69.
159 K.-S. Lin, A. K. Adhikari, C.-N. Ku, C.-L. Chiang and H. Kuo, *Int. J. Hydrogen Energy*, 2012, **37**, 13865–13871.
160 L. H. Wee, M. R. Lohe, N. Janssens, S. Kaskel and J. A. Martens, *J. Mater. Chem.*, 2012, **22**, 13742–13746.
161 R. A. Heintz, H. Zhao, X. Ouyang, G. Grandinetti, J. Cowen and K. R. Dunbar, *Inorg. Chem.*, 1999, **38**, 144–156.
162 B. S. Brunschwig, C. Creutz and N. Sutin, *Chem. Soc. Rev.*, 2002, **31**, 168–184.
163 N. A. Travlou, K. Singh, E. Rodriguez-Castellon and T. J. Bandosz, *J. Mater. Chem. A*, 2015, **3**, 11417–11429.
164 X.-W. Liu, T.-J. Sun, J.-L. Hu and S.-D. Wang, *J. Mater. Chem. A*, 2016, **4**, 3584–3616.
165 K. M. Choi, H. M. Jeong, J. H. Park, Y.-B. Zhang, J. K. Kang and O. M. Yaghi, *ACS Nano*, 2014, **8**, 7451–7457.
166 P. Srimuk, S. Luanwuthi, A. Krittayavathananon and M. Sawangphruk, *Electrochim. Acta*, 2015, **157**, 69–77.
167 W. Bao, Z. Zhang, Y. Qu, C. Zhou, X. Wang and J. Li, *J. Alloys Compd.*, 2014, **582**, 334–340.

168 K.-Y. Andrew Lin, F.-K. Hsu and W.-D. Lee, *J. Mater. Chem. A*, 2015, **3**, 9480–9490.
169 X. Wang, Q. Wang, Q. Wang, F. Gao, F. Gao, Y. Yang and H. Guo, *ACS Appl. Mater. Interfaces*, 2014, **6**, 11573–11580.
170 L. Shen, L. Huang, S. Liang, R. Liang, N. Qin and L. Wu, *RSC Adv.*, 2014, **4**, 2546–2549.
171 Z. Zhao, S. Wang, R. Liang, Z. Li, Z. Shi and G. Chen, *J. Mater. Chem. A*, 2014, **2**, 13509–13512.
172 C. Petit and T. J. Bandosz, *Dalton Trans.*, 2012, **41**, 4027–4035.
173 C. Petit and T. J. Bandosz, *J. Mater. Chem.*, 2009, **19**, 6521–6528.
174 L. Benoit, P. Camille and T. J. Bandosz, *ACS Appl. Mater. Interfaces*, 2010, **2**, 3606–3613.
175 C. Petit, B. Mendoza and T. J. Bandosz, *ChemPhysChem*, 2010, **11**, 3678–3684.
176 C. Petit and T. J. Bandosz, *Adv. Funct. Mater.*, 2011, **21**, 2108–2117.
177 C. Petit, L. Huang, J. Jagiello, J. Kenvin, K. E. Gubbins and T. J. Bandosz, *Langmuir*, 2011, **27**, 13043–13051.
178 Y. Zhao, M. Seredych, Q. Zhong and T. J. Bandosz, *ACS Appl. Mater. Interfaces*, 2013, **5**, 4951–4959.
179 Y. Zhao, M. Seredych, J. Jagiello, Q. Zhong and T. J. Bandosz, *Chem. Eng. J.*, 2014, **239**, 399–407.
180 S. Bashkova and T. J. Bandosz, *J. Colloid Interface Sci.*, 2014, **417**, 109–114.
181 P. Pachfule, B. K. Balan, S. Kurungot and R. Banerjee, *Chem. Commun.*, 2012, **48**, 2009–2011.
182 Y. Zhang, X. Bo, C. Luhana, H. Wang, M. Li and L. Guo, *Chem. Commun.*, 2013, **49**, 6885–6887.
183 Y. Zhang, A. Nsabimana, L. Zhu, X. Bo, C. Han, M. Li and L. Guo, *Talanta*, 2014, **129**, 55–62.
184 X. Lin, G. Gao, L. Zheng, Y. Chi and G. Chen, *Anal. Chem.*, 2014, **86**, 1223–1228.
185 L. Ge, L. Wang, V. Rudolph and Z. Zhu, *RSC Adv.*, 2013, **3**, 25360–25366.
186 Y. Yang, L. Ge, V. Rudolph and Z. Zhu, *Dalton Trans.*, 2014, **43**, 7028–7036.
187 O. Shekhah, H. Wang, M. Paradinas, C. Ocal, B. Schupbach, A. Terfort, D. Zacher, R. A. Fischer and C. Woll, *Nat. Mater.*, 2009, **8**, 481–484.
188 M. C. So, J. Shengye, S. Ho-Jin, G. P. Wiederrecht, O. K. Farha and J. T. Hupp, *J. Am. Chem. Soc.*, 2013, **135**, 15698–15701.
189 H. Aisheng, L. Qian, W. Nanyi, Z. Yaqiong and C. Jürgen, *J. Am. Chem. Soc.*, 2014, **136**, 14686–14689.
190 J. Zhao, B. Gong, W. T. Nunn, P. C. Lemaire, E. C. Stevens, F. I. Sidi, P. S. Williams, C. J. Oldham, H. J. Walls, S. D. Shepherd, M. A. Browe, G. W. Peterson, M. D. Losego and G. N. Parsons, *J. Mater. Chem. A*, 2015, **3**, 1458–1464.
191 S. Qiu, M. Xue and G. Zhu, *Chem. Soc. Rev.*, 2014, **43**, 6116–6140.
192 L. Dumee, L. He, M. Hill, B. Zhu, M. Duke, J. Schutz, F. She, H. Wang, S. Gray, P. Hodgson and L. Kong, *J. Mater. Chem. A*, 2013, **1**, 9208–9214.
193 Y. Yue, B. Guo, Z.-A. Qiao, P. F. Fulvio, J. Chen, A. J. Binder, C. Tian and S. Dai, *Microporous Mesoporous Mater.*, 2014, **198**, 139–143.
194 H. Jia, Z. Sun, D. Jiang and P. Du, *Chem. Mater.*, 2015, **27**, 4586–4593.
195 S. R. Ahrenholtz, C. C. Epley and A. J. Morris, *J. Am. Chem. Soc.*, 2014, **136**, 2464–2472.
196 I. Hijazi, T. Bourgeteau, R. Cornut, A. Morozan, A. Filoramo, J. Leroy, V. Derycke, B. Jousselme and S. Campidelli, *J. Am. Chem. Soc.*, 2014, **136**, 6348–6354.

197 F. Dai, W. Fan, J. Bi, P. Jiang, D. Liu, X. Zhang, H. Lin, C. Gong, R. Wang, L. Zhang and D. Sun, *Dalton Trans.*, 2016, **45**, 61–65.
198 S. Lin, C. S. Diercks, Y.-B. Zhang, N. Kornienko, E. M. Nichols, Y. Zhao, A. R. Paris, D. Kim, P. Yang, O. M. Yaghi and C. J. Chang, *Science*, 2015, **349**, 1208–1213.
199 N. Kornienko, Y. Zhao, C. S. Kley, C. Zhu, D. Kim, S. Lin, C. J. Chang, O. M. Yaghi and P. Yang, *J. Am. Chem. Soc.*, 2015, **137**, 14129–14135.
200 I. Hod, M. D. Sampson, P. Deria, C. P. Kubiak, O. K. Farha and J. T. Hupp, *ACS Catal.*, 2015, **5**, 6302–6309.
201 E. M. Miner, T. Fukushima, D. Sheberla, L. Sun, Y. Surendranath and M. Dinca, *Nat. Commun.*, 2016, **7**, 10942.
202 J. Mao, L. Yang, P. Yu, X. Wei and L. Mao, *Electrochem. Commun.*, 2012, **19**, 29–31.
203 L.-F. Song, C.-H. Jiang, C.-L. Jiao, J. Zhang, L.-X. Sun, F. Xu, W.-S. You, Z.-G. Wang and J.-J. Zhao, *Cryst. Growth Des.*, 2010, **10**, 5020–5023.
204 Y. Gong, H.-F. Shi, P.-G. Jiang, W. Hua and J.-H. Lin, *Cryst. Growth Des.*, 2014, **14**, 649–657.
205 Y. Gong, Z. Hao, J. Meng, H. Shi, P. Jiang, M. Zhang and J. Lin, *ChemPlusChem*, 2014, **79**, 266–277.
206 Q. Xu, H. Li, F. Yue, L. Chi and J. Wang, *New J. Chem.*, 2016, **40**, 3032–3035.
207 Y. Gong, T. Wu, P. G. Jiang, J. H. Lin and Y. X. Yang, *Inorg. Chem.*, 2012, **52**, 777–784.
208 C.-W. Kung, J. E. Mondloch, T. C. Wang, W. Bury, W. Hoffeditz, B. M. Klahr, R. C. Klet, M. J. Pellin, O. K. Farha and J. T. Hupp, *ACS Appl. Mater. Interfaces*, 2015, **7**, 28223–28230.
209 J. E. Mondloch, W. Bury, D. Fairen-Jimenez, S. Kwon, E. J. DeMarco, M. H. Weston, A. A. Sarjeant, S. T. Nguyen, P. C. Stair, R. Q. Snurr, O. K. Farha and J. T. Hupp, *J. Am. Chem. Soc.*, 2013, **135**, 10294–10297.
210 B. Nohra, H. El Moll, L. M. Rodriguez Albelo, P. Mialane, J. Marrot, C. Mellot-Draznieks, M. O'Keeffe, R. Ngo Biboum, J. Lemaire, B. Keita, L. Nadjo and A. Dolbecq, *J. Am. Chem. Soc.*, 2011, **133**, 13363–13374.
211 H. Wang, F. Yin, G. Li, B. Chen and Z. Wang, *Int. J. Hydrogen Energy*, 2014, **39**, 16179–16186.
212 L. Yang, S. Kinoshita, T. Yamada, S. Kanda, H. Kitagawa, M. Tokunaga, T. Ishimoto, T. Ogura, R. Nagumo, A. Miyamoto and M. Koyama, *Angew. Chem.*, 2010, **122**, 5476–5479.
213 B. Liu, J. H. Chen, C. H. Xiao, K. Z. Cui, L. Yang, H. L. Pang and Y. F. Kuang, *Energy Fuels*, 2007, **21**, 1365–1369.
214 Y. Pan, B. Yuan, Y. Li and D. He, *Chem. Commun.*, 2010, **46**, 2280–2282.
215 T. T. Dang, Y. Zhu, J. S. Y. Ngiam, S. C. Ghosh, A. Chen and A. M. Seayad, *ACS Catal.*, 2013, **3**, 1406–1410.
216 Q.-L. Zhu, J. Li and Q. Xu, *J. Am. Chem. Soc.*, 2013, **135**, 10210–10213.
217 J. Long, H. Liu, S. Wu, S. Liao and Y. Li, *ACS Catal.*, 2013, **3**, 647–654.
218 P. Miao, G. Li, G. Zhang and H. Lu, *J. Energy Chem.*, 2014, **23**, 507–512.
219 H. Wang, F. Yin, B. Chen and G. Li, *J. Mater. Chem. A*, 2015, **3**, 16168–16176.
220 K. Xi, S. Cao, X. Peng, C. Ducati, R. V. Kumar and A. K. Cheetham, *Chem. Commun.*, 2013, **49**, 2192–2194.
221 S. J. Yang, T. Kim, J. H. Im, Y. S. Kim, K. Lee, H. Jung and C. R. Park, *Chem. Mater.*, 2012, **24**, 464–470.
222 L. Li, Y. Wang, X. Gu, Q. Yang and X. Zhao, *Chem. – Asian J.*, 2016, **11**, 1913–1920.
223 A. Banerjee, K. K. Upadhyay, D. Puthusseri, V. Aravindan, S. Madhavi and S. Ogale, *Nanoscale*, 2014, **6**, 4387–4394.

224 S. Pandiaraj, H. B. Aiyappa, R. Banerjee and S. Kurungot, *Chem. Commun.*, 2014, **50**, 3363–3366.
225 B. Liu, H. Shioyama, T. Akita and Q. Xu, *J. Am. Chem. Soc.*, 2008, **130**, 5390–5391.
226 W. Chaikittisilp, K. Ariga and Y. Yamauchi, *J. Mater. Chem. A*, 2013, **1**, 14–19.
227 H. B. Wu, S. Wei, L. Zhang, R. Xu, H. H. Hng and X. W. D. Lou, *Chem. – Eur. J.*, 2013, **19**, 10804–10808.
228 C. O. Ania, M. Seredych, E. Rodriguez-Castellon and T. J. Bandosz, *Appl. Catal., B*, 2015, **163**, 424–435.
229 X. Wang, J. Zhou, H. Fu, W. Li, X. Fan, G. Xin, J. Zheng and X. Li, *J. Mater. Chem. A*, 2014, **2**, 14064–14070.
230 Z. Guo, J. K. Cheng, Z. Hu, M. Zhang, Q. Xu, Z. Kang and D. Zhao, *RSC Adv.*, 2014, **4**, 34221–34225.
231 J. Kim, N. D. McNamara and J. C. Hicks, *Appl. Catal., A*, 2016, **517**, 141–150.
232 F. Afsahi, H. Vinh-Thang, S. Mikhailenko and S. Kaliaguine, *J. Power Sources*, 2013, **239**, 415–423.
233 A. Banerjee, U. Singh, V. Aravindan, M. Srinivasan and S. Ogale, *Nano Energy*, 2013, **2**, 1158–1163.
234 J. Lu, W. Zhou, L. Wang, J. Jia, Y. Ke, L. Yang, K. Zhou, X. Liu, Z. Tang, L. Li and S. Chen, *ACS Catal.*, 2016, **6**, 1045–1053.
235 Y. Luo, L. A. Estudillo-Wong, L. Cavillo, G. Granozzi and N. Alonso-Vante, *J. Catal.*, 2016, **338**, 135–142.
236 S. Zhao, H. Yin, L. Du, L. He, K. Zhao, L. Chang, G. Yin, H. Zhao, S. Liu and Z. Tang, *ACS Nano*, 2014, **8**, 12660–12668.
237 J. Shui, C. Chen, L. Grabstanowicz, D. Zhao and D.-J. Liu, *Proc. Natl. Acad. Sci.*, 2015, **112**, 10629–10634.
238 S. Ma, G. A. Goenaga, A. V. Call and D. J. Liu, *Chem. – Eur. J.*, 2011, **17**, 2063–2067.
239 T. Wang, Q. Zhou, X. Wang, J. Zheng and X. Li, *J. Mater. Chem. A*, 2015, **3**, 16435–16439.
240 F. Zou, X. Hu, Z. Li, L. Qie, C. Hu, R. Zeng, Y. Jiang and Y. Huang, *Adv. Mater.*, 2014, **26**, 6622–6628.
241 T. Y. Ma, S. Dai, M. Jaroniec and S. Z. Qiao, *J. Am. Chem. Soc.*, 2014, **136**, 13925–13931.
242 S. Mao, Z. Wen, T. Huang, Y. Hou and J. Chen, *Energy Environ. Sci.*, 2014, **7**, 609–616.
243 J. Masa, W. Xia, I. Sinev, A. Zhao, Z. Sun, S. Grützke, P. Weide, M. Muhler and W. Schuhmann, *Angew. Chem., Int. Ed.*, 2014, **53**, 8508–8512.
244 H. Yue, Z. Shi, Q. Wang, Z. Cao, H. Dong, Y. Qiao, Y. Yin and S. Yang, *ACS Appl. Mater. Interfaces*, 2014, **6**, 17067–17074.
245 W. Chen, Z. Zhang, W. Bao, Y. Lai, J. Li, Y. Gan and J. Wang, *Electrochim. Acta*, 2014, **134**, 293–301.
246 F. Zheng, G. Xia, Y. Yang and Q. Chen, *Nanoscale*, 2015, **7**, 9637–9645.
247 D. Tian, X.-L. Zhou, Y.-H. Zhang, Z. Zhou and X.-H. Bu, *Inorg. Chem.*, 2015, **54**, 8159–8161.
248 B. You, N. Jiang, M. Sheng, W. S. Drisdell, J. Yano and Y. Sun, *ACS Catal.*, 2015, **5**, 7068–7076.
249 H. X. Zhong, J. Wang, Y. W. Zhang, W. L. Xu, W. Xing, D. Xu, Y. F. Zhang and X. B. Zhang, *Angew. Chem., Int. Ed.*, 2014, **53**, 14235–14239.
250 Q. Qu, J. Yun, Z. Wan, H. Zheng, T. Gao, M. Shen, J. Shao and H. Zheng, *RSC Adv.*, 2014, **4**, 64692–64697.
251 J. R. Warren, PhD thesis, 2012, University of Birmingham.
252 L. Vegard, *Z. Phys.*, 1921, **5**, 17–26.